WORLD VIEWS AND THE PROBLEM OF SYNTHESIS

EINSTEIN MEETS MAGRITTE: An Interdisciplinary Reflection on Science, Nature, Art, Human Action and Society

Series Editor

Diederik Aerts, Center Leo Apostel, Vrije Universiteit Brussel, Belgium

---

Volume 1 *Einstein Meets Magritte: An Interdisciplinary Reflection*
The White Book of 'Einstein Meets Magritte'
Edited by Diederik Aerts, Jan Broekaert and Ernest Mathijs

Volume 2 *Science and Art*
The Red Book of 'Einstein Meets Magritte'
Edited by Diederik Aerts, Ernest Mathijs and Bert Mosselmans

Volume 3 *Science, Technology, and Social Change*
The Orange Book of 'Einstein Meets Magritte'
Edited by Diederik Aerts, Serge Gutwirth, Sonja Smets and Luk Van Langenhove

Volume 4 *World Views and the Problem of Synthesis*
The Yellow Book of 'Einstein Meets Magritte'
Edited by Diederik Aerts, Hubert Van Belle and Jan Van der Veken

Volume 5 *A World in Transition; Humankind and Nature*
The Green Book of 'Einstein Meets Magritte'
Edited by Diederik Aerts, Jan Broekaert and Willy Weyns

Volume 6 *Metadebates on Science*
The Blue Book of 'Einstein Meets Magritte'
Edited by Gustaaf C. Cornelis, Sonja Smets, Jean Paul Van Bendegem

Volume 7 *Quantum Structures and the Nature of Reality*
The Indigo Book of 'Einstein Meets Magritte'
Edited by Diederik Aerts and Jarosław Pykacz

Volume 8 *The Evolution of Complexity*
The Violet Book of 'Einstein Meets Magritte'
Edited by Francis Heylighen, Johan Bollen and Alexander Riegler

---

VOLUME 4

World Views and the Problem of Synthesis

The Yellow Book of "Einstein Meets Magritte"

Edited by

Diederik Aerts, Hubert Van Belle and Jan Van der Veken
Brussels Free University,
VZW Worldviews,
Catholic University of Louvain

VUB UNIVERSITY PRESS

VRIJE UNIVERSITEIT BRUSSEL
BELGIUM

KLUWER ACADEMIC PUBLISHERS
DORDRECHT / BOSTON / LONDON

Library of congress Cataloging-in-Publication Data

Kluwer ISBN 0-7923-5760-4 (Volume 4)
ISBN 0-7923-5765-5 (Set)

VUB University Press ISBN 90-5487-229-2 (Volume 4)
ISBN 90-5487-234-9 (Set)

Published by Kluwer Academic Publishers,
P.O. Box 17, 3300 AA Dordrecht, The Netherlands.
and
VUB University Press,
Waversteenweg 1077, B-1160 Brussels, Belgium

Sold and distributed in North, Central and South America
by Kluwer Academic Publishers,
101 Philip Drive, Norwell, MA 02061, U.S.A.

In all other countries (except Belgium), sold and distributed
by Kluwer Academic Publishers,
P.O. Box 322, 3300 AH Dordrecht, The Netherlands.

Sold and distributed in Belgium
by VUB University Press,
Waversteenweg 1077, B-1160 Brussels, Belgium

Table of contents

DIEDERIK AERTS

THE GENERAL INTRODUCTION OF *EINSTEIN MEETS MAGRITTE*

The series of books 'Einstein meets Magritte' presented here originates from an international interdisciplinary conference with the same title, which took place in Brussels in spring 1995. On the eve of the third millennium, we assembled scientists and artists to reflect together on the deep nature of reality and the knowledge and skill humankind has gathered in this field. We had decided to call this meeting 'Einstein meets Magritte' because we believed that meaningful keys could be found at the place where the two meet. It is the way of the world that has made Einstein and Magritte into icons of our culture. The purpose of the conference was to reflect and debate without fear on the most profound and timeless questions.

On one of those evenings, when the talks and discussions were long and exhausting and the press were doing all they could to get Albert Einstein and René Magritte in front of the microphones and cameras, a few of my most loyal aides and myself succeeded in getting them safely and quietly to a taxi, which then carried us off into the Brussels night. We got out at Manneken Pis, since that was on Einstein's list, and we concealed ourselves among the many tourists who were coming and going, expressing their wonder in every language under the sun at the famous little statue. And one of us was taking pictures: Einstein and Magritte leaning against the railings, with us beside them, and one more, arm in arm, and then another in case the first was no good, when suddenly I felt a heavy slap on my shoulder:

"How you doing, mate?"

It was Jacky and his inseparable girlfriends Nicole and Sylvie, and everyone embraced everyone else. I introduced Albert and René, and interest was immediately shown, and I had my heart in my mouth, because Jacky was a painter, poet and urban philosopher. We walked together through the alleys of Brussels in dismal Belgian rain, over cobblestones that glistened in the street lamps.

When we had provided for the inner man with 'Rabbit in Beer' and 'Mussels with Fries', and finally a 'Dame Blanche' topped with warm chocolate sauce as apotheosis, Jacky enticed us to his house in the Rue Haute where we threw ourselves into deep, soft armchairs. Albert and René were offered the best places and as always Jacky told the story of his life and discussed his rightness, as he did repeatedly, with a confidence and suppleness that distinguished him so sharply from modern science. Albert listened enthralled and René was fascinated, and once more my heart was in my mouth, but Nicole winked reassuringly, and Sylvie brought us snacks on cushions of Brussels lace and sweet white wine in tall, old-fashioned crystal glasses.

The topic of discussion for the evening turned out to be 'the doubts of modern science'. In science there is not a single hypothesis for which one cannot find two groups of hard-working scientists, one of which can 'prove' the hypothesis while the other can 'prove' its negation. And the more fundamental and important the question is, the more clearly the situation turns out like this.

"It's crazy," maintained Jacky, "In fact science states that one doesn't know anything anymore."

"That's right," said Albert, "Truth is not a simple concept, and I believe that the history of science makes it clear how often erroneous hypotheses have been believed over the centuries."

"A good thing too," replied René, "Things can only happen as a result of the movement brought about by that constant doubt."

Meanwhile Sylvie came to join us and handed round pictures of the exhibitions of Jacky's paintings and poems. Jacky suddenly got very excited, as if something had inspired him, and he leapt up and vanished into his studio. A few minutes later he returned with his palette and brush poised. Before I could stop him he had started painting violently right at the spot where Albert and René were sitting. A large, gossamer-thin piece of Brussels lace gradually took shape and Albert and René vanished. Fortunately, my young assistants, Jacky's girlfriends and myself got away with just a few vicious daubs of paint in the face.

The series of eight volumes introduced here are not just the results of the conference, as would be the case with a record of the proceedings. The authors were invited to write with the events at the conference in the back of their mind, so that the books would form a second phase in the process of thought set in motion at the conference. A second phase more clearly crystallised than the self-organising forum that arose during the conference, but one which focuses on the same timeless questions and problems.

The whole ensemble was already streamlined at the conference into a number of main topics named after the colours of the rainbow - red, orange, yellow, green, blue, indigo and violet, as well as white, the synthesis of all colours. This order was maintained and led to eight separate books in the series.

Volume 1: Einstein *meets* Magritte: an Interdisciplinary Reflection
The White Book of *Einstein meets Magritte*

The white book contains more fully developed versions of the contributions made by the keynote speakers at the conference. So this white book covers various scientific topics. In his article, 'Basically, it's purely academic', John Ziman asks himself what 'basic research' really is in today's world. In his contribution, 'The manifest image and the scientific image', Bas Van Fraassen analyses the considerable differences between the theoretical scientific description of the world and the way it appears to us. He argues that most formulations of this problem may themselves be tendentious metaphysics, full of false contrasts, and that insistence on a radical separation between science and what we have apart from science, and on the impossibility of accommodating science without surrender, may be a way of either idolising or demonising science rather than understanding it. In the 'Microdynamics of incommensurability: philosophy of science meets science studies', Barbara Herrnstein-Smith examines the bemusing but instructive logical, rhetorical and cognitive dynamics of contemporary theoretical controversy about science. In his contribution 'Subjects, objects, data and values', Robert Pirsig proposes a radical integration of science and value that does no harm to either. It is argued that values can exist as a part of scientific data, but outside any subject or object. This argument opens a door to a 'metaphysics of value' that provides a fundamentally different but not unscientific way of understanding the world. Ilya Prigogine discusses in 'Einstein and Magritte: a study of creativity', the global transformation of a classical science which was based on certainties into a new science that takes possibilities as its basic concepts. Constantin Piron demonstrates in his contribution 'Quanta and relativity: two failed revolutions' that none of the two great revolutions in physics, quantum mechanics and relativity theory, have actually been digested by the physics community. He claims that the vast majority of physicists still cling to the idea of a non-existent void full of little particles, in the spirit of Leibniz or Descartes. Rom Harré reflects on the significance of the theory of relativity. In his article 'The redundancy of spacetime: relativity from Cusa to Einstein', he defends the hypothesis that relativity theory is best interpreted as a grammar for coordinating narratives told by different observers. In his contribution 'The stuff the world is made of: physics and reality', Diederik Aerts analyses the consequences of the recent advances in quantum mechanics, theoretically as well as experimentally, for the nature of reality. He analyses the deep conceptual paradoxes in the light of these recent data and tries to picture a coherent model of the world. In his contribution 'Dasein's brain: phenomenology meets cognitive science', Francisco Varela puts forward the hypothesis that the relation between brain processes and living human experience is the really hard problem of consciousness. He argues that science needs to be complemented by a deep scientific investigation of experience itself to move this major question beyond the sterile oppositions of dualism and reductionism. In his contribution 'What creativity in art and science tell us about how the brain must work' William Calvin defends the prospects for a mental Darwinism that operates on the milliseconds to minutes time scale, forming novel ideas and sentences never previously expressed. Adolf Grünbaum in his article 'The hermeneutic versus the scientific conception of psychoanalysis: an unsuccessful effort to chart a via media for the human sciences' argues that the so called 'hermeneutic' reconstruction of psychoanalytic theory and therapy proposed by Karl Jaspers, Paul Ricoeur and Jürgen Habermas fails both as a channel and as alleged prototype for the study of human nature. In his article 'Immortality, biology and computers', Zygmunt Bauman analyses the shift that postmodern society has provoked regarding the concept of immortality. He points out that strategies of collective and individual immortality have shifted from the modern deconstruction of death to a postmodern deconstruction of immortality, and points out that the possible consequences of this process need to be taken into consideration. Brian Arthur, in his article 'The end of certainty in economics', points out that our economy is very non-classical, meaning that it is based on essentially self-referential systems of beliefs about future economic conditions. He argues that our economy is inherently complex, subjective, ever-changing, and to an unavoidable degree ill-defined.

Volume 2: Science and Art
The Red Book of *Einstein meets Magritte*

And then Magritte comes in. Many obvious differences exist between science and art. But the *Science and Art* volume of this series addresses not only these differences but also the possibilities of crossing several of the gaps between science and art. Several contributions deal with sociological and philosophical elaborations of the similarities and differences between science and art, while others approach science from an artistic point of view and art from a scientific point of view. The volume also considers several approaches that attempt to go beyond the classical dichotomy between the two activities. In a special section, attention is paid to the particular role played by perception in both science and art as a regulator of human understanding. Together, these contributions strive for an intensive interaction between science and art, and to a consideration of them as converging rather than diverging. It is to be hoped that both science and art will benefit from this attempt.

Volume 3: Science, Technology and Social Change
The Orange Book of *Einstein meets Magritte*

The major subject of the orange book is that society as a whole is changing, due to changes in technology, economy and the changing strategies and discourses of social scientists. The collected articles in the orange stream discuss a range of specific societal problems related to the subject of social change, the topics of the articles range from the scale of for instance sociology of health and psychohistory to more specific social problems like for instance anorexia nervosa, art academies and the information superhighway. Although the authors approach different subject matters from dissimilar perspectives and work with various methods, all the papers are related to the theme of science, technology and social change. In the orange book the reader will find a lot of arguments and hints pertaining to questions like: To what exactly will this social change lead in the 21st century? What kind of society lies ahead? She/he will be confronted to a plethora of enriching conceptions of the relationships between social sciences and social changes.

Volume 4: World Views and the Problem of Synthesis
The Yellow Book of *Einstein meets Magritte*

A rapidly evolving world is seen to entail ideological, social, political, cultural and scientific fragmentation. Many cultures, subcultures and cultural fragments state their views assertively, while science progresses in increasingly narrowly defined areas of inquiry, widening not only the chasm between specialists and the layman, but also preventing specialists from having an overall view of their discipline. What are the motive forces behind this process of fragmentation, what are its effects? Are they truly inhospitable to the idea of synthesis, or do they call out, more urgently than ever before, for new forms of synthesis? What conditions would have to be met by contemporary synthesis? These and related questions will be addressed in the yellow book of Einstein meets Magritte.

Volume 5: A World in Transition; Humankind and Nature
The Green Book of *Einstein meets Magritte*

'A World in Transition. Humankind and Nature' is appropriately entitled after its aim for an intrinsic property of reality: change. Of major concern, in this era of transformation, is the extensive and profound interaction of humankind with nature. The global scaled, social and technological project of humankind definitely involves a myriad of changes of

the ecosphere. This book develops, from the call for an interdisciplinary synthesis and respect for plurality, acknowledging the evolving scientific truth, the need for an integrated but inevitably provisional world view. Contributors from different parts of the world focus on four modes of change: i) Social change and the individual condition, ii) Complex evolution and fundamental emergent transformations, iii) Ecological transformation and responsibility inquiries, iv) The economic-ecological and socio-technical equilibria. Primarily reflecting on the deep transformations of humankind and on the relationship between humans and nature it addresses major points of contemporary concern.

Volume 6: Metadebates
The Blue Book of *Einstein meets Magritte*

This book provides a meta-disciplinary reflection on science, nature, human action and society. It pertains to a dialogue between scientists, sociologists of science, historians and philosophers of science. It covers several topics: (1) the relation between science and philosophy, (2) new approaches to cognitive science, (3) reflections on classical thinking and contemporary science, (4) empirical epistemology, (5) epistemology of quantum mechanics. Indeed, quantum mechanics is a discipline which deserves and receives special attention here, for it still is a fascinating and intriguing discipline from a historiographical and philosophical point of view. This book does not only contain articles on a general level, it also provides new insights and bold, even provocative theories on the meta-level. That way, the reader gets acquainted with 'science in the making', sitting in the front row.

Volume 7: Quantum Structures and the Nature of Reality
The Indigo Book of *Einstein meets Magritte*

This book refers to the satellite symposium that was organised by the International Quantum Structure Association (IQSA) at Einstein meets Magritte. The IQSA is a society for the advancement and dissemination of theories about structures based on quantum mechanics in their physical, mathematical, philosophical, applied and interdisciplinary aspects. The book contains several contributions presenting different fields of research in quantum structures. A great effort has been made to present some of the more technical aspects of quantum structures for a wide audience. Some parts of the articles are explanatory, sketching the historical development of research into quantum structures, while other parts make an effort to analyse the way the study of quantum structures has contributed to an understanding of the nature of our reality.

Volume 8: The Evolution of Complexity
The Indigo Book of *Einstein meets Magritte*

The violet book collects the contributions that consider theories of evolution and self-organisation, on the one hand, and systems theory and cybernetics, on the other hand. Both can add to the development of an integrated world view. The basic idea is that evolution leads to the spontaneous emergence of systems of higher and higher complexity or "intelligence": from elementary particles, via atoms, molecules, living cells, multicellular organisms, plants, and animals to human beings, culture and society. This perspective makes it possible to unify knowledge from presently separate disciplines: physics, chemistry, biology, psychology, sociology, etc. The volume thus wishes to revive the transdisciplinary tradition of general systems theory by integrating the recently developed insights of the "complex adaptive systems" approach, pioneered among others by the Santa Fe Institute.

Even these books only signify a single phase in the ever-recurring process of thought and creation regarding the basic questions on the reality that surrounds us and our place in it.

Brussels,
July 17, 1998.

HUBERT VAN BELLE & JAN VAN DER VEKEN

WORLD VIEWS AND THE PROBLEM OF SYNTHESIS

1. FRAGMENTATION AND DISINTEGRATION

Our complex and rapidly evolving world is extremely fragmented in cultural, ideological, social, political and scientific areas. There remains little or no trace of cultural unity. We have to deal with a confusing mixture of many cultures, subcultures and cultural fragments. Fast-paced developments, increased mobility and the growing impact of the mass media have also disoriented many people. The world around them has changed so fast that they have become alienated from it and can no longer find their way. They no longer find anything to hold onto in the traditional ideological, social and political systems, which have lost much of their naturalness, credibility, attractiveness and influence. We are experiencing the end of the big dreams and the 'great narratives'. It seems that there are no longer clear and generally accepted views about the nature of reality and about man's task in the world. The pluriform society is not succeeding in formulating powerful and universal answers to the pressing and worldwide problems with which it is being confronted. Even science do not offer a satisfactory answer. The trend of differentiation in science is continuing unabatedly. The large amount of information that has been collected through specialized science does not form a coherent whole. The chasm between specialists and the layman is widening, and even specialists are no longer able to have an overall view of their discipline.

The rapid changes and the large-scale structures are leading to an increased alienation from the modern world. So many changes have taken place in such a short time that we no longer recognise ourselves in the world. The overview is lost and we become unsure. Questions about values and meaning are no longer given a clear and generally accepted answer by the collective systems from the past which once provided a meaning. Meaningful orientations and projects for the future are lacking or are no longer accepted. The general faith in progress has disappeared and important current problems seem insoluble. Many young people can no longer find a place for themselves, lack a project to which they can dedicate themselves and see no goal for which to live. For this reason, some turn against society or flee from reality.

For many, the world is becoming meaningless and hopeless. All that remains are a number of fragments that have little structure or coherence. When a satisfactory and coherent view of the whole is lacking, the individual is forced to try to find his own way and to make his own choices among the immense and confusing possibilities that are offered. Little time remains for fundamental questions and no unanimity develops around certain insights. Everything seems to end in a chaotic entanglement of contradictory ideas and actions that neutralise each other. This leads to paralysis and indecision. Although a uniformly-thinking society and one single imposed world view are neither attainable nor desirable, the members of a society need to be sufficiently in agreement about a number of points for the society to be viable. In the current fragmented world there is often a lack of the necessary minimum agreement required for urgently-needed common measures to be taken.

A group of scientists is unwilling to resign itself to this situation of fragmentation and disintegration and is calling for integrating world-view research. This interdisciplinary project presents itself as a challenge to the entire scientific community. These scientists wish to provide a framework in which the world views that are developing in the various fields of culture and science can confront each other. Many scientists who are unhappy with the disintegration are seeking universalising theories. From within their own area of specialisation they were driven to consider the world-view problem and try to form a picture of the whole. Most of them are of the opinion that humankind may not remain passive but that traversable paths to a better world need to be marked out once more. Reliable maps are essential, maps that correctly show the dangerous sand banks and the safe sea ways. The 'maps' that allow people to orient themselves in the world we call 'world views'. Many parts of existing world views have become unreliable or have fallen into disuse. The group is therefore of the opinion that a deliberate effort must be made to construct new world views. Not only must this be done individually and spontaneously, but also a concerted effort must be made in unifying research into world views. *Worldviews*, a Belgian non-profit organisation established in 1990, made a programme declaration for this ambitious project.

2. WORLD VIEWS

What are world views, and what is their purpose? For a comprehensive answer to these questions we refer to *World Views: From Fragmentation to Integration.* [1] in which the project of *Worldviews* is introduced. In this introduction we will limit ourselves to a brief summary. A world view can be defined as a coherent whole of concepts and propositions, enabling one to form a global picture of reality which can incorporate as many elements of one's experience as possible. It is the way in which people organize their understanding of the world. A world view offers some clarity regarding our place in the world, gives insight into the greater relationships, provides us with orientation and calls for possible action. With the help of a world view, we find our place in the universe, try to comprehend the evolution of the cosmos, and try to give stature to our task on earth. In world views, both societies and individuals find an answer to the underlying questions relative to being human and to the world in which we live. Nor can one evade some awareness of that which transcends us and raises questions about our destiny.

One can also characterise a world view concisely as a *model* of total reality viewed as a whole. So a world view is a *reference framework* which must include a place for all our manifold experiences of the world and ourselves. It is a symbolic representation system that should enable us to integrate everything we learn about the world and ourselves into a total concept. This view should provide a coherent picture of the world and should correspond with reality. To be useful in daily life, a world view must be sufficiently comprehensible and reliable. If it contradicts too many elements from our experience, it is no longer of any value to us. A world view throws light on the whole of reality as it is seen within a certain culture. World views help us to find our way in a complex environment and to act in a coherent manner. The questions of purpose, values and meaning that are central to the current crisis form the core of the world-view problem.

Although world views grow organically and historically, they can also be developed. The construction of world views is comparable to the work of cartographers in antiquity and in the Middle Ages. They mapped out the world on the basis of information from sailors, merchants and explorers. This information was often incomplete, inaccurate, contradictory and even fabricated. Gradually, however, the data were put together and a more reliable picture emerged. Analogously, the construction of world views can start from an inventory of existing world-view fragments. They can be found, for example, in philosophies of life, concepts of man and society, ideologies, philosophical systems and even in so-called 'common sense'. The available fragments can serve as a starting point in the construction of new world views. To form adequate world views, the fragmentary data must be integrated as much as possible. Science can make an important contribution. World views may not be in conflict with known scientific facts, but they do not coincide with them either. They must also incorporate systems that give values and meaning.

We continually speak of world views in the plural because a unique and monolithic world view — considering the immense complexity of reality — will remain an unattainable ideal. These world views illuminate various aspects of one and the same reality, and partially overlap each other. They must be as reliable as possible and they may not contradict each other. World views can be rightly compared to a set of maps showing the geological, political and economic situation, which are bundled together in one atlas. As with maps, world views should be in agreement in areas where they overlap. There are many maps but there is only one world.

3. THE WORLD-VIEWS DEBATE

After the publication of the *Worldviews* programme paper, a discussion has been going on in Belgium concerning the desirability of world-view construction. The spokesmen of the debate have been mainly Leo Apostel, the founder of *Worldviews* and Herman De Dijn, inspired by Wittgenstein's life-forms. Wittgensteinians tend to feel that there is no need for dialogue between the sciences and the humanities. To understand the world and to live in it are completely different interests, which should not be intermingled. World views are not constructed. They are found, and are part of our "Lebensform" (or form of life). It was the conviction of Leo Apostel, on the other hand, that today it was no longer possible to rely on the traditional world views, because they were scattered and fragmented. The reason is that mostly the scientific world view has not left the other aspects of civilisation untouched.

The underlying discussion can be traced back to a difference of opinion between Whitehead and Wittgenstein. In August 1913 Wittgenstein was a visitor at the Whitehead country house in Lockridge, Wittshire. There was never any meeting of minds. Three things about Wittgenstein annoyed Whitehead intensely. (1) He was passionately certain of the correctness of his opinions. (2) He kept insisting that you *must not* ask certain philosophical questions. (3) Instead of trying to bring science and philosophy together, he drove them farther apart, by making philosophy a very special kind of linguistic activity [2]. Leo Apostel clearly sided with Whitehead. Because of the prevailing influence of the scientific outlook on reality, there is a tendency to substantialize or radicalize this outlook on reality. Positivism is a case in point. Although Leo Apostel respected the knowledge acquired by modern science, he thought that it should not be separated from other approaches to the real, as there are: esthetics, ethics, religion. Whitehead in *Science and the Modern World* [3] states: "In every age each of these topics suggest a view of the world. We may ask ourselves whether the scientific mentality of the modern world in the immediate past is not a successful example of such provincial limitation.

Philosophy, in one of its functions, is the critic of cosmologies. It is its function to harmonise, re-fashion and justify divergent intuitions as to the nature of things. It has to insist on the scrutiny of ultimate ideas, and on the evidence in shaping our cosmological scheme. Its business is to render explicit, and — so far as may be — efficient, a process which otherwise is unconsciously performed without rational tests."

4. THE *WORLDVIEWS*-PROJECT

After a year of intense study and discussion the *Worldviews*-group published in 1991 the Dutch version of the programme declaration. This consensus text has been translated in English and also published [1]. Starting point was the fact that the disintegration of the great world-explanatory systems of Modernity (of what has been called by Heidegger Weltbild, Cfr *Die Zeit des Weltbildes*) has left our world without orientation. However, Heidegger limits this "Weltbild" to the modern times, and interprets is as the endeavour to dominate the world through understanding.

The understanding of "world view" by Leo Apostel is at the same time broader and more modest. Every period has its own world view. Without a coherent world view there is no culture. Our time is characterized by the collapse of a unitarian view of the world. But we do not want to restore one single world view. We simply want to understand what is going on, and to do the best we can to think about the world as coherently as feasible, taking into account every bit of trustworthy experience we can get.

In the programme paper the *Worldviews* -project is defined as the endeavour to form "a coherent collection of concepts and theorems that must allow us to construct a global image of the world, and in this way to understand as many elements of our experience as possible." [4] The Whiteheadean overtones of this definition are evident (and is somewhat lost in the English edition of the programme paper). The *Worldviews* -project has a *theoretical* and a *practical* aspect.

The Theoretical Aspect

A. Knowledge of Reality
First we have to focus on the question how do we get to *know* reality as a whole ("world" has the broad meaning of all-encompassing framework of our lives. This is the phenomenological meaning of the world, as in the Husserlian "Lebenswelt".) What kinds of knowledge are there? Is one knowledge privileged over against the other? Is all knowledge nothing more than a strategy to dominate our "Umwelt" (environment), or should we hope that our best knowledge expresses somehow the real features of the world ? Leo Apostel, for sure, was more a critical realist than a Kantian idealist. Here a confrontation with contemporary philosophy of science is required.

B. An Explanatory Model of Reality
Sciences does not just want to describe, but also to explain. The explanatory value of scientific theories has always been a hot issue in our discussions. Leo Apostel was rather optimistic about the explanatory value of science. He was looking for the most general structures of the real which could allow us somehow to grasp not just "how things are" but "*why* things are the way they are". Apostel thought that symmetry and asymmetry was somehow a last clue to the nature of the real. Without symmetry and asymmetry, at least at some limited corners of this universe, life would not be possible. The question of reversibility versus irreversibility is an aspect of this problematic.

A challenging insight of Leo Apostel was that the last explanatory causes of the universe are not efficient causes, but formal causes. His education as a logician may have brought him to this insight. As far as this is concerned, he sided with Leibniz, and was enthousiastic about research in "mathematical metaphysics" by a colleague, Wim Mielants.

C. A Construction of a Model of the World

The *Worldviews*-group has studied the different metaphors that are used to talk about the total system of the universe. Should it be conceived as an organism, a living body, or rather as a purposeless mechanism? The *Worldviews*-group identified the traditional oppositions (mechanic versus organic, objective versus subjective, macro versus micro) as resulting from the way different disciplines focus on different levels of reality. It may be that at some levels of reality one description is more adequate than another. There has been in the group an overall feeling that reductive schemes are less interesting than more holistic ones.

The Practical Consequences of Having or Not Having a Coherent World View

World views will differ depending on how they structure their model of values (axiology). The question of meaning and the way in which we judge and attach value to global reality are key topics in this aspect of the program.

An interesting research project has been elaborated by the Law department of the KU Leuven, and more especially the department of criminology. Is there a link between the disintegration of world views and delinquency? The word delinquency itself seems to suggest such a link. Delinquent behaviour is behaviour which is un-bound, which has lost its contact with the supporting structures, such as community, nature, culture which give some solidity to the acting person.

Another interesting aspect of this problematic has been labeled by Apostel, first as "a-theistic" and later as "non-theistic religiosity". Religion is derived from religare, to connect, to bind. It was Apostel's conviction that some deep seated awareness of being related to a larger all-encompassing Whole is a requirement for a healthy and meaningful life. The way to live out this relatedness to All Encompassing Reality could according to his view, not be conceptualized. Therefore he felt closer to Zen-Buddhism than to traditional theism. The Center of Metaphysics and Philosophical Anthropology in Leuven encourages rather a panentheistic understanding of the Ultimate than a traditional onto-theological approach. For that reason there appeared to be some structural similarities with Leo Apostel's non-theistic religiosity. As far as the conceptualisation however is concerned, Apostel opted for what technically could be called a negative a/theology.

A last aspect of the common project, which has not retained too much attention of the group until now, but which was vital according to Leo Apostel, is the praxiological implication of having or not having a coherent world view. We are involved in the world not only by knowing, valuing or feeling but also by acting. There is a general tendency to neglect the sciences of action when searching for an integrative world view. Those praxiological approaches are called "applied sciences", although this does not imply that they merely apply knowledge which has been acquired theoretically. A global theory of the world can only be built in interaction with a general praxiology.

5. SURVEY OF THE CONTRIBUTIONS

The different contributions to "World views and the Problem of Synthesis" can for the greater part be divided into four groups. Section 1 contains the contributions of the members of the *Worldviews*-group and are presented as an homage to Leo Apostel, who died on August 10, 1995. We are honored to include one of his last presentations. These contributions introduce the *Worldviews*-project and deal with the philosophy and thinking of Leo Apostel. Section 2 offers papers presented by the International Research Community "Integrating World views." Two immediately related papers are also included. In this section contributions are brought together which give a survey of important questions related to interdisciplinary approaches, global understanding of reality and world-view research. Besides, the position of social sciences and the "social scientization" of all knowledge is discussed. An attempt to develop a world view based on quantum physics and a layered structure of reality is also presented. The papers of section 1 and 2 should provide the reader with a general feeling of what the *Worldviews*-project is all about.

Section 3 and 4 contain papers as they were read at the *Einstein Meets Magritte* conference. They present a broad spectrum of research within the general context of the *Worldviews*-project. Section 3 deals with epistemological question and discusses fundamental problems about unity in science. It presents several attempts to overcome the Kantian dichotomy and to reconcile the different theories, views and beliefs. A variety of methodologies critically combined together in a "multimethodology" is proposed as a general framework to reconcile different paradigms and as a basis for real-world interventions. The idea of complementarity, first suggested by Niels Bohr in quantum physics, is also introduced as a principle to reconcile different visions which contradict each other but have their value. In section 4 new ways in interpreting science and society are tried out. In a first group of papers generalizations of physical theories and the limitations a such an extrapolation are discussed. It is shown that mechanics in and of itself is not mechanistic. Synergetics is considered as a bridge between east and west as well as a step towards a synthesis of natural sciences and humanities. A second group of papers is methodology oriented and deals with polyscopic modeling techniques and their application to create "high-level information". In a last group of contributions attention is paid to philosophical theories and their practical consequences for real life problems such as mental health and the prevention of delinquency. A biological model to explain social development and change in computer science is also proposed. The disparity of the contributions reflect the wide scope of the *Worldviews*-project. We hope, nevertheless that this publication will be a stimulus for the ongoing and new international world-view research. Indeed, the *Worldviews*-project can be considered as a challenge for the scientific world as a whole.

6. REFERENCES

[1] **Aerts, D., Van Belle, H., Van der Veken, J. et al.**, *World Views: From Fragmentation to Integration*, Brussels, VUBPRESS, 1994.

[2] **Lowe, V.**, *Whitehead, A. N., The Man and His Work. Volume II: 1910-1947*, New York, John Hopkins University Press, 1990, p. 278.

[3] **Whitehead, A. N.**, *Science and the Modern World*, New York, The Free Press, 1967, p. vii.

[4] **Aerts, D., Van Belle, H., Van der Veken, J. et al.**, *World Views: From Fragmentation to Integration*, Brussels, VUBPRESS, 1994, p. 17.

JAN VAN DER VEKEN

IN SEARCH OF AN INTEGRATED VIEW ON THE WHOLE OF REALITY: INTRODUCING THE WORLDVIEW PROJECT

A "World View" is "a house for the Spirit" (F. Lange). In a house, there are many rooms. They do not all have the same function. A living-room is normally not a study room, and a bathroom has its own finality. Nevertheless, the rooms of a house are connected with one another: they are integrated into a whole, because the same family is living in it. Why should it be different with the family of mankind, living on earth?

Sometimes there is a tendency in architecture to "integrate" the whole living space into one big multi-functional living room. This encourages communication, but has also some inconveniences: the different functions are not recognized in their particularity: sometimes there is too much interference. A possible reaction is to isolate the rooms almost completely.

Once I read on a poster :"A house must be clean enough to be healthy, and dirty enough to be happy." The same can be said about the house of the spirit. Too much rigidity makes it an unfit place to dwell. Not enough integration makes it unlivable. That is the reason why "we" are in search of an integrated view on the whole of reality.

LET ME FIRST SAY SOMETHING ABOUT THE "WE"

The *Worldviews* group started under the impulse of Professor Leo Apostel, a well known logician and philosopher, teaching at the State University of Ghent and the Vrije Universiteit Brussels. The group includes Diederik Aerts, (Physics, University of Brussels), Diderik Batens (Philosophy and Logic, University of Ghent), Bart De Moor, (Engineering Sciences, University of Leuven), Staf Hellemans, (Sociology, University of Utrecht), Edel Maex, (Psychiatry and Psychotherapy, General Hospital Middelheim, Antwerp and Helen Dowling Institute, Rotterdam), Hubert Van Belle, (Engineering Sciences, Bombardier Transportation), Marie-Claire Van de Velde (Research Council, University of Ghent) and Jan Van der Veken, (Philosophy and Theology, University of Leuven). In 1990, the group presented a consensus paper to the public, under the title *World Views. From Fragmentation to Integration.* This is a proposal, a kind of methodology for world view-construction, followed by a series of possible projects. It is not an exposition of new facts or theories, but an invitation to look at known facts and theories from another point of view, with a new, different, integrated purpose. This consensus paper has been translated into English, and is published by the VUBPress, Brussels, 1994. The group has also published a second book, *Cirkelen om de wereld: concrete invullingen van het wereldbeeldenproject* (Perspectives on the World: an Interdisciplinary Reflection). This book offers attempts by the various members of the group to say what can be said about a common and shared world view from the point of view of their specific disciplines. The group has also established contacts with research institutes all over the world.

What do we mean in this context by *world*? World is surely not to be identified with planet earth. We do not live on a planet. The world is our "Lebenswelt," as Husserl has beautifully put it. The "world" in which we live is the broader totality of all that is meaningful for us. In this sense, we can talk about the world of medieval man, or about the world of the Eskimo. It is clear that all those "worlds" may be very different from one another. Hence the question arises: is it meaningful to talk about a common world view, or should we rather accept the fact that there is a great number of incommensurable world views?

A *world view* is a symbolic system of representation. A clue is taken from a definite domain of reality (e.g., our own body, or a clockwork) and is extended to somehow represent the whole of reality. This is the reason why all language about our world is metaphorical (in the literal sense of that word): we borrow meanings and concepts applicable to one domain of our experience and transfer them to other domains of existence, and sometimes extend them towards the whole. New metaphors appear as a result of important cultural changes. In the Middle Ages the world was seen as Gods handiwork. In the 17th century, the world was conceived as a mechanism. According to Heidegger's *Die Zeit des Weltbildes*, the 17th and 18th century can be characterised as the attempt to encompass the whole into one single "Bild" or "Image." Such a guiding image or "root metaphor" almost irresistibly determines the way we conceive every domain of reality. When the world is conceived as a mechanism, God is conceived as a cosmic watchmaker. Today we are more inclined to use organistic metaphors. The whole universe is rather conceived as an evolutionary whole, as an unfolding story, or perhaps, as a play or as a work of art. Images of the world are at the same time images of "man." Cosmos and anthropos are intimately related. So we can see, depending upon our world view, man as the crown on Gods work, as a very intricate robot, or as a "magnificent accident" of evolution. Images of man and images of the world are so intimately related, that we can talk as well about world views, "Weltauffassung," as conceptions of "man." Intercultural dialogue before all is dialogue between different world views.

Integration presupposes a certain communication between the different spheres of interest. The various human interests "suggest cosmologies and are influenced by them" (A.N. Whitehead in *Science and the Modern World*). Those different interests are, a.o., science, esthetics, ethics, religion. In a culture in which a definite interest is predominant, that approach has a tendency to occupy the whole sphere of meaning, and to inspire our total vision on reality. Religious fundamentalism is a case in point. When a religious outlook on reality is all-pervasive in a certain society, different human interests, such as politics, morals, the organization of the community are "solved" from one, all-encompassing religious point of view. Positivism is another example. Impressed by the evident success of science, the scientific outlook at reality has a tendency to extend itself to every sphere of our existence. "Philosophy, according to Whitehead, in one of its functions, is the critic of cosmologies. It is its function to harmonise, re-fashion, and justify divergent intuitions as to the nature of things." This means that the world view project is in principle anti-reductionistic. I consider this to be one of the most salient features of our project: although it recognizes fully the impact of contemporary science on our world view, it is in principle open to other human interests, such as ethics and religion. We do not consider these different interests as completely isolated "language games," as some Wittgensteinians would have it. Science, ethics and religion speak different languages, to be sure, but the speaker is the same. Only an educated speaker can speak and hold together different languages in the right circumstances. This is probably the main "aim of education." The integration of the different perspectives is very well expressed by the German word "Bildung." "Bildung" is always "in eins bilden": it is a way to foster the communication of the images (Bilden) which arise on the different levels of existence.

If the integration of the different perspectives on reality is no longer successful, fragmentation or disintegration is the result. Hence the subtitle of our consensus paper reads: *From Fragmentation to Integration*. The tentative title of the English translation of our second book, which is an attempt to start the construction of an adequate, contemporary world view from different perspectives, is: *Fragments of Integration*.

The quest for an integrated view on the whole of reality contains an empirical and a theoretical moment. There are no isolated facts. "Seeing" is always "seeing as" (Wittgenstein, *Philosophical Investigations*). It is evident that our project is very congenial with what Whitehead conceived to be the task of speculative philosophy. His was "the endeavour to frame a coherent system of general ideas in terms of which every element of our experience can be understood." As a method, Whitehead suggests: "The true method of discovery is like the flight of an aeroplane. It starts from the ground of particular observation; it makes a flight in the thin air of imaginative generalisation; and it again lands for renewed observation, rendered acute by rational interpretation." Our aeroplane is probably flying on lesser heights than Whitehead's, and for that reason we hope to keep a closer contact with reality than Whitehead's impressive speculative system. Our project is more descriptive, and tries to "map" all we know about our world today. Our aim is not so much to arrive at one, single "map" of the all-encompassing universe. That might well be impossible, or unconvincing. It is rather an endeavour to encourage a dialogue, not just between the sciences and the humanities, or between the two cultures, to use P.C. Snow's expression, but also between fragments of the same culture, be it scientific or humane. Specialisation makes communication not only difficult between different disciplines, but often between the representatives of subdivisions within the same discipline. The world view project can be characterised as truly interdisciplinary. We try not to talk so much *about* interdisciplinarity as to talk to one another, from the point of view of our disciplines and about our common concern: an integrated view on the whole of reality.

World view construction is a form of "cosmic mapmaking." "Map" is itself a metaphor, but a useful one. Different maps serve different purposes. Some maps may be more adequate than others. Even a "distorted map" may be useful for some specific purposes. An inspiring example is the well known "Mercator projection." It is fully inadequate as a representation of the relative surfaces of the different countries, but it allows navigation, because the angle that a ship has to follow is exactly the same as the one that can be measured on the map.

Hence, it will be important to map the different realms of our world, and to show how they are related.

In reflecting upon the problem of building a coherent and adequate scheme which allows to "think together" the main domains of our world, it has occurred to me that Karl Popper's distinction between World 1, World 2 and World 3 is most helpful. John Eccles has suggested to call the first world World 1, etc. Popper has followed this suggestion. In *Facing Reality*, John Eccles has drawn some diagrams, visualising the relationships between World 1, World 2, and World 3. Popper was very pleased with those diagrams, and of course anyone is entitled to draw one's own diagrams.

World 1 is the material universe, the entire cosmos, the inorganic and the organic world. Our bodies and our brains belong to World 1, as well as human artifacts and works of art as material structures.

World 2 is the psychic world: the world of our experiences and our self-awareness, the world of our emotions and free decisions. A very important part of World 2, without which we would not be ourselves, and lack any sense of identity, is our memory. It has its "seat" in some specific parts of our brain.

World 3 is the World of culture: it includes our languages as well as our scientific theories, and of course, our world views. We could say that a very important part of World 3 is the memory of mankind. Past achievements of the human race are stored in our libraries, our museums, and even in prehistoric caves. Popper acknowledges that it was most difficult for him to convince his audience about the objective or independent status of World 3. Of course, World 3 cannot exist without some material bearer, which belongs to World 1: the books in our libraries, the paintings, sculptures in our museums. It seems that elements of World 3 may somehow survive, even if no actual interpreter or representative of World 2 is around. An inspiring example given by Popper is the so called language linear B, some form of pre-Homeric Greek. It was recognized as a language, even if no one on earth was able to decipher it. Once it was interpreted, it came "back to life." The same could be said about the hieroglyphs. They are works of art in their own right: they may be admired even if not understood.

The problems which arise in World 1, World 2 and World 3 are very different; yet, they cannot be tackled in isolation. When asking the question; how is World 1 related to World 2 and World 3, again and again we discovered that all the problems arising in the philosophy of science are also problems arising within the world view project: how is consciousness related to the cosmic evolution? (Emergence of World 2 out of World 1). In which sense the categories of World 3 (our scientific theories) are applicable to World 1? That our conceptual schemes can be applied to World 1 is far from evident. Eugen Wigner has written an article with the inspired and inspiring title: "The unreasonable effectiveness of mathematics in physics." Indeed, why should we take it for granted that our ways of thinking can keep in contact with a universe which seems to belong to an entirely different ontological realm? "That mathematics works is the greatest mystery of all," according to A. Einstein, who could not help philosophizing about science, about the universe and even about God. What are the implications of the fact that the world can be known? Plato, the Stoics, Leibniz and so many others devoted their best efforts to thinking about this far from obvious problem. There are plenty of analogies in contemporary philosophy of science. Why are the initial conditions of this universe and the laws of nature the way they are? If they were even slightly different, no "anthropos" or no "observer" would be there to wonder about the universe and his or her own existence. Does this throw light upon the Universe itself? According to the highly debatable "cosmic anthropic principle," it does. It seems as if the observer is part and parcel of the very structure of the universe itself. What is debatable about the Weak Anthropic Principle (WAP) is not what is says, but its explicative power. After all, it does not say more than the following: the totality of reality must be such that we are possible, given the fact that there is an observer. But that can also be said about flies and about computers. Maybe, it just states the obvious. The WAP does not say, as some interpreters wrongly read it, that an observer *must* exist.

All those problems can be illumined by asking them in the language of the relationships between World 1, World 2 and World 3.

Concerning World 1 we would like to know why the initial conditions and the four forces governing this cosmic epoch are so incredibly fine tuned, that the problem arises whether man is somehow intended right from the beginning. At least the initial conditions have to be such that man *could* arise, given the fact that we are living in a universe with an observer. But that seems very different from the stronger assertion (the Strong Anthropic Principle) that man *must* arise.

Concerning World 2 is seems obvious that consciousness somehow arose out of the biological evolution. The very intricate system of the brain seems at least to be a condition of possibility of conscious life on earth. The very fact that World 2 exists throws some light upon the Universe which seems unexpectedly "hospitable to life." Is man to be conceived as a "magnificent accident" (according to the title of the book of Wim Kayzer, who brought together six or seven renowned scientists to discuss about their world view? If an accident, why so magnificent? And if so magnificent, can it be an accident?

World 3 is possible because we are the only beings capable of representing our world symbolically. World 3 is truly a human creation, although, of course, the conditions of possibility for such an achievement have to be present somehow in reality itself. Merleau-Ponty has borrowed the expression "l'esprit sauvage" from Claude Lévi-Strauss to point to that ontological dimension, which allows sooner or later for the emergence of the spoken logos. To work on World 3 is truly building "a house for the spirit."

The fact that World 2 mediates necessarily between World 1 and World 3 accounts for what can be called "the Kantian bias": all our world views, all our theories, all our endeavours to make sense out of our world are necessarily framed by the kind of concepts we use to talk about our world. Here the issue of realism versus anti-realism arises. Are our representative schemes merely a kind of network that we throw over World 1 or should we say, in Hume's (or Philo's) words, "That the cause or causes of order in the universe probably bear some remote analogy to human intelligence" (*Dialogues concerning Natural Religion*, 609)? Jacques Monod rejects that analogy. His main point is that the "alliance" between man and the cosmos came to an end: "L'ancienne alliance est rompue." We are finding ourselves in a cosmos which has not willed us, and which at some time will go on without us. It seems that a new awareness is growing in certain scientific circles, that man cannot just be a result of chance. The original and more inspiring title of the influential book of Ilya Prigogine and Isabelle Stengers (who by the way is very attracted by Whitehead's philosophy) is *La nouvelle alliance. Métamorphose de la science.* The main idea of the book is that given a new and transformed view on science, it is no longer necessary to exclude that there might after all be an alliance between man and the universe. The emergence of order, and of more complex structures such as the brain, are no longer to be conceived as a matter of chance, a local accident in a universe which as a whole is governed by the law of entropy. At least at one place and possibly also at other places an emergence of order is no longer to be conceived as an anomaly. Contemporary science throws some light upon the emergence of *Order out of Chaos* (the title of the English translation of Prigogine's book). Problems such as autopoiesis, self-organization etc. arose again and again in our discussions. The conceptual status of such notions is far from clear. But at least we are confronted here with serious questions, where a dialogue between science and philosophy seems possible and required. At least it can be said that the emergence of novelty and more order is not an anomaly, and in this sense it is somehow expected to happen in a universe which lives long enough to allow for such an extraordinary event to happen. We live in a "layered" universe, where one domain presupposes the other. Life, e.g., presupposes the presence of carbon—all life we know of is "carbon based life." Carbon is a heavy molecule, which can arise in no universe younger than ten billion years. So it seems almost certain that a smaller, (i.e., a younger universe), could not contain us (or an observer). This universe seems in a most wonderful way "calibrated for life." Intelligent life, which makes possible World 3, can be seen in the line of complexification, which may seem a "law of nature," as universal as the law of entropy.

Christian de Duve puts it this way:

> When I do listen to music, when I am walking around in an art gallery, when I admire the pure lines of a gothic cathedral, when I do read a poem or a scientific article, when I see my grand children playing or when I simply reflect upon the fact that I can do all these things, including the fact that I can reflect upon my ability to do all those things, then it is impossible for me the conceive the universe of which I am a part as not bound, by its very nature, to give birth somewhere, at some time, may be at different places and at different times, to beings capable to appreciate its beauty, to experience love, to search for truth and to fathom the mystery. For that reason I will probably be counted amongst the romanticists. So be it."

One of my professors once said: "Metaphysics is an illness. Happily enough it is without a cure."

It seems to me that all this allows us to understand the predicament of philosophy in this post-modern times and to ask age old questions in a new and more informed way.

WHAT HAS MADE GREEK OR WESTERN PHILOSOPHY "SPECULATION CONTROLLED BY REASON"

Judeo-Christianism has adjusted the insights of the Bible and of Greek cosmology in an impressive synthesis, which has dominated the West for more than fifteen centuries. The metaphor of the two books is a way to integrate the world view of the Bible and the philosophical world view of the age. Modernity can be understood as the endeavour to arrive at a definitive and fully rational world view. Heidegger calls modernity "Die Zeit des Weltbildes." He gives, however, a somewhat distorted view of what a world view is. According to his reading of the German "in Bild bringen" he interprets the project itself of constructing a world view as an act of intellectual imperialism. No wonder that a fully rational, unique and definitive world view which would bring every element of our experience within the boundaries of reason is no longer possible today. We live in the time of the end of "the great Narratives" (J.F. Lyotard). It seems fashionable to replace the great Narratives by small narratives: isolated islands of meaning within a sea of contingency. It could be the case, however, that in spite of this evident fragmentation of our culture, some new and "worldwide" world view is in the making. The insights of relativity theory, of quantum mechanics, of genetics, together with the new ecological awareness are determining in such a profound way our outlook on reality, that a new great narrative may be in the making.

The two extremes to avoid are a new brand of positivism which sees science as the only solution of all human woes and does away with all kinds of metaphysics, world views, ideologies or philosophy. The other extreme is an unenlightened recourse to the tradition, as a lifeline for those drowning in an ocean of relativism.

To quote Whitehead's end-remark in *Symbolism: Its Meaning and Effect*:

> "Those societies which cannot combine reverence to their symbols with freedom of revision must ultimately decay, either from anarchy, or from the slow atrophy of a life stifled by useless shadows."

A last and important feature of the world view project is that it does not intend to be purely theoretical, philosophical our scientific. It is also directed towards offering reasonable guidelines for action. Hence it must be possible to draw, by way of conclusion, at least some lines leading to responsible action.

1. The impact of world-problems will increase immensely. The problems which are the result of the use and abuse of technology in the West will arise on a worldwide scale. There will be a need for a nation-transcending action, e.g., in order to tackle problems such as ecological problems, the threat of damage to the ozone layer, the explosion of the world population, etc. Such problems cannot be addressed without at least some shared presuppositions.

2. Hence the necessity of serious information, and of a battle against simplifications. This necessity for objective information is connected to the empirical side of the world view project. Responsible action is not possible if we are ignorant about the objective elements of the situation and of the impact of our action in the long run.

3. Isolated insights are not up to the task. Alain Badiou puts it as follows: "After all, the task of philosophy is to draw a map of what we can really think, and then to conceive for that map a way to figure out how these different domains of reason or of truth [art, science, politics, love] are interrelated ("reconnaître les lieux de pensée, ou de vérité -"l'art, la science, la politique et l'amour"). "En somme, la tâche de la philosophie consiste à dresser la carte de ce qui relève effectivement de la pensée, puis à confectionner pour cette carte un cadre permettant de concevoir comment s'agence le passage de l'un à l'autre").

4. Courage will be needed to go on "naming" our individual and collective ideals. They are not entirely different from those of the Enlightenment, even after the end of "the great Narratives." The recognition of human rights may well be the requirement of an ethical minimum without which a contemporary outlook on reality would not be morally acceptable.

5. The economy is not a natural process, governed by an "invisible hand." It clearly belongs to World 3: it is a culturally determined way to use or abuse the resources of World 1. Without the resources of World 1 the representatives of World 2 cannot survive. There is a need for a worldwide solidarity (World 3) to take into account the needs of all human beings, without deteriorating in an irreversible way the sustainability of the planet (World 1). It would be foolish not to take into account, in any decision, the known impact upon the environment. A new way of coping in a responsible way with the limited resources of the globe is an element of all sound ethics. A new and responsible risk assessment is necessary. The first ethical imperative for the coming decades seems to be the reduction of suffering, caused by war and technological wrongs.

6. Different world views should not be seen as contradicting one another completely. In this respect world views are different from ideologies (although there are also obvious similarities). World views (as we conceive them) are in principle open to rational criticism, and should not contradict the known facts. World views may be partly overlapping, so that a responsible common action is still possible. Only then our cosmos is going to be also in the future a cosmos in which life is possible, also for the future generations.

AFFILIATION

Jan Van der Veken
Worldviews Group
Higher Institute of Philosophy
University of Leuven
Belgium

HUBERT VAN BELLE

SYMMETRY AND SYMMETRY-BREAKING IN THE PHILOSOPHY AND THINKING OF LEO APOSTEL

Leo Apostel was fascinated by the opposition between symmetries and the breaking of symmetries ('symmetry-breaking'). In his contribution to *Perspectives on the world* [1], entitled "Symmetry and symmetry-breaking: ontology in science" (An Outline of a whole), he considers this polarity to be the key to the understanding of the nature of global reality. According to him, the recognition of symmetries and symmetry-breakings or asymmetries as characteristics of totality offers unifying as well as specifying possibilities. He considered the 'unifying' pole of this opposition to be of great importance to the construction of world views, a project in which he was enthusiastically involved. Since Leo Apostel believed that the symmetries and asymmetries can be found on each of the levels of reality, he considered them to be fundamental characteristics of the world and as a final explanation. The aesthetical aspect of this explanation complied with his deepest desires.

SYMMETRIES AND ASYMMETRIES

In nature numerous symmetries appear to exist in often unexpected shapes. Symmetries of entities are transformations which, in certain respects, leave these entities invariant. Symmetries of a body are (rigid) movements of that body which leave it externally unaltered. A sphere is invariant for an arbitrary rotation around each one of its possible axes. A cube, too, appears to remain unaltered when it makes a 90-degree rotation around one of its axes perpendicular to the six faces. Perfect crystals look the same when they are rotated over certain angles or when they are looked at from certain angles. The human body is largely symmetrical for reflection of the left and right halves. It goes without saying that a human body has less symmetries than a sphere. The perfect symmetry of the sphere is broken up in reality.

Not only objects but also phenomena, processes and laws can display symmetries. A periodic wave is invariant to a time shift over one or more periods. The equations of motion of a pendulum are invariant to a reversal of time. Also of great importance is the conclusion that the laws of nature are invariant to the orientation of the measuring equipment, the place of measurement, the moment (time) of the experiment and the (constant) velocity of the laboratory. In other words, the laws of nature are not dependent on place and time. Without this remarkable characteristic of reality an inductive generalisation would be impossible. From the four rudimentary principles of invariance the basic laws of classic pre-quantum mechanics can be deduced: the conservation of momentum, the conservation of energy and the laws of the special theory of relativity. Thus, the bodies in the universe move according to laws which express the conditions for inductive knowledge!

In the real world perfect symmetries are rare. Even crystals tend to display a certain asymmetry. There also exist many irregular bodies with which each motion produces a different image. This also counts for phenomena and processes, especially for irreversible processes. These processes are not invariant to reversion of time. If a videotape of a smoking chimney is rewound, then one can see a phenomenon which does not answer to the laws of nature and to reality. Irreversibility can be brought into connection with the spontaneous increase of entropy of a closed system and the arrow of time. Consequently, the history of the universe is considered to be a succession of symmetry-breakings which, starting from a completely symmetrical situation, lead to an increasing asymmetry. If the possibility of spontaneous symmetry-breaking could be deduced from the nature of the symmetries displayed by the laws, then "time" could be deduced from "being". Time is then considered as an ordered series of successive symmetry-breakings. The irreversible processes "carry" time. Also causal processes can be considered as asymmetrical. If "a" is the cause of "b" then "b" cannot be the cause of "a": Experience also teaches us that the consequences cannot precede their causes. We cannot change the past, but we can (partially) change the future. Furthermore, science does not accept the future to account for the past.

BEING, SYSTEM AND CAUSE

Leo Apostel starts his argument from the observation that certain relations remain unaltered in reality, but that we are also surrounded by continuous change. According to him, the entirety of reality is characterised by invariability and change, being and becoming, characteristics which are strongly interconnected, but which also differ fundamentally. Their opposition and their alliance would unite the whole of nature. As a result, he asks himself what invariants and changes are and makes the connection with symmetries and symmetry-breakings. This leads to the claim that reality can be characterised by certain symmetries and by certain symmetry-breakings. He also asks himself what relation symmetry and asymmetry have to each other. Are they side by side, are they deducible from each other or do they originate from common ground? Leo Apostel chooses the last option, an ontological deduction. "Everything that is exists." It should therefore be possible to deduce the most general characteristics of totality (the symmetries on the one hand and the symmetry-breakings on the other) from the essence of existence as such, "being".

In doing this, he also makes a link from symmetry and asymmetry to system and cause. In the existing, invariants as well as irreversible processes occur. According to him, "being" also implies "system" as well as "cause". Everything that exists has an internal autonomy, an inner structure which renders it into a system and keeps it invariant in different contexts. Furthermore, it is systematically interconnected with other existents so that alterations of one cause alterations of the other. Systems are determined by symmetries, causes by asymmetric relationships. Causes influence the outside due to "forces" and generate consequences. In this "story" two families of opposing concepts appear. These are the family of "causality, symmetry-breaking, change, process and force" on the one hand, and "system, symmetry, structure, invariancy, constancy" on the other hand. Within this ontological vision, both families need to be generated by a common "entity", the notion of "being".

If this line of thought is correct, then it should be possible to conclude from the ontological point of departure why these and no other invariants and irreversible processes occur. This has to be at least feasible for the major types of invariants and irreversible processes. The deduction is fruitful if we can show that what we see has to be present in every existing totality. For example, each universe has to display gauge symmetries. These local symmetries allow for us to randomly "change" parts of a system, yet retaining the system in its global form. Without the existence of gauge symmetries the existence of systems as such would not be possible since only gauge symmetries preserve the global structure under local transformations. In other words, it is claimed that the whole shows itself in such a way that it cannot show itself any differently.

In all this, Leo Apostel does not limit himself to inorganic nature but also looks for examples in life and in humanity. In doing this he tries to bridge the gap between the world of the natural sciences and the world of the social sciences. He perceives, among others, symmetries and antisymmetries in the structure of DNA and symmetries in family relationships. Every kind of system is characterized by a number of symmetries and every process is characterized by a number of symmetry-breakings. The totality is both a system and a process. According to Leo Apostel, the basic symmetry-breakings must imply the basic symmetries and *vice versa*, the basic symmetries the basic symmetry-breakings. The link between the two follows from the mutual relation between system and causality. The coherence of the components of a system and the retention of its structure should follow from the causal interaction of the components. So the symmetries (systemhood) have to be deduced from the symmetry-breakings (i.e., the causal interactions). On the other hand an object can only work on and affect another object when both belong temporarily or enduringly to a common system. So the symmetry-breakings (causation) have to be deduced from the symmetries (systemhood). Leo Apostel assumes then that the symmetry-breakings that occur are those that maximise the degree of symmetry of the whole, and that the symmetries that occur maximalise the degree of asymmetry of the whole. Symmetry-breakings maximalise the global invariance. Symmetries maximalise the global process-innovation, emergence or creativity.

MEANING AND VALUE

His approach to the concepts of meaning and value is remarkable. He claims meaning to be that which gives our global personal existence and/or our collective existence a globally positive value. He describes value as something that at the same time includes a reference, displays an orientation, encloses a call for action (i.e., has an asymmetrical structure) and yet also perpetuates itself and "deserves" to be by what it is (i.e., has a symmetrical structure). Consequently, characteristic for value is the unity between the co-occurring symmetry and symmetry-breaking characteristics. Such a mutual implication has been realised in totality as well. Indeed, "being" simultaneously involves system-being and cause-being, while systems are determined by symmetries and causality by symmetry-breaking. As a result, total being has a positive value. This is also the case for our existence, which is part of "being" and which is simultaneously capable of reflecting the totality of that "being". The positive meaning of totality does not exclude the fact that certain parts of it have a negative value. One should not lose sight of the tragedy, the pain and the evil. Our value-experience encloses a revolt against that part of the existing reality.

The possibility to deduce laws of nature from qualitative principles like invariants inspires philosophers who are searching for an ontological basis. Leo Apostel was looking for an answer to the question concerning the ultimate structure which shows the totality and the fundamental characteristics of evolution. He found this answer in symmetries and symmetry-breakings. These also offered him a basis for an action theory which on the one hand takes into account the traditions and on the other hand offers the possibility to breach these traditions. On the 24th of July 1995 he wrote the following in a rather mysterious letter:

It appears to me that I should write a text in which I should pose "Time and Ethics" as the central theme (promises as typically human; and also the breaking of promises in order to keep other—deeper—promises; loyalty and remembrance, anticipation as a condition for life in time and transcending time; and the religious life as an attempt to, situated in time, experience time as a whole, without intervention and with surrender).

He further explains this by referring to a recent period of meditation in Tiltenberg:

Everybody is struggling with one or more personal problems which are close to his kernel. In my case this definitely is time, variability, the slipping by, transitoriness and, beside this, the (impossible) desire to bind time, to bridge gaps. This theme fulfilled my meditation. What should one do if nothing that one can do helps? Now you know that I actually wanted to keep my Zen separated from the construction of world views. However, at the end of the sesehin it seemed as clear as day to me that this ethical-existential problem was at the centre of my "Perspectives"-article. Considering symmetry (stability) and symmetry-breaking (becoming) as mutual conditions for each other = keeping time, and yet moving along with the stream.

AFFILIATION

Hubert Van Belle
Worldviews Group
Belgium

REFERENCES

[1] **Apostel, L.**, "Symmetry and symmetry breaking: ontology in science". in: **The Worldviews Group**, *Perspectives on the world*, Brussels, VUBPRESS, 1995, pp. 175-217.

LEO APOSTEL

WHY NOT NOTHING?

1. THE BELIEF IN EXPLANATION

"Why" questions (which express a need to find explanations) are asked in countless situations. Why were there patches of mist in the early universe? Why did the Roman Empire decline and fall? Why do bees prefer a hexagonal honeycomb? Why did the scientific revolution take place in Europe? Why do physical constants have the values they do?

We seek an explanation for all facts, events, processes, structures, things or systems that we consider to be real. Naively and spontaneously, we believe that reality is rational, and that there is an explanation for everything. Optimistic individuals assume that mankind will find all the explanations; less optimistic thinkers assume that we will come close to finding the answers, while some people (who have become suspicious as a result of failure) resign themselves to the fact that mankind, faced with insurmountable obstacles (which they define in different ways), will not and cannot find certain explanations (even though they exist). The radical belief in the explicability of reality is expressed in the principle of sufficient reason. "For everything that exists, there exists a sufficient reason, which explains why it exists".

The key concepts of this belief are "explanation", "existence" and "reason". Important chapters in ontology, epistemology and logic provide an analysis, albeit a controversial one, of these three concepts. In this article we shall not concern ourselves primarily with the analysis of the three basic concepts. They will come up at various stages, but we shall not worry about them here. We shall concentrate our attention on the question formulated in the next section.

2. THE MOST UNIVERSAL FACT AND THE THEORY OF NON-EXISTENT OBJECTS

The most universal thing we know is that "Something exists". This fact does not refer to the existence of a particular thing or person, but to the existence of something. It is not the fact that stars, living beings or conscious human beings exist that interests us if we emphasise this point, but purely and simply that something exists. Moreover, we are not saying anything specific about that something; we are not attributing it with any property or relation, we are merely saying that it is real, that it exists.

A short, symbolic translation conveys the extreme universality of the meaning that this fact describes.

(1) (Ex) E!x

x is a variable for individuals.

Ex is an individual quantor (an operator who chooses an element from a given domain).

E! is the name that logicians who concern themselves explicitly with existence, have given to "existence".

(1) means "There is an x, whereby this x has the property that it exists".

We cannot convey this most universal fact if we cannot also say:

(2) (Ex) $\sim$ (E!x)

(2) means "There is an x, whereby this x does not have the property that it exists".

In our way of thinking, we can rise above existence. The domain in which the quantor (Ex) chooses his objects is broader than that of real entities. We can talk about fictitious things and processes. Some of these are possible, others are impossible; some of these are individual, others abstract. The logician Alexius von Meinong, who thought most clearly about "fictions", suggests that in contexts such as the one mentioned above the quantor "Ex" should be allowed to choose from the domain of conceivable or even imaginable objects, which he calls "subsistent", while actual, real objects are "existent". This proposition implies that we can construct a theory of non-real objects alongside a theory of real objects. Several contemporaries, including Charles Parsons and Richard Routley, have done that [1], [2], [3]. This is worth mentioning because without these constructs we would trivialise the most universal fact "Something exists" since, if the variable x can only vary among real individuals, it follows that all of its particular values exist. One immediately has to conclude that there is no individual who does not exist: all individuals exist.

(3) (Ax) E!x

(because the statement (Ex) $\sim$ E!x becomes a contradiction according to this interpretation of the individual variable x; x only varies over existing objects.)

It is interesting to note down some axioms that might characterise "E!".

1. What exists, is consistent.

2. What exists, is complete (in the following meaning of the word: for each pair of predicates of the form (P, -P) that can belong to something, that something either has P or $\sim$ P.

3. If something exists, all its constituent parts exist too.

4. If something has external relationships, it exists.

Two debatable but interesting axioms are:

5. Nothing exists with necessity.

6. It is necessary that something exists.

To our knowledge no theory has been expanded that examines the implications of these axioms.

It is not necessary to go into any more detail here about the logic of non-existent objects. We should emphasise, however, that this apparently academic observation about non-existent objects is consistent with a fundamental characteristic of our thinking. We understand reality on the basis of possibility and necessity. Only by distancing ourselves from the facts, by recognising that other facts could have occurred instead (i.e., by surrounding the facts with a "cloud" of possibilities), but did not in fact occur, because what we encounter in reality becomes relatively necessary as a result of other facts that also exist in reality, only then can we understand and explain. Only by comparing with what could have been, but is not, and thus gaining insight into what must be necessary, can we understand what is.

3. THE EXPLANATION OF THE MOST UNIVERSAL FACT

If we apply the principle of sufficient reason to the most universal fact, we have to assume that sufficient reason exists for the most universal fact. In other words, something explains why something exists. To put it another way, one can and must ask "Why does something exist?" (i.e., Why does anything exist?), and there is an answer to this question.

If one asks "Why does something exist?", one expresses his surprise about existence as such, and one does that because one thinks that an alternative was possible! In this case the alternative can only be: "It is false that something exists". Hence nothing exists.

(4) $\sim$ [(Ex)E!x] or (equivalent to 4):

(4') (Ax) $\sim$ E!x

The title of our article "Why not nothing?" uses one and four together and asks for the explanation of it. "Why is (1) true?" (i.e., "Why does something exist?") and "Why is (4) false?" (i.e., "Why does nothing not exist?").

If "Something exists" is true, it of course follows directly from this that "nothing exists" is false. Suppose, however, that it was proved that "nothing exists" is false. Does it simply follow from this that "something exists" is true? Only if one assumes the principle of the excluded third in the following form.

(5) (Ex)E!x or $\sim$ [(Ex)E!x]

It is known that intuitionist logic does not apply this principle of the excluded third limitlessly. More particularly, for reasons that do not concern us here the principle of the excluded third is not applied to infinite, non-constructively defined sets. It is obvious that the set of conceivable entities from which the existential quantor chooses, is not constructive and infinite. Hence intuitionism will not subscribe to the principle in form (5). It follows from this that we have to keep asking the two subquestions since no answer to the first question can be deduced from an answer to the second, at least in intuitionist logic. It is not certain, however, that we should apply this logic here. The invalidity of (5) in one of the important logical systems does present an argument for continuing to deal with the two subquestions separately.

Three basic concepts crop up in the first part of the complex question: "why" (explanation), "existence" (reality) and "something" (an individual variable). Basic concepts also feature in the second part: "why", "existence", "not" (denial, negation) and "something". The term "nothing" can be reduced to negation (as demonstrated in 4). The second part of the universal "why" refers to all the concepts to which the first part refers and, in addition, to negation.

One might think that one could choose to answer the question "Why does something exist?" without bothering about the more complex question "Why does nothing not exist?". But we are only asking for explanations in the awareness that the facts that we are surprised at could have been different. We are only asking why something exists at all because we believe that it was possible that nothing existed. In fact our question concerning the explanation of the existence of "something" is based on three convictions:

(1) (Ex)E!x: something exists (5)

(6) M[$\sim$ (Ex)E!x]: it is possible that nothing exists.

(7) If V denotes "explanation" and "q" denotes an arbitrary judgement, then the third conviction is the one that postulates the question

(Eq)q is true as is qV(Ex)E!x and $\sim$ (Ax) $\sim$ E!x.

4. THE MOST UNIVERSAL "WHY" QUESTION AND THE PRINCIPLE OF SUFFICIENT REASON

Anyone wishing to confirm the principle of sufficient reason must necessarily ask and analyse the question in (3) and seek an answer to it. The philosopher who first posed the question with great force, G.W. Leibniz, was—as we know—a confirmed believer in the principle of sufficient reason.

It is important to understand that one can have reasons for asking the question about the explanation of the most universal fact without subscribing to the principle of sufficient reason. One can ask and answer a specific (even, as here, an extreme, ultimate, ultra-universal) "why" question without claiming that all "Why questions" have to be answerable. Having made this reservation, we shall presuppose the principle of sufficient reason in the remainder of this discussion.

5. THE MOST UNIVERSAL QUESTION POSED BY LEIBNIZ, AND HIS ANSWERS

The question "Why something and why not nothing?", however natural and legitimate it may be, was clearly articulated for the first time in philosophical literature in 1697 by G.W. Leibniz [4]. Leibniz's basic text can be divided into two parts: the first three sections and sections 8-17 on the one hand, and sections 4, 5 and 6 on the other. In most of the text the explanation of the existence of something is given by means of a very conventional argument for the existence of God.

1. The world—whether finite or infinite in time—is made up of entities, the nature of which, their essence, does not imply their own existence.

2. One cannot conclude, a fortiori, the existence of the totality of the world as it is, or the existence of something, from a finite or infinite set of such entities.

3. Since there must be a reason for the world, that reason can only be found in an essence outside the world, whose very essence implies existence, and which must be the sufficient reason for the world (section 3).

In the shorter part of the text another explanation is given, which we would like to quote in full here since it is the first explicit explanation of the most universal fact that we have come across.

"Par celà même qu'il existe quelque chose plutôt que rien il y a, dans les choses possibles, c'est-à-dire dans sa possibilité même ou essence, quelque exigence d'existence—l'essence tend elle-même à l'existence. D'où il s'ensuit de plus que tous les possibles—tendent à l'existence d'un droit égal selon leur quantité d'essence, ou de réalité; selon le degré de perception qu'ils enveloppent; la perfection n'est rien d'autre, en effet, que la quantité d'essence" (section 4).

Section 5: "Par là on comprend très manifestement que parmi les infinies combinaisons de possibles et séries possibles, celle qui existe est celle par laquelle la plus grande quantité d'essence ou de possibilité est amenée à l'existence" [4, p. 340].

Like many philosophers in the Middle Ages, Leibniz distinguishes essence from being (the "being as it is" from "being"). He assumes that there are essences (which he equates with possibilities), in other words he does not pose the more fundamental questions "Why do possibilities exist?" and "Why do essences exist?". He assumes that they are affirmatively resolved.

He attributes these "possibilities" or "essences" with a "tendency" towards existence. Furthermore, he assumes that these tendencies can differ in intensity and that the intensity depends on the degree of perfection of the possibilities or "essences".

We know from other texts that for Leibniz the degree of perfection corresponds with a function of complexity and integration. We can conclude from this construct that for Leibniz the explanation for "Something exists" must be found in the fact that for at least some possibilities or essences, the tendency towards existence, alone or in combination with one another, is strong enough to be realised.

We have quoted these two parts of Leibniz's pioneering efforts because there is still an echo of his second answer to the most universal explanation question in the works of Robert Nozick, John Leslie and Nicholas Rescher [5], [6], [7]. First, however, we would like to demonstrate that the answer to the question cannot be accepted by the argument for the existence of God.

Either the existence of the world follows necessarily from the existence of God, or not. If the world necessarily follows from God, this world is part of His essence and must be deduced from that essence. In this case, we shall only have an explanation for the most universal fact if we can really deduce God's being from His essence (1), and if we can deduce the essence of the world from God's essence (2).

This dual task has never been undertaken by anyone. In any case, it is in conflict with every religious experience. Only if it is brought to a successful conclusion would the question "Why does God exist?" (a definition of God as a necessary essence provides no explanation either of His necessity or of His existence) be answered, and consequently also the question "Why does something exist?". If, on the other hand, the world cannot be deduced from God with necessity, then the existence of God does not provide any explanation for what that world is, nor for the fact that He exists. The fact that Leibniz refers to God is based on non-cogent reasoning. He proceeds from the fact that even an infinite set of essences, whose being is not implied by their nature, would not be able to explain themselves. What is more, he advances from the premise that no part of the world explains its own existence or that of the totality, to the conclusion that no property or structure of properties of the whole world would be able to explain the existence of that world or of its parts. This transition is not justified by any logical rules. We shall see that in contemporary philosophy it is precisely this last theory that will be tested.

Taking into consideration that the explanation of the most universal fact through God does not hold water, if one refers only to Leibniz's first schema, then we can see:

1. that several degrees of existence have to be introduced into such a schema. Possibilities that do not exist strive to exist with varying degrees of force. These degrees of being (which will be infinite in number according to Leibniz's own metaphysics) demand an explanation time and time again. "Why does something exist?" is divided up into an infinite number of sub-questions (in the form: Why does something exist to the nth degree?)

2. This multiplication of degrees of being does help us understand the distinction between "being" and "essence", however. The distinction seems conceptually correct to us: the question "Why is there something and not nothing?" differs fundamentally from the question "Why is the world thus and not otherwise?". Certainly these questions, which have almost the same degree of universality (the second is, however, more specific), can be dealt with in relation to one another (one can give priority to the question about existence, or to the question about essence, or else treat them both equally). As soon as one distinguishes between "being" and "essence", in reality the following question arises: what is the essence of being as being and what is the being of essence as essence? These questions can and must be answered; they lead not to a vicious infinite regressus, but once the idea of degrees of being has been established, one can deal with them more easily than if one treats the distinction between being and not being as discontinuous. These comments are not intended as real arguments in favour of postulates; their sole object is the development of Leibniz's first schema (of which—as mentioned—we shall find echoes).

3. As Rougier [8] demonstrates, a number of scholastics who subscribed to the real distinction between essence and existence have nevertheless proposed multiple ways-of-being that culminate in the assertion that

there are as many essences-of-being as there are essences. This multiplication of the concept of being would not prevent Leibniz from applying his first explanation schema, which depends only on the strongest tendency towards existence of possibilities.

6. OPPOSITION TO THE MOST UNIVERSAL EXPLANATION QUESTION

Why does something exist? Why not nothing? Why do possibilities exist? These questions are quite legitimate, but we should realise that they are of an extreme, an ultimate nature. The answers that we might put forward will be strange and unusual. That is why these answers can and will repel us and leave us unsatisfied. This cannot and should not prevent us from taking these answers seriously. How fundamental the questions may be, until now they have rarely been dealt with constructively. The vast majority of philosophers have dismissed them as meaningless. The tiny minority who regarded them as acceptable have treated them as unsolvable mysteries. The first group includes Kant, all neo-Kantians and all positivists or neo-positivists. The second group includes Heidegger and Wittgenstein and all their followers (few in number) who still raise the most universal explanation question [9], [10].

We, on the other hand, regard the questions as acceptable and—in principle—as partially solvable. The reason why we think they are acceptable is clear from what we have said: we should seek the explanation for both the most universal fact and the most specific fact. Our curiosity and our faith in the comprehensibility of reality compels us to do so. The history of civilisation has shown time and again that anyone who tries to place limits on human thinking and prohibit particular problems, has been proven by history as mistakes. Comte thinks that one cannot think positively about the chemical composition of stars and then helium is discovered in sunlight by spectroanalysis. Kant declares that one cannot talk about the finiteness or infinity of the world without contradictions and then Einstein demonstrates in 1915 that it is an empirically solvable problem. Accepting that limits be placed on human thinking a priori and that particular aspects of reality are essentially unknowable, presupposes a static, conservative attitude towards thinking and leads to an abdication of the will to knowledge that can have potentially self-destructive consequences. We think that the belief in reason and science is fundamentally inspired by the belief in ontology and metaphysics. We therefore accept the questions, but "the proof of the pudding is in the eating" as they say. The only way we can show that they are solvable is by putting forward partial solutions and comparing them. In this way we hope to contribute to the research tradition.

Before proceeding we can at least refute two pseudo-proofs that are constantly put forward when our questions are asked.

1. First Absurdity argument

Anyone seeking an explanation for a fact must have an explanation rule and an explanation basis. We wish to know why something exists. As explanation we can only offer another something (which also exists) or, conversely, something that does not exist. In the first case, however, the "existence" of the explanation basis must be assumed. Then we no longer give a universal answer to the question about the explanation of the most universal fact. In the second case we have to deduce the existence of something from absolute nothingness. This involves us in a contradiction.

2. Second Absurdity argument

The sufficient reason for a fact is its cause. The cause of the fact that "something exists" cannot, however, be found since the cause itself must already exist.

Our answers to these two arguments are as follows.

The second argument is based on the identity of "reason" and "cause". The explanation for each partial fact is indeed its cause (together with the forces that this cause exerts and the laws according to which these operate). But the "reason" for either the total fact (the world) or the most universal fact (the existence of something) cannot be a cause since the total fact already encompasses all causes, and since the most universal fact is not an occurrence that is localised in space and time (one that can only have causes). In order to pose our questions, we therefore have to provide an analysis of "explanation" and "reason" that distinguishes them from "cause". This is a high price to pay, but an affordable one. In every respect we find here a reason for the "strangeness" that will be characteristic of every explanation of the most universal or the total fact. The refutation of the first absurdity argument forces us to come to even stranger conclusions.

If we deduce "something exists" from properties of "the existent" itself, we have to explain something by means of an internal property or structure of a property of it. A few possible examples will clarify this point: we deduce from the fact that laws apply (1) or from the fact that explanations exist (2) or from the fact that judgements are true (3) that "something exists". This is analogous to Kant's deduction of concrete ethical rules from the formal conditions for every code of ethics. There are a great many other possible bases. We can also assert that it follows from the existence of a world that it cannot be empty (and hence must encompass existence) (4). The list is not an exhaustive one and should be put into some sort of order. Explanations like this are exceptional and almost circular, but not impossible.

If we deduce "something exists" from "nothing", we should not define it in such a way that it cannot bring forth anything other than—or even opposite to—itself.

The strategies we are forced to adopt are—and this also fits in with our exceptional question—themselves exceptional. We have to actually apply them to see whether and to what extent they can work.

7. FIRST PROPOSITION: "NOTHING EXISTS" IS LOGICALLY IMPOSSIBLE

Note: we do not assume that this proposition will work, but it is the first one that is put forward.

If nothing exists, then all the models of all possible theories are empty, including the models of all logical theories. This has all kinds of strange consequences!

A few examples will illustrate this point:

(8) (Ex)E!x implies (Ax)E!x

(9) (Ex)E!x implies (Ex)-E!x

(10) (Ex)E!x implies [(Ex)E!x]

(8) is true if "implies" is used in the sense of material implication since (Ex)E!x is always false. (9) is true for the material implication since (Ex) $\sim$ E!x is true. (10) is true since the antecedent is false and the consequent is true. These consequences occur if one uses "the material implication". If we assume that reality (sic) is not only actually empty but also necessarily empty, if we state that the antecedents of 8, 9 and 10 are "necessarily" true, then the paradoxical properties continue to apply (for the most "strict" implications).

Paradoxes also occur in systems for relevant implication. To summarise, if we say that a theory is contradictory if it has no model, it follows from the fact that "nothing exists" that no models exist and that, in that sense, all theories are contradictory. If we demand that in some "mysterious" way or other logical laws are universally valid, we have to preclude the postulate that all models are empty. This would imply that "something exists".

The scope that we can attribute to the logical impossibility that "nothing exists" depends, however, on the status that we accord to logic. We can choose from various systems of logic. Some of these will not include the paradoxes of material and strict implication that have been used here to demonstrate the necessity that "something exists". The "logical" approach only prompts us, in an open quest, to search for the logical paradoxes that in a given system of logic result from (Ex)E!x being false. If a system of logic is based on conditions for efficient problem-solving, or on conditions for optimum or contextually efficient dialogue, the logical impossibility of nothingness follows from the fact that there is no search and no dialogue in nothingness. This trivial insight does not satisfy our search for "Why not nothing?".

Nevertheless, the attempt to prove that "Nothing exists" is logically impossible suggests a more general strategy: we choose a set of laws, for which we have reason to believe they apply in all possible worlds. These laws do not have to be logical laws. Then we demonstrate that they do not apply in an empty world (for which "something exists" does not apply). We deduce from this that no possible world can be empty.

Some modern physicists have attempted to prove that the laws of nature are logically necessary. Geoffrey F. Chew [13] says that "nature is as it is because this is the only possible nature consistent with itself". This programme has not been successful so far but if one were to succeed in implementing it, one would—by showing that the world should be thus and not otherwise—have immediately demonstrated that it must be, and not, not-be.

8. SECOND PROPOSITION: "NOTHING EXISTS" IS TOTALLY INEXPLICABLE

If the principle of sufficient reason applies in all possible worlds, then every true judgement in a possible world must be explicable. In an empty world the true judgement that "nothing exists" certainly applies. This true judgement must be explained. But how?

1. "Nothing exists" does not explain itself.

2. Other facts that might be considered for the explanation of "Nothing exists" are not found in an empty world.

3. The structure of an empty world does not itself exist either and cannot be used as the basis for an explanation.

Hence, in an empty world, "Nothing exists" is a brutal and inexplicable fact. If one accepts the principle of sufficient reason, an empty world is an irrational world.

And we can go even further. Robert Nozick has proposed an infinite series of ways to weaken the principle of sufficient reason. SR1 would be: every fact has either a sufficient reason, or a sufficient reason for the meta-fact that it has no sufficient reason. SR2 then takes the following form: every fact has an SR, or an SR1 for the meta-fact that it has no SR, or an SR3 for the meta-fact that it has no SR2.

This series can be extrapolated to infinity.

It now appears than not one of this infinite number of ways of weakening an SR principle can apply in an empty world. Does this satisfy us? Everything depends on the status we accord to the principle of sufficient reason.

To give one example of this dependency: does the validity of the SR principle itself have a sufficient reason? If it does, then the incompatibility of the SR principle with an empty world is a powerful argument. If it does not, then this incompatibility is merely an incompatibility between two brutal facts.

To summarise, in an empty universe nothing can be explained, not even that it is empty. The simple demand that in a world it must be possible to explain something precludes the empty world. This is a much weaker demand than the full principle of sufficient reason, and hence a much stronger preclusion.

9. THIRD PROPOSITION: "NOTHING EXISTS" IS INFINITELY IMPROBABLE

In section 8 we applied the principle of sufficient reason. Its negative counterpart is the principle of insufficient reason. We are familiar with this from the fundamentals of the probability theory. "If n possibilities are given and there is no sufficient reason to choose one of these in preference to another, then they are all equally probable". If "e_1 or e_2 ... or e_n" has a probability factor of 1 (or is even necessary), and if there is no SR to make e_1 preferable to e_2, all e_1 will have a probability of 1/n. We shall apply this to the set of possible worlds. In the most unfavourable case for the refutation of "Nothing exists", we place the empty worlds (which we shall call L) on one side and the not empty worlds (which we shall call B) on the other side. We consider the two sets as a unity and assume there is no reason to choose empty in preference to not empty, or not empty in preference to empty. Assuming that there have to be "worlds", then L has a probability of 1/2 and B too. This is an extreme case, however. There is only one way of being "empty", whereas there are many ways of being not empty. We can therefore assume that there are 1, 2, 3, 4, ... n, ∞ different non-empty worlds. In each of these assumptions, assuming that we have no reason to suppose that one of these worlds is more probable than another, the principle of insufficient reason determines that the probability of L always decreases (from 1/2 to 1/n) and that the probability of B always increases (from 1/2 to 1/3+1/3 to 1/4+1/4+1/4 etc.). If an infinite number of possible worlds exist, the probability of L will approach the limit zero. We shall now apply the principle of insufficient reason to the totality of all these distributions of probabilities over possible worlds. In one distribution L has a probability of 1/2. In all the other distributions L has a lower probability, and in an infinite number of distributions the probability of L is arbitrarily close to the limit zero.

All this follows if we can apply the principle of insufficient reason both to the individual 2, 3, ... n possible worlds and to the assumptions about the number of possible worlds, and if our ignorance or ontological indifference (or both) are so great that there is no known and/or existing reason to choose one world in preference to another. In concrete terms, all the foregoing follows from the fact that being can offer more variety than nothingness.

We do not have to make all the assumptions mentioned if we simply demand: the probability of a world increases with its diversity and decreases with its uniformity or monotony. It follows immediately that an empty world is the least probable world. We can even provide a "probabilistic" justification for "something exists" in another way. Let us assume that (in stark contrast to the postulate - OU (organic unity) and the principle of sufficient reason) the real world is as contingent as possible. Every fact can exist, but can also not exist, and the facts are so few in number and are as loosely interconnected as possible. One has to think that such a world is the result of a coincidence. Once such an assumption is accepted, it quickly emerges that the empty set is such an exceptional and special set that the chances of obtaining it, as the result of a coincidence, as a model of reality are either zero or negligible.

A cosmology that relies on coincidence and chaos permits the conclusion that "something exists" just as well as a cosmology that relies on determinism and order. If we compare this third proposition with the previous one, we can see in section 8 that the empty world has a minimum degree of unity and in section 9 that the empty world has a minimum degree of diversity. This immediately takes us back to Leibniz, who equated the degree of perfection of a possibility with the extent to which it combined maximum unity with maximum diversity. It follows from sections 8 and 9 that the existence of something is necessary to give reality the necessary unity (8) and diversity (9). Without introducing concepts such as "perfection" and "the striving of possibilities for self-realisation", the results achieved are not identical to those of Leibniz, though they are related.

Here a warning is in order. It is much too early to claim victory. In sections 8 and 9 it was deduced that "something exists" must be true from the fact that something has to be explained (or that everything has to be explained) and from the fact that something that can be realised in many ways has more chance of being realised than something that can be realised in just a few ways. However, we would only be satisfied with these arguments if the premises on which they are based were both ultimate (i.e., no further explanation needed) and self-explanatory and fruitful (i.e., with a plethora of relevant consequences for the rest of reality). The task therefore remains to demonstrate that the demand for explanation, the principle of sufficient reason and the principle of sufficient diversity are ultimate, self-explanatory and fruitful.

10. FOURTH PROPOSITION: "NOTHING EXISTS" CANNOT BE RECONCILED WITH THE ORGANIC UNITY OF A WORLD

Our question is related to the question: can a possible world be empty? So far we have not asked ourselves what meaning is given to "world". Let us now rectify this omission. A world is a set of objects that are interconnected through relations in the following manner:

a. All objects are systematically connected with other objects: the facts that correspond with true judgements about these objects are deducible from one another.

b. A world stands on its own: it constitutes its own reason and can be explained on the basis of itself.

c. A world is complete: no objects can be added or removed without contradiction.

A world is an organic unity if all facts in this world are explained by the way in which they are connected with the rest. If a world is an organic unity, no facts can exist that are more fundamental than others: there are no axioms or consequences, no fundamental facts or superficial facts. All facts presuppose one another and can function as premises and consequences in turn.

Every world is necessarily an organic unity since only an organic unity can explain itself. If the facts were interconnected (i.e., have the property a) on the basis of a hierarchy of facts, some of which are fundamental and others deduced, then the fundamental facts that explain the rest would themselves remain without explanation, thus rendering the whole system inexplicable. This contradicts the idea of the autonomous existence of a world.

If possible objects arrange themselves in sets that have the property of being "worlds", these sets also have the property of being "organic unities". Why would these sets of possibilities arrange themselves in this way? The fact "The world w is an organic unity" must, if one applies it to itself, also be explained by its relations with the other facts of w. That fact "fits in perfectly" with the other facts of w because it connects all facts—and also itself—perfectly with all other facts. The principle "the world w is an organic unity" is thus a principle that explains itself according to the schema.

(ZV): All laws with a property C are true. We shall call this judgement P.

Judgement P has the property C.

Hence P is true.

Applied to OU:

All facts that are perfectly interconnected in OU are real.

This is fact F.

Fact F is connected perfectly with the other facts.

Hence F is real.

All possible worlds are organic unities because isolated possibilities could not be in autonomous existence, since they could not have any explanation. An organic unity cannot be an empty world since in an empty world all facts are isolated and cannot be interconnected. It follows necessarily from the analysis of the concepts "world" and "organic unity" that no possible world can be empty. This fourth proposition is an elaborated version of the second proposition (which partially overcomes the arbitrary nature of the principle of sufficient reason).

Two extremely important consequences follow from the concept "Organic Unity":

1. The non-hierarchical nature of OUs underpins an invariance principle: let V be the explanatory relation in a world w. The facts of w are expressed in relation to V (i.e., $V(f_1\ f_2\ f_3 \dots f_n)$: they explain one another in that order). Let us perform a permutation on $f_1 \dots f_n$. If the world w is not hierarchically arranged, V remains valid for all permutations of the f series. The organic unity of a world implies the validity of invariance principles in that world.

2. The demand to explain everything, and still find a self-explanation for every world, implies—as we have seen—that principles that have themselves as special cases must be true. This means that an invariance "$B_1 V B_2$" is as such that the permutation "$B_2 V B_1$" leaves the principle valid.

Robert Nozick [5, p. 142] notes that "symmetry is a mark not of the truth of an explanation, but of our understanding a phenomenon. This requires further investigation". On p. 135 he states: "It will not be arbitrary that this principle (leaving aside for the moment which principle he is referring to) holds if it satisfies some deep invariance principle". In the preceding analysis we have demonstrated why symmetry and invariance are of such crucial importance in answering the most universal question about the "why" of the most universal fact.

The relationship between this "consistency theory of reality" and the idealistic "consistency theory of the truth" cannot be denied. The traditional objection to these consistency theories of the truth has been that very different sets of judgements could be consistent. Thus, here, very different sets of possibilities could be organic unities.

We either have to reconcile ourselves to this radical plurality and confirm that they all exist (1), or introduce one radical contingency: the selection of the organic totality that exists (2), or else deny the radical plurality in order to differentiate degrees of organic totality and kinds and thereby introduce a maximisation principle (the real world is the most organic of all possible worlds) and/or a qualitative selection (the kind of OU in the real world explains why it exists and the alternative possible worlds do not).

Since a degree and form of symmetry and invariance corresponds with every degree and form of organic totality, a programme is thus outlined for the study of forms of symmetry and invariance.

11. FIFTH PROPOSITION: "SOMETHING EXISTS" BECAUSE ALL POSSIBLE WORLDS EXIST

Foregoing propositions have already referred to the symmetry and invariance under transformations within a world. It is an obvious step to ask oneself whether the set of all possible worlds does not also satisfy invariance and symmetry principles. One of the most radical invariances that is so often cited is the "principle of fecundity or plenitude". "Existence" is invariant when each possible world is replaced with another. "The principle of fecundity is an invariance principle" [5, p. 130]. All possible worlds exist. If one can prove that a non-empty world is a possible world, it follows that this non-empty world exists, whereas, if one can prove that an empty world is a possible world, it follows that it exists as well. One can clearly link invariance theory with modal logic.

The theory of Saul Kripke is well known. Possible worlds are connected with themselves and with one another through "accessibility relations". The expression "$w_1 R w_2$" means that w_2 is accessible (we shall leave aside the meaning of accessibility) from w1. Kripke has constructed his elegant semantics for modal logic out of various postulates for the accessibility relation. If one assumes that R is simultaneously symmetrical, transitive and reflexive, then R is an equivalence relation. It follows that R is invariant under whichever permutation of possible worlds. The properties of R are expressed in Lewis' S5 system. This structure (which is generated by a permutation group) conveys the same basic intuition as the plenitude or fecundity principle. The existent world has no precedence over the possible worlds.

In view of the previous sections, we can conclude that empty worlds are not possible. Does it follow from this that at least one non-empty world is possible (and hence (!) is real if we accept the principle of plenitude)? Only if we apply the principle of the excluded third to possible worlds (we have realised that this is not without difficulties).

One can also cast doubt on the cogency of the preceding arguments and make a radical new start. Our basic assumption is that existence has no priority over non-existence, nor does non-existence have any priority over existence. Neither existence nor non-existence are the "normal", privileged state from which the other deviates. For all possibilities it holds that their existence should not be given any preference over their non-existence. The principle of insufficient reason deduces from this that they will both occur and not occur. It follows from this complete discontinuity or rift between being and essence that all possibilities will be realised. This discontinuity also implies that every possibility can be realised, and hence is realised (in accordance with the principle of plenitude). Not all possibilities are "compossibilia", however. Why should we accept the principle of plenitude? Robert Nozick puts forward two arguments.

1. In our existent world, laws are invariant under a number of transformations (think of the GRT!). Thus one can also think of "existence" as invariant under a number of transformations (of laws and of individual world histories). The most radical invariance is one that regards all possible legal systems and world histories as existing. This argument is very weak. The fact that within one world properties remain invariant under transformation provides only very feeble reasons to suspect that properties (such as "existence") would remain invariant during the transition from one world to another. Even if this were the case for some properties, it does not follow that this would hold for "existence".

2. The second argument attempts to apply the principle of plenitude to itself. One treats a union of possible worlds as a possible world. If one permits oneself such a "move", a meta world in which all possible worlds are real is itself a possible world. If one accepts the principle, one must ipso facto accept that this meta world exists. Robert Nozick [5, p. 131] understands that this cannot be: a meta world in which only two possible

worlds exist is also a possible world (according to this view) and hence must also exist. The earlier conclusion and this last statement give rise to a contradiction. Hence, one cannot regard unions of possible worlds as a possible world (this actually follows from the fact that these unions are not organic unities).

We therefore have to reject the two arguments and can only deduce the principle of plenitude from the total contingency of each existence, the rift between essence and being mentioned earlier, combined with the principle of insufficient reason.

This basis is not completely solid either. Clearly we need a more profound analysis of "existence" or "being". We shall return to this point later. If we pursue this analysis, we have to direct our attention primarily to the trans-world invariants.

12. SIXTH PROPOSITION: "SOMETHING EXISTS" BECAUSE IT IS BETTER THAT SOMETHING EXISTS THAN THAT NOTHING EXISTS

We have repeatedly referred to "possible entities" and "possible worlds". Universal laws apply in every realised world for the relations between the entities realised in such a world. However, there are also possible entities outside the realised world. One can (!) assume that these entities do not remain arbitrary, but rather obey a higher order of laws, which we (with Nicholas Rescher) call protolaws, and which convey necessary conditions that entities have to fulfil in order to exist (not laws governing beings but laws governing the process of coming into being). The idea assumes that all possible worlds share a certain structure, one that is not logically necessary, but that has to be realised in every possible world that can be realised. It is not apparent why this assumption should be true, but it is certainly not absurd. This common structure, which is ontologically necessary, will certainly also be realised in the "real" world.

If one could prove that every possible world that has this structure is not empty, it follows that the real world is not empty either. Generally speaking, a plurality of worlds will therefore exist (but not all). It is now no longer based on the principle of plenitude, but on a hypothetical property of the hypothetical protolaws [7, p. 31].

Rescher owes us two arguments: 1) proof of the existence of protolaws and 2) proof that all possible worlds that fulfil the protolaws are not empty. This does not reduce his approach to nothing, however: the idea of a structure of possibilities (a) and of the ontological implication of that structure (b) is implicit in all the previous propositions. Rescher does not provide the proof required. Instead he says: "The most plausible candidates for protolaws ... are the fundamental principles of physical nature ... the basic cosmic equations (say the field equations of general relativity" [7, p. 33]. The next page seems to contain the fundamental equations of quantum electrodynamics.

These laws, however, have been discovered to be fundamental laws of the actually realised nature. If they were to be protolaws, it must be possible to demonstrate that they are conditions for existence (and not just properties of particular being). In order to demonstrate this, one would at least have to prove their invariance under a number of transformations that are ontologically justified. The invariance programme comes up once again. Yet the ontological justification, which is so necessary, is not even described in broad outline. If we—quod non—could successfully perform this task, we would also have to prove that "the fundamental equations themselves admit of no empty solutions—that any solution that satisfies them must incorporate the sorts of singularities we call "things"" [7, p. 34].

If we were to succeed in this attempt, the fact that "something exists" would follow from the protolaws. Why should we accept the protolaws? In view of our earlier propositions, we would think that the protolaws are either conditions of the possibility of explanation and/or of regularity, or guarantees of a maximisation of explanation and regularity, or conditions for organic unity, or conditions for the evolution of existence as a random process. We therefore have no shortage of candidates to explain any protolaws.

Rescher goes in a different direction, directly inspired by Leibniz. The protolaws owe their privileged status to the fact that they give the world a global value, greater than the value that it would have in any system that does not fulfil the protolaws. The values from which the global quantity is maximised are not aesthetic or ethical or vital values [7, p. 48]. They are intrinsic properties of legal systems and of worlds. Rescher [7, p. 49] mentions: simplicity, uniformity, systematicity, elegance and harmony. This raises further questions: 1) Why do these very properties define the value of a legal system or of a world? 2) How should these properties be defined specifically? 3) If these properties are relatively independent of one another, how should the multi-criteria optimisation problem of their global maximisation be solved? We have to note in criticism that:

1. The value series of Rescher is totally unsystematic and less subtle than even that of Leibniz.

2. If one can solve the cosmic value problem, one can also talk about the meaning of the universe as such.

3. The concepts of simplicity, uniformity, harmony and elegance can all be linked with certain symmetries and invariances. Hence in an "axiarchic" system we again encounter the question of invariance, though in another form.

Rescher does not even make any attempt to answer these questions. They form just one part of the research project he proposes. The fundamental problem he does wish to address is: why are the protolaws such that they "give the highest possible value to the world"? His answer (which is borrowed from Robert Nozick, without accrediting it to him) is that the "optimisation" principle (as we shall call it) can be considered as a special case in its own right. The universal fact "Reality maximises its value" is true because the very truth of this principle maximises the value of reality. Incredibly, Rescher refutes his own principle in note 13 [7, p. 53] by rightly observing that a principle of "minimisation" can equally well be deduced from itself: "reality minimises its value" is true because the fact that this principle is true does indeed minimise the value of reality. The empirical choice between the two "axiological" meta-principles he proposes is not practicable.

If one has an objective value theory for systems (one that can be defended), one can perhaps prove that particular structures lend a maximum of simplicity, elegance and harmony to a system. One may even be able to prove that a particular body of laws is necessary to realise that maximum. A non-empty world is assumed by this maximisation of cosmic values.

To summarise, something exists because existence has more value than non-existence (not for us, but in itself).

Values are not regarded as causes here, nor is an evaluating essence presented as the cause of the world. Values merely "explain" why the laws have the form they do. This is only possible, however—since the self-explanation of the evaluation principle has clearly failed—if we can furnish an independent argument for "The value of a system implies its existence". No such argument exists. On p. 167 of *Universes* [14] Leslie has this to say: "The notion that an ethical requirement or set of requirements could itself be creatively effective may supply a philosophically tidy answer to the question why there is something rather than nothing". Mackie and Smart have taken this idea seriously; it is not intrinsically impossible that values have a causal efficiency if they are objectively real. However, they both reject the objectivity of values.

13. SEVENTH PROPOSITION: "SOMETHING EXISTS" BECAUSE NON-EXISTENCE CAN GENERATE EXISTENCE

The seventh proposition (together with the eighth and last, which we shall examine below) differs from the six previous propositions (and is of a more speculative nature because of that difference) in that the seventh proposition implicitly (and the eighth explicitly) postulates one or more theories about the meaning of "being" and "nothing". It is natural to use a definition of "being" and "nothing" in the search for an answer to the question "Why does something exist and not nothing?". So far we have used information about "explanation" (why), "world" and "something". However, we have not taken the properties of "being" and "nothing" as a basis. We were afraid that our starting point would become more uncertain and debatable than it is now. We should not let this fear hold us back, though. First, we shall put forward a dilemma argument. If "nothing" is the natural state that requires no explanation (whereas all deviations from "nothing" must be explained), it can be deduced that "something exists", and if "existence" is the natural state, all deviations from which have to be explained, it can also be deduced that "something exists". Hence, if a natural state can be thought of that requires no explanation, then we know that "something exists".

We shall proceed on the assumption that existence boils down to exerting a Force 1 and non-existence to exerting a Force 2. Force 1 is a force that produces objects, while Force 2 is a force that eliminates objects. The idea of considering "existence" as a "force" comes as some surprise. Yet existence as a static property or relation is indeed a more familiar, though unsatisfactory hypothesis. Our intuition about "existence" is (we think) that existence is an activity, whereby beings make a difference in the rest of their world, exert influence on other beings. That is why it is normal to consider existence as an active relation, a force. It is more difficult to understand why we would define non-existence as a negative force. Can one not interpret "non-existence" as a complete absence, the antipode of existence? A negative definition, however, never grasps the essence of what is being defined, just its relation to something else. If one defines "nothing" in a completely negative way, the assertion that "nothing exists" is logically impossible because it is synonymous with "something that exists can only be defined by its relation with something that does not exist" (namely—in this case—existence). This would be a rather trivial proof of the assertion we are seeking to establish. We choose not to solve the problem verbally. We then have to positively define "non-existence", however. The only definition that corresponds with the definition of existence is the one proposed: something that is not, is the object of or the bearer of a negative force. We shall postulate a large number of positive and negative forces.

If "nothing" is the normal state, the negative forces will dominate completely. But if we assume that they can also interact, it is possible that they will cancel one another out. If the elimination of a sufficient quantity of negative forces, or all negative forces, produces a positive force, nothing will necessarily generate existence.

This metaphysical adventure story might (!) be analogous to the fluctuations of the quantum void, which some physicists think were responsible for the Big Bang. The assumptions on which it is based are 1) non-existence = object and subject have negative forces; 2) negative forces can interact; 3) constellations may exist in which they cancel one another out; 4) the complete elimination of two or n negative forces produces a positive force. In order to generate existence out of nothing, one must therefore assume a large number of hypotheses for which there is no obvious proof. The mechanism was first devised by Robert Nozick [5, p. 23] (by clear analogy with "two negatives make a positive"), but the argument for negative forces and the analysis of the four assumptions needed to make them "positive" are our own work. Not that we take it all very seriously, but the positive definition of non-existence seems to point us in the direction of reality.

If "existence" is the normal state, then it is immediately clear that "something exists". In this assumption, however, all possible objects are normally real and all possible worlds are real. The principle of plenitude between worlds and within each world is true, and is only restricted by the requirements of non-contradictoriness. We know that the principle of plenitude is an invariance principle.

One might wonder whether the complete elimination of all negative forces does not imply the principle of plenitude too (if the number of positive forces produced as a result of the elimination of negative forces is proportional to the number that was eliminated). Let us repeat that we do not take this seventh proposition completely seriously. Critics of Nozick have called it pure nonsense. We nevertheless wanted to discuss it because it alludes to a natural definition of existence and non-existence (that is often forgotten), and because this is the only theory where existence arises out of nothing.

Why should "existence" or "non-existence" be given priority, however? The most common answer refers to the fact that the onus of proof falls on the person who confirms something. This quasi-legal argument can have no ontological significance, however. If we dig further we can see that the decision to opt for "normality" of existence or non-existence depends on the ontology one supports.

One can say that "nothing" is simpler, more uniform, more symmetrical, more stable and more invariant than "existence" and that "existence" therefore has to be explained from "nothing". One might also think that every existence assumes an "act of existence", a positive effort of force, an energy charge, which is not assumed by nothing; that is why one would conclude that "non-existence" is normal.

The priority of nothing can therefore be defended on the basis of a number of fundamental options. If one has the opposite fundamental options and if one attributes more productivity, variability, asymmetry and diversity to "existence", then one can defend the priority of existence on the basis of these options and set oneself the task of understanding why something is not (our problem then becomes "Why not everything?" and is no longer "Why not nothing?").

These two sets of "fundamental ontological values" lead to two opposing priorities and to two incompatible ways of dealing with "Why not nothing?".

Is the decision to opt for the first or the second ontology completely arbitrary? If so, then our problem becomes entirely illusory. We are convinced that this is not the case. We can deduce which set of criteria should be given priority from the definitions of "existence" we are considering (as a result of the eighth proposition). There is insufficient time and space to explore this question here.

14. EIGHTH PROPOSITION: "SOMETHING EXISTS" BECAUSE "NOTHING EXISTS" IS INCOMPATIBLE WITH EVERY DEFINITION OF EXISTENCE

If "Nothing exists", then reality is empty. Either there are no possible objects or possible worlds, or all possible worlds are empty. In that case it is true that "Nothing exists" and there is an external, autonomous reality that possesses the real property that it contains no objects, properties or relations. It seems absurd to say that reality is empty and that this characterises "reality". Yet we have to speak in such terms. "There is no reality", a phrase that a complete sceptic could defend, is not the same as "Nothing exists". What we would like to explore in this eighth proposition is whether it might be possible for objective reality to contain nothing.

We know of no method to run through all possible criteria for "existence", but the set that we quote below (from Routley) is fairly exhaustive. We realise that arguments can be brought against each of these criteria. Our intention here is not to analyse the accuracy of the criteria, but to use them to examine whether a completely empty real world is possible.

1. Holistic criteria: something exists if it is related to the whole (the world).

2. Spatio-temporal criteria: something exists if it occupies a place in space and time.

Variations of this are: something exists if it occupies a place in space (1), if it takes place at a moment in time (2), if it is located in the vicinity of or at a particular distance from a place in space or a moment in time.

3. Intensional criteria: something exists if it can be observed in principle.

4. Causal criteria: something exists if it has a direct (or indirect) causal relation with an entity (arbitrarily or specified).

5. Completeness criteria: something exists if it is consistent and complete in every respect.

6. Genetic criteria: something exists if it has acquired properties through extensional interactions [3, pp. 697–730].

In view of these six criteria, how can we interpret the assertion "Nothing exists"?

One set of interpretations is the following:

1. Nothing is related to the whole.
2. Space and time are empty.
3. Nothing can be observed.
4. Nothing has a causal relation with something else.
5. Nothing is consistently and completely defined.
6. Nothing has acquired properties through extensional interaction.

A second set of interpretations is the following:

1'. There is no whole.
2'. Space and time do not exist.
3'. Observation does not exist.
4'. There are no causal relations.
5'. Definition does not exist.
6'. Genesis does not occur.

The two sets of interpretations are not identical but are in fact related. It is sufficient to test one of them to demonstrate that "Nothing exists" can be refuted. For some views about space and time an empty space and an empty time are impossible. With these views, 2 implies 2' and an "empty" reality (which, according to criterion 2, would always be spatio-temporal) cannot occur. If nothing is related to the whole, the whole has no parts and 1 implies 1'. A total reality to which nothing is related cannot, therefore, occur. According to criterion 1, "Nothing exists" is also impossible. Criterion 3, on the other hand, does not permit any decision to be made: if in the event that nothing is observable and observation does not exist, a reality that is completely unobservable is certainly possible. The only way that criterion 3 can yield anything useful for our purposes is to generalise "observability": something can be observed if it can produce partially structurally the same effect in its environment. It depends on properties of the causal relation as to whether a world in which something like this is not possible can be conceived. We do not think it can. This reduces the ontological discussion based on 3 to an ontological discussion based on 4. If nothing is causally related to something else, it cannot be related to the whole either (and criterion 1 intervenes), while, within reality, nothing explains or can be explained. If nothing is consistently and completely defined, it is either inconsistent (and hence impossible), or incompletely defined (and hence the whole of reality is undefined in its interrelations). Empty reality then has no autonomous status. If nothing has acquired properties through extensional interaction, then nothing has properties (which, as in the previous case, leaves reality undefined), or all properties are internally acquired (which completely fragments reality and reduces it to chaos, which is impossible), or else all properties are intensional (which makes reality fictitious).

We may conclude that, based on a more concrete realisation of the concept of existence, it is impossible that "Nothing exists" can be true. The question therefore remains unresolved. We have seen that an empty reality cannot be. We are now faced with more profound questions: why does connection with the whole, space-time, partial isomorphism, causality, complete definition (individuality) and partial external genesis occur in that reality? The question "Why not nothing?" is no longer inaccessible. It seems to give rise to a number of research projects as a result of the various propositions that have been put forward.

In particular, these research projects often seem to be related to the invariance problem. Space-time is defined by its invariance group (remember Klein's Erlanger programme). A whole is defined by the permutations of its parts in which it is invariant. Observability is defined by the possibility of producing partial isomorphous representations. Causal relations are defined by the invariants of interactions. Completeness is defined by properties of objects that are invariant in all changes of context. Genesis is defined as a combination of causality and completeness.

It is not our intention here to use the invariant theory to provide an "explanation" for the fact that reality comes under the six criteria. Let us simply note that the other propositions put forward to answer the question "Why not nothing?" nearly all refer to the invariant problem too [12].

Note: The present text demonstrates how difficult it is to find constructive answers to the question "Why does something exist?". However, an infinite series of even more difficult and more profound questions exists. It is not logically necessary that something (an object, a world) is ontologically possible. "Why is something possible?" is an even more difficult question to answer than "Why does something exist?". One asks such a question if one believes that not all objects, about which one can talk or think, are possible.

(10) [(Ax)M!x] (where M! means ontologically possible, by analogy with E!, which expresses ontologically real).

In order to answer this question, one will probably have to talk about objects that may conceivably be possible, but are not possible. Questions can also be asked about their degree of being. In order to answer this question, one has to climb higher up the ladder of possibilities. An infinite number of questions are then generated. Can we overcome these regressions or not? We can express them in slightly more concrete terms at least. It is acceptable to state that only for an object that is possible is there any point in saying that it exists.

$(E!x \text{ or } \sim E!x) \rightarrow Mx$. If there is a property P that means that what P has is possible $(Px \rightarrow Mx)$, then there is also a property that must have an x in order to exist or not to exist. $(E!x \text{ or } \sim E!x) \rightleftarrows Px$. The questions that follow "Why does something exist?" are:

1. Are there objects that do not have P? (i.e., are there objects that can neither be, nor not be?)

2. Why should one believe that?

3. How and why does P arise with a much larger set of objects that do not have P?

4. What are the properties of objects that do not have P? Do they form a system? Do they have a structure?

5. Is the boundary between E! and $\sim$ E! blurred? Is the boundary between P and not-P blurred? Is the boundary of a possible set of all objects O blurred?

Since one has an infinite series of modalities, one can repeat these questions with other fundamental properties P', P", P''' (etc.). The question is for how many of these questions a fixed meaning can be established. In this article we have tried—prompted by Nozick and Routley (and to a lesser extent by Rescher)—to put forward a few fixed meanings for "Why not nothing?". It is not clear whether we can proceed further and—if so—how we should do so. However, it is certainly necessary to put forward and describe in broad outline the "ladder of possibilities". Nozick, who, to our knowledge, is the only person who has looked beyond ("Beyond", pp. 150–158) the level of "Why not nothing?", resorts to mystical experience to say something about objects that have "not-P". How important mystical experience may be, this change of tack cuts off the progression of possibilities and one simply switches from one linguistic game (metaphysics) to another (religion) without a legitimate reason.

15. "WHY NOT NOTHING?" AND HEIDEGGER

In his "Introduction to Metaphysics", Martin Heidegger (we shall quote the 1958 French translation of his "Einführung in die Metaphysik", 1952, Ch. 1 "La question fondamentale de la métaphysique) put forward the question that we have dealt with here as being the fundamental question of metaphysics. Yet not a single solution is proposed in the almost fifty pages that he devotes to this question (neither in the remainder of his work, in which this question constantly crops up in various forms). He points to the importance of the question (the most universal and the most profound), to the human significance of the question, to the need to develop the question by exposing its presuppositions (there is being), by indicating what is asked (the beings) and by emphasising that one cannot confine oneself to "Why is something?", but should positively stress the contrast "Why is there something and not nothing?". Yet he rejects the tools that could help him to approach the question by affirming that (p. 35) it is illogical and unscientific to talk about nothingness. One cannot and should not use logic and science to seek the solution because they postulate a special answer to the question, which—according to Heidegger at any rate—cancels out the question. For Heidegger the question can only really be asked if one knows what "being" means. "Mais l'être reste introuvable, presque comme le néant ou finalement tout à fait de la même façon" (p. 47). Nietzsche ensures that the word "existence" has no meaning and Heidegger quotes him to ensure that the "experience of being" is lost in the course of the history of Western civilisation. It is difficult to imagine a greater contrast than the one between our attempts to answer the question without assuming a definition of being (the first six propositions), or by giving an explicit definition of being and nothing (the last two propositions), and his emphasis on the meaninglessness of the "being concept" and the loss of the "being experience". We want to approach the answer to the question that Heidegger considered unsolvable using means that are familiar to Carnap, who rejects Heidegger's question as absurd.

Heidegger's emphasis on the "difference" between "being" and "beings" makes him break down the question into further questions: "What is being?", "Why being and not nothing?", "Why beings and not nothingness?". This question becomes a historical and political question for him. The future of the relationship of mankind with the total being and with being as such are at stake.

To a large extent we agree with Heidegger. Mankind lives in the qualitative and quantitative totality of beings, in the system of the universe, from house to galaxy cluster, from quarks to cultures. Will mankind concentrate on the totality of beings, on the exploration of it, the experience of it, reflection over it and contact with it, or will it get wrapped up in technical control and regional analysis? Heidegger affirms that concentrating on the totality of beings also assumes a grasp of "being" (we might say: of the essence of being). We agree with this too. The wish to grasp being does in fact imply the question "Why not nothing?". We cannot radically concentrate on being without wanting to know why it has arisen out of nothing (p. 40), why it does not revert into nothingness, why it does not stop, why it does not collapse. What explains the stability of totality and being and what explains the appearance of that being in, opposite and (perhaps) from nothingness? For us, as for Heidegger, these questions are inextricably linked with our will to concentrate on the totality, to orient ourselves in it, to be open to it and to experience it and understand it. We can live in and through that will or we can refuse to concentrate on the whole. Once again we agree with Heidegger that these questions are historical, and for two reasons. 1) At every moment in history the totality is given to us in a historical manner through the content of science and technology and through the integration of these (or not) by philosophers. 2) At every moment in history our societies and cultures have to find different, practical answers to the question of how one should live in order to be able to and want to ask the question about the totality of beings, about the essence of being, about the reason for beings and being (and hence also about their and our future). For Heidegger, however, this question cannot be approached scientifically because it is a historical—political and historical—philosophical question. For us, on the other hand, it can and must be approached scientifically because it is also of a historical nature (in the two respects we have just mentioned). We should like to make a few observations about the historical, political and ethical dimensions of the eight propositions put forward in answer to the radical question.

At first sight the political atmosphere of Heidegger's work (the reference to the fate of the planet Earth, but the destruction of the Earth, the destruction of a "... Europe" by the techno-scientific giants Russia and the United States, the flight of the gods and the massification of society) resembles a passistic—agrarian—elite description of a European who sees his own continent occupy a central position and wants to offer it up as the—unlikely—saviour of the planet. The connection between the most universal question and these concerns is not clear. Unless "Why not nothing?" should also be interpreted as "Why not ignore mankind as a whole in practice?". Why not disregard the universe as a whole (starting with the Earth as a whole)? Why not break out of the autonomy and the 'differentness' of the outside world in order to merely control it and/or make a model of it? "Why not nothing?" can produce answers that, because they reveal the reason for being, also call upon us in our being to form an organisation of all beings! How can that happen?

At first sight one does not suspect any relation between ethical and political problems, historical trends and the seven propositions. Yet something is noticeable:

1. Robert Nozick has established his reputation on the basis of his "Anarchy, State and Utopia", a lucid defence of a libertarian minimal State.

2. Richard Routley is a militant environmentalist and adherent of an analytical eco-philosophy.

3. Leslie, who was a source of inspiration to Rescher (together with Nozick), is a Christian neo-Platonist.

The authors whose work we are continuing, re-classifying, changing and combining are definitely not ideologically neutral. It is therefore not necessarily an impossible task to seek the ethical-political and historical relevance of their work. A few contrasts immediately strike us: if "something exists" because all possible worlds exist (principle of plenitude), then the reason for existence is the absolute equivalence of all possible worlds. If this "egalitarianism" explains existence, the meaning and value of that existence can only be found in the equivalence of cultures, ethics and political systems. Diversity, divergence and multiformity thereby increase in standing.

On the other hand, if the reason for existence is found in the organic unity of those possible worlds that are realised, its meaning and value will also lie in the consensus, the communication, the solidarity of mankind, culture and ethical system. The fact that this organic unity is, therefore, completely unhierarchical also has an egalitarian message in this case. This first contrast between two "degrees of being" therefore also underpins an ethical-political contrast.

Other contrasts also have axiological significance. If "something is" because it is logically impossible for it not to be, we have to seek meaning and value in logical coherence and consistency. If, on the other hand, "something is" because a process of coincidence will almost certainly produce "being" (the empty world is an extreme, isolated exception in the set of possibilities), then taking risks, gambling, taking advantage of coincidence and living coincidentally would appear to be most consistent with the nature of being.

If something exists because nothing cannot explain itself (our second proposition), the universe, as a self-explanatory system characterised by maximum consistency, is an incentive to trust in the comprehensibility (not in the controllability) of the world and, at the same time, an incentive to lead a personal and collective life oriented towards maximum openness for that reality (conservation, exploration and contemplation of reality in its entirety as the supreme goal).

If, on the other hand, something exists because it is intrinsically better that something should exist, generally speaking the maximum realisation of cosmic values will inspire ethics, politics and culture (the commitment will be more active in this case and less exploratory).

We should not forget that several of the seven answers are reconcilable with one another, which means that the ethical-political options associated with them can also be associated with one another.

There are still two answers that require an ethical-political interpretation. If "something exists" because nothingness necessarily produces being, then negation, revolt, denial itself can be thought of as creative. This consequently leads to commitment to revolt. If "something exists" because it follows from the internal meaning of "existence" (as completeness, productivity, relation to the whole, spatio-temporal localisation, genesis) that this "existence" must be realised, then the values associated with these different meanings of existence will also give meaning to it. Anyone who refuses to ask the question about the reason for being will thereby close himself off from the totality of being and from the objectivity of the values. This leads almost inevitably to fragmentation, relativism and subjectivism. Anyone who considers the question to be definitively unanswerable or definitively answered, condemns himself to either a positive or negative mysticism (Wittgenstein and Heidegger) or to a dogmatic, authoritarian system of ethics and politics. Only if one can see the various answers as never-ending research topics, will one safeguard at the same time the objectivity of interpretation and also the openness of ethical-political commitment. We can see, therefore, that Heidegger's assertion about the ethical-political significance of the question about the reason for being also appears to be true if one attempts to give a number of interim answers and not (as Heidegger does) avoid an answer.

Can we also understand the historical context of the various answers? We believe that the history of science and socio-political history make certain answers more or less probable and fruitful.

One general aspect strikes us: we are relating the treatment of the fundamental question "Why not nothing?" to a theory of possible worlds. This option is an obvious one in view of the rapid development of the modality theory in the twentieth century. Yet in that same twentieth century, science (quantum mechanics) has made the idea of a multitude of possible worlds (Everett interpretation) topical, and supplied technology with another version of "possible worlds" through the virtual worlds of space communication and through space travel, while the unification of the planet Earth has demonstrated the multiplicity of the culture worlds. The socio-cultural as well as the scientific situation in which mankind now finds itself gives the question "Why not nothing?" its own concrete content. The content of this question will change in the future because the meaning of the "why", of the "negation" and of the "being" concept will change. However, in all previous periods the most universal question has been the most radical question (and will remain so in all subsequent periods). The refusal to ask it or the impotence to attempt a solution, just like the kind of solution on which the entities that do that are working, fundamentally determines both the world view and the action (however strange the question has always been and will always remain).

16. CONCLUSION

We would like a rational kind of metaphysics that has the broadest possible empirical base and forms an open research tradition, but one that still asks the radical metaphysical questions. That is why we wanted to link symmetry and invariant metaphysics, which we put forward elsewhere, with the radical question of Leibniz by specifying, classifying and associating answers to that question of Nozick, Leslie and Rescher, and to show how they almost all lead to research projects that focus on the symmetry and invariance approach.

It is too early to say which of the eight propositions we reject and which ones we accept. We are more sceptical about the first, third, fifth, sixth and seventh propositions than about the second, fourth and eighth. We would like to combine 2, 4 and 8. All of the propositions, however, need to be developed further and—we suspect—can be developed further.

We hope to link them in a more concrete way with various branches of science in the publication we are planning entitled "Natuurfilosofie" (Natural Philosophy).

AFFILIATION

Leo Apostel (1925-1995)
Emeritus in Philosophy and Logic
University of Ghent
Belgium

REFERENCES

Note: This text was originally written in Dutch by Leo Apostel in February 1995. When he died a few months later, he hadn't had the chance to publish it. We are grateful to be able to reproduce its translation here.

[1] **von Meinong, A.** (ed.), *Untersuchungen zur Gegenstandstheorie und Psychologie*, Barthe, Leipzig, 1904.

[2] **Parsons, T.**, *Nonexistent Objects*, Yale University Press, 1980.

[3] **Routley, R.**, *Exploring Memory's Jungle*, Departmental Monograph, Philosophy Department, Research School of Social Sciences, Australian National University, Canberra, 1980.

[4] **Leibniz, G.W.**, "De rerum originatione radicali", *Leibniz's Philosophische Schriften*, 7 volumes, Vol. **VII**. The French translation is quoted here, published by Lucy Prenant, Leibniz, Oeuvres, Autier, Montaigne.

[5] **Nozick, R.**, *Philosophical Explanations*, Harvard University Press, 1981.

[6] **Leslie, J.**, *Value and Existence*, Oxford University Press, 1979.

[7] **Rescher, N.**, *The Riddle of Existence*, University Press of America, 1984.

[8] **Rougier, L.**, *La Scolastique et le Thomisme*, Paris, no date.

[9] **Wittgenstein, L.**, *Tractatus Logico-Philosophicus*, London, 1922.

[10] **Heidegger, M.**, *Was ist Metaphysik*, Niemeyer Verlag, 1958. **Heidegger, M.** *Einführung in die Metaphysik*, 1952.

[11] We are indebted to Robert Nozick for this kind of reasoning, which explains "ultimate" (statements that cannot be further explained) on the basis of itself (as special cases of itself). A fundamental circularity cannot be avoided in answers to "ultimate" questions. That circularity is not necessarily vicious, though. Peter Aczel's "Non well founded sets" (CSLI Lecture Notes, no. 14, 1988), quoted by John Barurse and John Etchemendy, presents hyper-sets that may be elements of themselves. The structure of a self-explanatory universe is that of a hyper-set.

[12] **Apostel, L.**, "Symmetry and symmetry breaking: ontology in science". in: **The Worldviews Group**, *Perspectives on the world*, Brussels, VUBPRESS, 1995, pp. 175-219.

[13] **Chew, G.F.**, "Bootstrap: A Scientific Idea", *Science*, vol. **II**, 1968, pp. 762-765

[14] **Leslie, J.**, *Universes*, Routledge, London, 1989.

DIDERIK BATENS & WIM CHRISTIAENS

LEO APOSTEL'S WORLD VIEWS PROGRAM IN THE PERSPECTIVE OF HIS CAUSAL ONTOLOGY A CRITICAL APPRAISAL

1. AIM OF THIS PAPER

Leo Apostel was the main instigator of the *worldviews* group and, as long as he lived, the driving force behind it. Although his philosophy cannot be reduced to this program, the latter clearly was of utmost importance to him.

In the present paper we[1] first try to spell out why this was so. We try to clarify the way in which Apostel's philosophy brought him to the world views program, the aims he wanted to achieve with it, and the sense it had for him. To this end we shall first discuss—in sections 2 to 5—Apostel's ontology (or scientific metaphysics), and show that the development of *a world view* is a natural outcome of it. We shall depict the deep unity of this ontology, the key role played by causality, and the hard objectivistic approach to values and norms. This part of our essay is mainly descriptive and explanatory.

We plan, however, to do more. More precisely we shall discuss four systematic questions which we briefly announce in the subsequent paragraphs. There we shall necessarily be interpretative and argumentative. We shall argue for some positions that Apostel never explicitly subscribed to, we shall not always agree with him, and occasionally we shall even run counter to his understanding of the world view program and its motivation.

The first question concerns the justification of the intrinsic pluralism of the world views program. Apostel wanted a multiplicity of world views developed. This clearly cannot be justified by his ontology. We shall argue that the clue to this puzzle is Apostel's epistemological stand, that this stand is basically correct, and that the aim at a multiplicity of world views is an unavoidable consequence of it. The central point here will be that world views are necessarily *hypothetical*.

The second question concerns the contents of a world view and its

[1] The research for the present paper was supported by grant from the Flemish Minister competent for Science and Technology (contract Caw96/54c).

place within a knowledge system. There are two points we want to clarify here. Many people consider a world view as the set containing the most basic ontological principles and they assign it a unifying function in that it is contained in all 'theories' of the knowledge system to which it belongs—in this sense it was also used by Kuhn [18]. For Apostel, a world view is a *system*, and it contains not only descriptive statements in the strict sense, but also *values and norms*. Moreover, we shall argue that world views do not play the unifying function that is usually ascribed to them.

Leo Apostel was a die-hard realist. On several occasions, he explicitly subscribed to epistemological realism, usually of the converging sort. This leads us to the third question, viz. whether the world views program presupposes such realism. We shall argue that the answer to this question is negative. But as the world views program is clearly incompatible with some epistemological positions, we shall try to point out its main epistemological presuppositions.

Our fourth and final question concerns the 'objectivity' of values and norms. By considering values and norms as integrating parts of world views, Apostel denies their being merely subjective. In section 5, we describe his extreme objectivistic position on them. In section 9, we shall argue that the world views program does not presuppose this position. What it does presuppose, however, cannot be clarified in a few words. For one thing, we shall point out that "objective" is a misleading term that should be replaced by "justifiable". One reason for this, among several, is that the first term is misleading—and possibly Apostel was misled by it. Basically, we shall argue that the world views program presupposes justifiable values and norms, but that the latter need not exists independently of humans.

In (our final) section 10, we consider the general question: Why world views? This question has been answered at length in several publications of the worldviews group, most explicitly in [9], very briefly summarized in the introduction to The Worldviews Group [21]. In an attempt to strengthen the case, we shall reconsider the question in terms of our discussion.

2. CAUSALITY, SCIENTIFIC METAPHYSICS, AND WORLD VIEWS

Over different periods, some separated by dramatic changes in his personal life, Apostel's interests have turned in different directions. Sometimes he was convinced to have taken a radically new start. His work concerns a variety of topics. Results are scattered over a potpourri of books and journals. In [10], looking back at a lifetime of academic work in philosophical logic, epistemology, and ethics, Leo Apostel reports a strong sensation of an implosion. The different parts of his work reveal a unity of which he had not been aware for many years—as if, unknowingly, he had been working all the time on the same embracing project. The chief unifying force was causality:

> "The concept of cause itself became the center of my world view. The reason is as complex as it is evident. Either the universe is an unorganized aggregate and as such ununderstandable (its existence would not be intelligible) or events and processes within this universe are glued together by a causal nexus, allowing one event or process to produce others." [10, p. 96]

The following explanation is in several respects typical for Apostel. Suppose that, on the one hand, you deny the existence of an omnipotent God (or super-natural force) who links one event, state, or process to another event, state, process (his atheism). Suppose, on the other hand, that you view the world (systems, events, states, processes—for example fires, societies, planets, human beings, ideas, values, etc.) as real and knowable (his scientific realism). The central question then has to be: How does one system, event, state (etc.) bring about another system, event, state (etc.)?

Apostel's theory of causality constitutes his answer to this question. Without it, one can neither understand his epistemology nor his ethical theory. His theory of causality is, moreover, at the heart of his metaphysics which depicts the universe as radically different from both the theistic universe and the mechanistic universe. It is a scientific metaphysics in at least two senses. First, this metaphysics should not only be compatible with the sciences, it should also derive its fundamental principles from the sciences (at least in all the domains in which elaborate sciences are available). Second, this metaphysics should be built in a scientific way. It should not be a starting point or presupposition; but a result, and more specifically the result of a rational endeavour.

Apostel's interest in causality goes back at least to his Ph.D.. thesis, defended in 1953, at the summit of the positivistic period. Major further steps are his 1963 program for a scientific metaphysics, and his 1965 program for a moral science. In the two 1974 volumes, he presents his (radically anti-epistemological) theory of causality and his epistemology (that is based on the former; knowledge is the result of causal interactions between a knowing subject and reality). From 1985 on, his main aims were the elaboration of a scientific metaphysics and the world views program. As early as 1963, Apostel writes:

> "we need the universe as an intelligible whole, as an emotionally significant totality, and as an organized field of action. These three needs, distinct from each other, all demand the development of a metaphysical system. They all demand also the development of a metaphysical system that has the collective, progressive, empirical, and rigorous character of the scientific system." [2, p. 11]

Where this clearly foreshadows the world views program, causality was not integrated in his metaphysics in those days. In reply to Armand Phalet, he writes: "He is completely correct when he summons me to apply to my metaphysical papers my work on causation. I have been a 'man of little faith' by trying to keep these two unpopular undertakings separate from each other" [7, p. 304]. Whence the effect of experiencing the implosion. Let us stress that the latter did not result from Apostel's relating the different aspects of his work, but from the insight that there was a single program behind it.

At Apostel's death, some three books were left unpublished. The younger author (WC) of the present paper prepares their publication. Needless to say, we only point to some highlights for our topic. Apostel's bibliography is far longer than the present paper and we cannot relate here all those books and papers to causality, metaphysics, and world views.

3. A CAUSALITY BASED METAPHYSICS

The basic entities in Apostel's ontology are singular causes. Individual causes are more fundamental than both individuals and laws.

The classical empiricist reduces singular causal facts to natural laws and (in a different sense) reduces the latter to empirical generalizations. The causal realist Leo Apostel reverses this view, assigning a secondary ontic status to natural laws.

> "Pour tout épistémologue qui ne croit pas que des universels régissent et dominent le réel particulier, il est évidemment clair que la causalité doit être le fondement de la légalité et non inversement." [4, p. 424]

In [10, p. 98] Apostel and Van Dooren relate this way of thinking to Harré, Bhaskar and Fisk. One might add Cartwright and Leszek Nowak. Remark that several of these, including Apostel, share a strong Marxist inspiration.

For Apostel, causality is the (ontological) criterium for existence: something exists iff it exerts a causal force [7, p. 141]; and existence requires unity: "if something is not a unity, it cannot exert a causal force" [*ibid.*]. The notion of a unity or system is itself dependent on causality: "in order for an a to produce a modification in a b it must actualize causal dispositions that use all the parts, if there are any, of this a" [*ibid.*]. If something cannot manifest itself as a causal force or cannot be transformed by a causal force, then it cannot be a unity or system:

> "An individual object exists iff it has [a] causal relation with at least some different objects and events, and if some of the properties of this individual have been acquired by causal interactions in the course of its genesis." [7, p. 212]

An individual is a concrete system. In [7, p. 142], Apostel presents his definition of an individual. He refers to three criteria, all of which are in terms of causality: (i) an individual cannot be divided in such a way that the aggregate of its parts has the same causal powers as the original; (ii) it is completely defined in all possible respects or along all possible dimensions (a is an individual iff, with respect to all individuals b and all causal interactions I, it is determined whether it can exert I on b); (iii) an individual is unique (no two different individuals have the same causal capacities). Thus, an individual is basically a Gestalt of powers:

> "I suggest that anything has real being that is so constituted as to possess any sort of power either to affect anything else or to be affected, in however small a degree, by the most insignificant agent, though it be only once. I am proposing as a mark to distinguish real things that they are nothing but power." [22, p. 992]

In recent approaches, possible worlds are taken as basic (unanalyzed) entities and are then used to define necessity and possibility, counterfactuals, causality, determinism, etc. For Apostel, possible words should be analyzed on the basis of his ontology, leading to an objective interpretation of the possible worlds semantics. It was typical for Apostel's approach to logic that he further analyses the semantic entities that are commonly taken as basic. His action logic is another typical example of this procedure. This made his position rather typical. Where most logicians consider semantics as an uninterpreted game that leads to the desired effects, Apostel sides with those that require an ontological interpretation of the basic entities of semantics. But where most of the latter are Platonists like David Lewis, Apostel wants the interpretation to be objective and concrete.

As causes are basic, possible worlds are 'becoming'. This allows for a strong accessibility relation dependent on bifurcation points in this becoming. But remark that the indeterminacy is not dependent on our knowledge, but on laws and causes that are 'really true'. "La notion d'indétermination réelle du devenir est donc impliquée par la notion du réellement ou ontologiquement possible" [4, p. 146]. Apostel implicitly anticipates Earman's [17] definition of determinism, but by reducing it, through the analysis of possible worlds, to causality. The effect on his analysis of counterfactuals is similar:

> "les connections causales réelles [...] sont suffisantes et nécessaires pour expliquer l'existence de lois nomologiques (subjonctives fortes) mais aussi d'une telle façon qu'en général (bien que ce ne soit sans exception) elles ne sont connaissables qu'à partir de la vérification de ces lois nomologiques fortes." [4, p. 424]

The title of Apostel's [4], "Matière et Forme", refers precisely to this approach and this position. The content (particular causes) determines the form (scientific laws), and reality is a constant interaction of content and form.

4. THE ANALYSIS OF CAUSALITY

Having explained the central role of causality, we now should (finally) address the question: What is a cause for Apostel? The answer is to be found in the first volume of his [4]. After presenting approaches in terms of, first, temporal logic and, next, of the logic of counterfactuals, Apostel [4, pp. 165–170] offers his basic analysis of causality. He does so by devising an ontologically interpreted semantics. The basic idea is that causes *produce* their effects. In this sense there is a structural likeness between causal production and human action.

The definition itself is too complex and too technical to be summarized here. We can only vaguely give a feel of it. Consider the structural relations that obtain between an actor, the instruments used by the actor, the materials that are transformed in the action, and the result of the action. We say that a causes b if there are two sets c and d such that the aforementioned structural relation holds between a, c, d, and b (in that order). The essential components of the formal analysis are: a set, a relation R defined on this set, and an ordering relation O defined by R; the two relations lead to a hierarchical ordering of the set.

Remark that no intentionality or teleology is involved in the structural relations. If there were, the definition would become anthropomorphic. The present definition is 'actomorphic'. But it still is an ontological definition of causality, *not* an epistemological criterion for it.

The definition of causality is related to other approaches. It is interesting to see how Apostel tries to incorporate these, for example by showing that they grasp some aspects of his approach. The following quotation is suggestive enough in itself:

> "It is interesting to notice that my 'production' relation introduces a type of logical non-linearity: a produces b if and only if (1) a is a strong (subjunctive conditional) necessary condition of b, (2) a is a strong necessary condition of other strong necessary conditions of b, and (3) a is a strong necessary condition of the fact that other events x, y are strong necessary conditions of b." [7, p. 96]

The definition of causality is also related to epistemology in two different respects. First, it determines (epistemological) criteria for identifying causes. But, second, it also determines epistemology in general, as humans belong to the causally ordered universe. Apostel even goes a step further in claiming, already in [4, vol. 2, pp. 233–450], that the causal properties of the domain about which we are thinking co-determine the mathematics, statistics and logic we should use in that domain.

5. INTEGRATING ETHICS

There is a strong analogy between the way in which Apostel characterizes the relation between the sciences and the aspects of reality they intend to capture, and the way in which he characterizes the relation between ethics and values. The last version of his ethical position is presented in his [8]. It can be characterized as a non-reductionistic naturalism that involves three central claims. First, moral statements can be true. Next, they cannot be identified with any non-moral statements. Nevertheless, there are rich and regular relations between the former and the latter. Apostel refers to an ontology that sees the world as composed of different layers which cannot be reduced to one another, but are nevertheless built on and from each other.

Given such an ontology, it is possible that statements that refer to different layers of reality imply each other (in being co-extensive). But this does not entail that they have the same meaning or that they can be reduced to one another.

The clarify his position, Apostel refers to the distinction between primary properties (such as form or mass) and secondary properties (such as colour, sound, and smell), and to the way in which the latter may be seen as real, viz. as complex structures of processes. He then introduces ternary qualities, such as the cruelty of an act or the injustice of the social order. These he sees as properties of concrete structures or *Gestalte*.

The result is a very hard objectivity in ethical matters. Apostel continues the parallel with non-ethical statements all the way. For example, he argues that, in our experience, we are confronted with (outside) information that allows for the characterization of objects as morally positive or negative. As any other form of experience, this one is not unproblematic. In order to detect the information and to interpret it in the correct way, humans need to be rational in a sense that also pertains to emotions. But the information (the ternary properties) belongs to reality, even if we do not perceive those ternary properties.

6. THE MULTIPLICITY OF WORLD VIEWS

We saw that Apostel's ontological position led to the necessity of constructing a world view. In a sense, a world view is just an ontology (in Apostel's sense) in disguise. We shall not comment on the reasons why the former term is liable to attract more interest in the cultural climate of the last decennia.

Leo Apostel, given his very specific and extreme ontological views, could be expected to attempt the formulation of a (single) world view. Actually, Apostel's work in scientific metaphysics—he devoted much time on this during the last twenty years of his life and left three fascinating unpublished manuscripts—is just offering a single position. Why then, did he devise a program that aims at a multiplicity of world views?

Our claim is that the answer lies in Apostel's theory of knowledge. It is difficult to argue for this from his writings. During the last years of his life, Leo Apostel stated regularly that he wanted to do ontology, not epistemology, because he considered the former as primary and more basic.

A careful reader of Apostel's work will have seen that it displays a constant tension between ontology and epistemology—sometimes he simply seems to contradict himself. Except for the later, unpublished works, where he basically tries to derive a metaphysics from the sciences, Apostel seems to ground ontology on epistemology. Yet, in discussing epistemology, he sneers at the usual philosopher's epistemology, arguing that epistemology should be derived from ontology. More than twenty-five years ago, the older of the present authors (DB) discussed the matter with Leo, but never obtained a satisfactory answer. Nevertheless, Apostel's position is coherent and strong. We somewhat reconstruct his position here, in terms of Batens [13]—bits and pieces of this are to be found in Batens [11], [12], [14]—but we should warn the reader that Leo Apostel considered the rationality defended there as too relativistic.

Apostel viewed human beings as integral parts of the world. In order to understand them, we have to look at the world and at the basic causal mechanisms that govern it. Such realistic outlook on human beings involves the rejection of any absolute knowledge: all human knowledge is hypothetical. Given their properties, these subsystems of the world have no direct and flawless access whatsoever to their environment, and they certainly do not have at their disposal any *via regia*. It follows that their knowledge, even their best knowledge, is always subject to revision.

World views are no exception: they are at best hypothetical. Only by the confrontation of a multiplicity of world views do we make a chance to arrive at more justified world views.

The hypothetical character of world views does not only depend on the hypothetical character of our present best scientific theories, from which they are derived in as far as this is possible. That hypothetical character *also* hinges on the fact that even our present best scientific theories are far from sufficient to construct integrated world views. Values and norms (in all relevant domains) are at best very *indirectly* derivable from the sciences—and this "indirectly" means that we cannot even rely on the ideology that lingers behind the sciences, but that we have to recur to more speculative and ideology-laden constructions. (We offer a more positive approach to the justification of values and norms in section 9.) It follows that world views have to be subjected to critical examination, and only the confrontation of a multiplicity of world views is capable of delivering this.

Let us try to summarize. For Apostel, philosophy is impossible unless the world views program is realized, i.e., unless it is based on an ontology (in his sense). Decennia ago, he argued that both metaphysics and ethics can be constructed in a scientific way—see [2] and [3]. But these papers may be misread as merely *deriving* metaphysics and ethics from the sciences—a strategy that is clearly bound to fail if the premisses are to be the *present* sciences. In this sense, world views present a way out.

Indeed, even given the present situation of the sciences, they may be confronted with each other, they may be subjected to the (rather weak) test that is provided by their confrontation.

Apostel had outspoken and extreme positions and worked hard to devise a world view in which these positions are built into a unity that is compatible with our present best scientific theories. But the epistemological implications of his ontology forced him to consider his own world view as provisional, temporary, and hypothetical. In this sense, epistemology is central for his ontology, even if the former was basically derived from the latter.

7. WORLD VIEWS AND THEIR PLACE IN KNOWLEDGE SYSTEMS

We saw that a world view contains not only descriptive statements, but also evaluative and normative ones; and we hinted at some of Apostel's reasons for this decision. A different and independent reason is related to their function. A world view should provide us with the basic guidelines for experiencing the world, understanding it, and acting in it. This is clearly impossible without values and norms.

Part of the analytic philosophical tradition separates knowledge from (non-cognitive) goals. In its extreme forms, it contends that only knowledge can be justified, whereas goals (and values and norms) are 'subjective' and not a valid object of either discussion or justification. More sophisticated forms of the *neutral knowledge position* recognize the import of *cognitive* values, norms, and goals, but assign them a rather special status by which they are independent of other values, etc. This position does not deny that non-cognitive goals and values are required for actions. However, it maintains that the aim to obtain objective knowledge (sometimes the need for such knowledge) is shared by all humans, that cognitive values and norms may be derived from that goal (respectively need), that experience and understanding can be justified, along with knowledge, in a way that is independent of non-cognitive goals and values, and that knowledge may be combined with almost *any* set of values and goals in order to generate actions that (attempt to) realize these values and goals. (In this sense, the title of Carnap [16] is quite significant: "Probability as a guide in life.")

Apostel rejected this position, and we are all on his side in this respect. Many arguments against the neutral knowledge position are well known. Even if all humans were to share the aim to obtain objective knowledge (or the need for it) they clearly disagree about the domains and aspects to which this knowledge should pertain. Also, during the last twenty years, philosophers of science avoided the term "objective" because of its misleading character, and replaced it by such predicates as "justified", "rational", and "scientific". However, it is false that all humans strive for scientific knowledge, and they heavily disagree on the meaning of "rational" and "justified". There are deeper arguments against the neutral knowledge position, not in the least because cognitive values and norms were subject to radical changes, even in the history of the sciences since the seventeenth century.

We do not spell out these arguments, because they are not as essential as they might seem to be. More fundamental is the question whether statements about (non-cognitive) values and norms may be justified—in section 9 we shall argue that they are. If they are, then world views *should* contain such statements. Even if such statements were not suitable for justification, they still should belong to world views in view of their function. In the latter case, however, the world views program becomes much more complicated, as we have to spell out which parts should be compatible with the results of our best scientific theories, may justly be criticized, may be confronted to the corresponding parts of other world views, etc., and which parts should be kept carefully apart from such discussion. Moreover, world views would be seriously crippled with respect to their chief function (as described in the first paragraph of the present section). Such a view is radically at odds with Apostel's, and he never made any attempt to work it out.

For Apostel, a world view is a *system* of descriptive, evaluative and normative statements. Its chief function is to guide our experience, understanding, and actions, not only in order to further the integration of our knowledge systems, but especially in order to integrate our lives. A different function, viz. to integrate our knowledge system, is subsidiary to the former and a side-effect of it.

We now come to the question about the place of world views in knowledge systems. Two common concepts of world views are mistaken. The first identifies a world view with the knowledge system (of a person or group) to which it belongs. Under this interpretation, the world views program aims at bringing together all available knowledge and to integrating it into a coherent and systematic unity. Needless to say, individuals and groups are wholly unable to adhere to a world view in this sense; it simply is too embracing to be captured by humans. By the same token, it cannot possibly fulfil the function of guiding the experience, understanding and actions of individuals or groups.

The second concept sees a world view as central to a knowledge system in a deductive sense: a world view would be common to all parts of the knowledge system of a person or a group. Put differently, a world view is assigned a unifying function with respect to a knowledge system in that (i) each part (theory) of the knowledge system is compatible with the world view and (ii) the most basic components of any such part belong to the world view. A theory from a knowledge system may then be obtained by first taking the relevant statements from the world view, next adding some specific but less central statements, and finally making the deductive closure of the result. This concept is that common, and at the same time that mistaken, that we have to spell out what is wrong with it.

The concept derives from the mistaken belief that a (decent) knowledge system is a monolithic, coherent and integrated entity. This belief is mistaken because it is unrealistic: actual knowledge systems are not structured in this way, and happily so, for thus structured systems would rule out the characteristic dynamics of the sciences. Actual knowledge systems are rather loose connections of more or less integrated subsystems, with two kinds of relations between some such systems. We cannot defend or even spell this out here but refer to Batens [13] and [14]. A world view is a (hopefully rather integrated) subsystem of a knowledge system, characterized by its chief function described above.

As we said some paragraphs ago, world views have also the function to integrate knowledge systems. This does not entail that a world view is a kind a common divisor of a previously unified knowledge system—if it were, then presumably no human person (or group) would ever have a world view at his or her disposal. But then, what might be the integrating function of a world view with respect to a knowledge system?

Actually the matter may be explained quite simply, provided we proceed in two steps. Suppose that a person adheres to a given world view. As the latter guides the experience, understanding, and actions of the person, it will have immediate effects on the future evolution of the person's knowledge system. If the person is confronted with a new domain or with problems in an old domain, the world view will impose constraints (in Nickles' sense) on the solution: it will suggest some solutions and rule out others; it will entail requirements on possible solutions; it will classify some knowledge elements (evidence and theoretical statements) as relevant and others as irrelevant; it will hint to certain heuristic rules by associating the problem with other problems solved in the past. Needless to say, all this will have an integrating effect on the knowledge system.

There is, however, a second step. World views are never themselves unproblematic. Obviously, they may be drastically incomplete—one of Apostel's central complaints was that people fail to have explicit and integrated world views. But even where they are present, they may be inconsistent or incoherent, and they may be subject to external difficulties: false or unjustifiable claims or presuppositions. The matter is complicated by the fact that internal and external problems with world views often become apparent while, guided by one's world view, one tries to solve a problem that is itself external to it. In this sense, the 'application' of world views causes an internal dynamics, along with the external dynamics that derives from the confrontation with other world views (as described in section 6).

Summarizing the second part of the present section: world views are not all-embracing and they are not central to a knowledge system in the sense that they would integrate it into a coherent and monolithic system. The central character of world views resides in the fact that they have the effect to increase the integration of knowledge systems. Even this, however, is only a side-effect of the fact that they are central in a different way—and there resides their strength—viz., in that they direct our experience, understanding, and actions.

8. DOES THE WORLD VIEWS PROGRAM PRESUPPOSE REALISM?

At the psychological level, Apostel's die-hard realism was an effect of his reaction to positivism, the prevailing forceful philosophy from the time of his youth. His also then outspoken rejection of positivism incited him to study with, among others, Carnap—and until Apostel's death, there was no philosopher he admired more. His rejecting the reduction of the world to sensations, drove him to Marx and Freud as attempts to understand and fight injustice and neurosis. Presumably, his reaction was overdone, just as, in his later years, his 'disinterest' for epistemology and his concentration on the *content* of physical and human sciences. The older of the authors of the present essay (DB) had many discussions with Leo Apostel on the status of world views. Time and again, Apostel expressed his conviction that all knowledge is fallible and hence at best hypothetical. Being an ontological *and* epistemological realist, Apostel realized very well that even our most justified claims about reality (at whatever level of generality) are fallible.

A more important question is the extent to which the world views program presupposes realism. In the previous sections, we have explained that world views are subject to criticism and change. The result might seem paradoxical: we have to use world views as a guide to our lives, but at the same time we know that they are only hypothetical, that they might be (and presumably are) mistaken.

The positivists resolved this paradox by restricting all knowledge to knowledge *about* our (possible) sensations. As "our" refers to all humans, the paradox is merely displaced. Indeed, if the world beyond *our* sensations is inaccessible to us, it is hard to see how sensations that are not *mine* might be accessible to me. A realist might easily locate other human beings in the world; how a positivist might do so is not very clear, as is well-known.

Returning to the question, we claim that the world views program does not require epistemological realism, but excludes positivism. In both cases, the reasons are epistemological, but they are different. Let us start with positivism.

We saw that a world view requires statements about the world, including statements about goals and values (and possibly norms). For all brands of positivism known to us, however, the latter statements are out. Positivism simply looses its gist if it tries to make room for goals and values. That Schlick wanted to make room for them is not a convincing argument as long as no one demonstrates that his philosophy (in this respect) may be embedded in the program of either Mach or Carnap. The crucial tenet of positivism is that all knowledge should be ultimately reducible to sensations, and hence that theoretical statements cannot have a truth value. There is no way in which statements about goals and values may be given a truthvalue in view of this position.

The rejection of positivism by no means entails siding with epistemological realism. At the end of the twentieth century, this should be utterly obvious (but unfortunately it is not to epistemological realists). The denial of positivism involves the claim that statements about reality make sense. Subscribing to this claim is necessary to engage in the world views program. But there are many ways in which this claim may be subscribed to. For example, in their [15], Batens and Meheus defend a position that rejects both positivism and epistemological realism. Roughly, they argue that, on the one hand, humans need a world view in order to sensibly interpret the world and organize their lives, and that scientific creativity presupposes interpretations, but that, on the other hand, the positivists were basically right in claiming that no theoretical statement can ever be justified in any absolute or final way by a set of experiential data. This position rules out (especially convergent) epistemological realism, but leaves ample room for world views; and so do most present-day positions. (Also Rescher's [19] position entails a rejection of positivism that does not lead to epistemological realism.)

Some readers may still expect an explanation of the paradox that merely hypothetical world views are supposed to guide our lives. This paradox, however, arises only within a foundational framework. The discovery of such a framework was a respectable aim of the mainstream tradition in western philosophy. It was respectable because it was a demanding view on rationality, but it also has proved a (complete) failure. Two roads are open after the failure became apparent. On the one hand, one might still try to reach the apparently unreachable aim. This is respectable, but all odds are against it. By present insights, not too many people should engage in this direction. On the other hand, one might try to work out an alternative. This is an extremely demanding enterprise, because one has to start almost from scratch, and the resulting requirements on rationality (apart from being heavily under discussion) turn out to be drastically different from the traditional ones—see Batens [12], [13], [14]. We do not see a (respectable) alternative. Some surviving positivists, for example, admitting that the old story is not correct, claim that it is *roughly* correct. This position might be heuristically justifiable, and as such useful, but is not a challenge to other positions.

9. DOES THE WORLD VIEW PROGRAM PRESUPPOSE HARD OBJECTIVITY IN VALUE MATTERS?

Our comments on Apostel's value-objectivism—see section 5—are similar to those made in the previous paragraph. So, we can be rather brief here.

We claimed that the world views program is not compatible with an extreme positivistic position on values and norms. An adherent of such a position could still develop a world view that takes the arbitrary character of values into account. He or she might even adopt some moral or aesthetic principles provided no claim on justification is made. But nothing would prevent such a person to drastically change his or her world view at any point in time. Such a world view, or at least the aspects under discussion, would not be suitable to be compared to others; confrontation might lead to a change, but such a change could never justifiably be considered as an improvement. According to an extreme positivistic position, there is nothing available to guide our actions in a justifiable way—we argued before that the same obtains for experience and understanding.

All this does not entail that the construction of integrated world views presupposes Apostel's hard objectivism. It is merely required that statements about values and norms may be seen as justifiable, and that rational discussion about them makes sense and might lead to progress. In this respect, the new concept of rationality that has been emerging during the last decennia is extremely important. As this relative rationality is replacing the older foundational one, it not only becomes apparent that scientific activities are less governed by strict rules than one used to think, but at the same time it becomes clear that the type of rationality that characterizes the sciences may be justly applied to values and norms. In both cases, we cannot start from scratch, in both cases we may have good reasons to replace certain convictions by others, and in both cases the outcome is sometimes justly identified with progress.

Let us for a minute turn to cognitive values and norms. If humans are considered as integral parts of the world, the justification of those norms and values may be provided by properties of humans and by properties of the world. In view of those properties, we can argue that, in order for humans to obtain knowledge about the world (themselves included), some methods are available whereas others are not, some methods are more suitable than others, some cognitive values are to be preferred over others. Of course, such justifications are never built on the world itself, but always on human hypotheses about the world. This rules out the foundation of knowledge (and of cognitive values and norms), but it does not rule out justification in view of our present best insights. The dynamics of descriptive knowledge may change our methodological insights. The latter change will force us to reassess and perhaps to modify our descriptive knowledge. This mechanism involves the well-known circularity, but as long as we are able to see those changes as progressive, we may consider the circle as healthy. It is precisely this mechanism that governs present-day philosophy of science, where methodological discussions are largely based on present historical and systematic knowledge. It is this mechanism that explains the superiority of contemporary philosophy of science over the speculative and explication approaches of the past.

There is no reason why this situation should not be taken as a model for ethics and aesthetics. If the values from those domains are seen as relational, as determined by properties of humans and their environment, they may be justified along the same lines as cognitive values. It is correct that far less relevant factual knowledge is available outside the cognitive domain. It is equally correct that hardly any coherent and detailed theories have been proposed—moral theories are just sets of vague principles that are not even seen as unexceptionable by anyone (apart from some fundamentalists). Also, it will be hard to find funding for such research in the present economic and political circumstances. This clearly makes the enterprise much more difficult than in the case of epistemology, but there is no impediment in principle. Here lies, it seems to us, a further and independent merit of the world views program. Integrated world views will offer the opportunity to start the discussion and confrontation, and might at least clear the field for hypothetical moral theories. In this way, they will facilitate the realization of the older program of Apostel's [3], directed at the construction of a moral science.

The present section should not be read as a criticism on Apostel's ethical position. One of us does not share the hard moral objectivism, and wanted to emphasise that the position expounded in the previous paragraph has nothing subjectivistic. Apostel's much more radical objectivism is not a prerequisite for the world views program. His position is the more courageous one and, if it proves successful, it will lead to fascinating results that are beyond the capabilities of less extreme positions. In the plurality of systems that the world views program is directed to, it suggests a road that certainly deserves to be followed.

10. WHY WORLD VIEWS?

Part of the motivation behind the world views program is that the absence of world views has a disorienting effect on humans. It seems to us that their actual absence at this historical point is at least in part caused by the fact that the positivistic outlook on the sciences became an integral part of our culture, for example through science handbooks. The import of circumscribing philosophical and cultural nonsense is unquestionable. But any such circumscription involves the danger that large parts of our actual experiencing, understanding and acting are seen as outside the realm of rationality, and hence are left in the hands of obscure and dangerous political, religious and ideological 'leaders' and movements. Please do not misunderstand this. We are not pleading to engage in the obscure. On the contrary, we are urging that real and unavoidable aspects of our lives be brought into the clear. In the same vein, we do not want to return to the old and worn out ideologies, but we plea for world views that are integrated with our best knowledge, and that might help to extend our knowledge to new domains in the future. As Apostel phrased it:

> "This dream is again to provide our culture with a picture of the universe that would bring all of us together as actors in a shared environment; this scientific, and continuously corrected and modified 'grand picture' would be the forever evolving 'myth of the twenty-first century'. I claim that a rational and humanly meaningful image of the cosmos is a social necessity and, for this reason, I try to replace 'metaphysics' as pure speculation by an evolving scientific metaphysics." [7, p. 314]

In this sense, world views are not only necessary, they are also possible. Indeed, there are methods for devising integrated world views, and they may be criticized internally, but also externally by confronting them with other subsystems of our knowledge systems (most prominently with the sciences) as well as with each other. The latter requires that a plurality of world views is projected, for only such plurality can lead to confrontation and hence to real progress.

We have argued that several aspects of Leo Apostel's philosophy were a part of his (partial) world view, but are not prerequisites for the world views program. We should at least avoid a misunderstanding here. Indeed, we claim that many aspects of his philosophy are essential to the program, and that it is no means an accident that precisely he came to this program. The most eminent example is the idea of a lawlike and causal world of which humans form an integral part, and its consequences for ontology and epistemology. This idea leads to the necessity of world views: without them, human experience, understanding and actions are arbitrary, but the same idea also provides us with the means to construct world views and to submit them to a justificatory endeavour. We are causally linked to the world, by experience and action; but we are also able to formulate hypotheses about these causal connections. We construct highly sophisticated (hypothetical) theories about the world; but we are also able to formulate hypotheses about the effects of these theories on our experiencing the world, on our understanding of the world, and on our actions in the world. Given the kind of beings we are, we cannot understand ourselves without seeking an understanding of the rest of the world, just as our understanding of non-human aspects of the world presupposes knowledge about ourselves. Irrespective of one's personal interests, only an integrated world view will do.

The old ideal of knowledge (at least in the western culture) was the correct and complete unified theory. This ideal has proved unrealistic because it is radically beyond the capabilities of humans. It cannot even be seen as a gradually approachable target located at the end of times. The dissipation of the old ideal leaves us with a concept of a knowledge system that consists of scattered bits and pieces all of which are merely descriptive. Such knowledge systems are unfit to guide our lives. World views are an alternative to both the old ideal and the present scattered conception. In contradistinction to the old ideal, they are within the reach of humans. They are not all-embracing. They are only subsystems of knowledge systems. They should be integrated, but do not require that the knowledge system to which they belong is monolithic. In contradistinction to the scattered conception, world views are able to guide our experience, understanding and actions.

AFFILIATION

Diderik Batens and Wim Christiaens
Center for Logic and Philosophy of Science
University of Ghent
Belgium

REFERENCES

[1] **Apostel, L.**, *La loi et les causes*, Unpublished PhD. thesis (Université Libre de Bruxelles), 1953.

[2] **Apostel, L.**, "Can Metaphysics be a Science?", *Studia Philosophica Gandensia*, **1**, 1963, pp. 7–95.

[3] **Apostel, L.**, "Pluralistische grondslagen van de moraalwetenschap", in: *Problemen rond de moraalwetenschap*, Gent, RUG, 1965, pp. 69–126.

[4] **Apostel, L.**, *Matière et Forme* (2 volumes), Gent, Communication and Cognition, 1974.

[5] **Apostel, L.**, *Logique et dialectique*, Gent, Communication and Cognition, 1979.

[6] **Apostel, L.**, "Some remarks on ontology", in: **Agassi, J. & Cohen, R.S.** (eds.), *Scientific Philosophy Today*, Dordrecht, Reidel, 1981, pp. 1–44.

[7] **Apostel, L.**, *The Philosophy of Leo Apostel. Vol. 3: The Philosopher Replies*, (edited by F. Vandamme & R. Pinxten), Gent, Communication and Cognition, 1985.

[8] **Apostel, L.**, *Wereldbeelden. Ontologie en ethiek*, Antwerpen, Humanistisch Verbond, 1994.

[9] **Apostel, L. & Jan Van der Veken, J.**, *Wereldbeelden. Van fragmentering naar integratie*, Kapellen, DNB/Pelckmans, 1991.

[10] **Apostel, L. & Van Dooren, I.**, *The Philosophy of Leo Apostel. Vol. 1: A Life History*, (edited by F. Vandamme & R. Pinxten), Gent, Communication and Cognition, 1985 (this volume contains the bibliography of Leo Apostel up to 1989).

[11] **Batens, D.**, "Incommensurability is not a threat to the rationality of science or to the anti-dogmatic tradition", *Philosophica*, **32**, 1984, pp. 117-132.

[12] **Batens, D.**, "Meaning, acceptance, and dialectics", in: **Pitt, J.C.** (ed.), *Change and Progress in Modern Science*, 1985, pp. 333-360.

[13] **Batens, D.**, *Menselijke kennis. Pleidooi voor een bruikbare rationaliteit*, Leuven/Apeldoorn, Garant, 1992.

[14] **Batens, D.**, "Do we need a hierarchical model of science?", in: John Earman (ed.), *Inference, Explanation, and Other Frustrations. Essays in the Philosophy of Science*, 1992, pp. 199-215.

[15] **Batens, D. & Meheus, J.**, "In-world realism *vs.* reflective realism", in: **Douven, I. and Horsten, L.** (eds.), *Realism in the Sciences*, Leuven, Universitaire Pers, 1996, pp. 35–53.

[16] **Carnap, R.**, "Probability as a guide in life", *Journal of Philosophy*, **44**, 1947, pp. 141–148.

[17] **Earman, J.**, *A Primer on Determinism*, Dordrecht, Reidel (now Kluwer), 1986.

[18] **Kuhn, T.S.**, *The Structure of Scientific Revolutions*, 1962. [=*International Encyclopedia of Unified Science* vol. **II**, nr. 2], Chicago, The University of Chicago Press (2nd enlarged ed. 1970).

[19] **Rescher, N.**, *Scientific Realism. A Critical Reappraisal*, Dordrecht, Reidel, 1987.

[20] **Vandamme, F. & Pinxten, R.** (eds.), *The Philosophy of Leo Apostel*, Gent, Communication and Cognition, 1989.

[21] The Worldviews Group, *Perspectives on the World. An Interdisciplinary Reflection*, Brussels, VUBPRESS, 1995. (translation of *Cirkelen om de wereld. Concrete invullingen van het wereldbeelden-project* Kapellen, Pelckmans, 1994).

[22] **Plato**, *The Sophist*, Hamilton, Cairns Edition, quoted in [7].

IMMANUEL WALLERSTEIN

"THE STRUCTURES OF KNOWLEDGE, OR HOW MANY WAYS MAY WE KNOW?"

The Report of the Gulbenkian Commission bears the title, *Open the Social Sciences*. The title bears witness to the sense of the Commission that the social sciences have become closed off, or have closed themselves off, from a full understanding of social reality, and that the methods which the social sciences had historically developed in order to pursue this understanding may themselves today be obstacles to this very understanding. Let me try to summarize what I think the Report says about the past 200 years, and then turn to what this implies for what we should now do.

The Commission saw the enterprise of the social sciences as an historical construction, institutionalized primarily in the period 1850-1945. We emphasized that this construction was therefore quite recent, and that the way in which social science was constructed was neither inevitable nor unchangeable. We tried to explain what elements in the nineteenth-century world led those who constructed this edifice to make the decisions that were made concerning the distinctions that were created between a named list of "disciplines." We sought to outline the underlying logic that accounted for why the multiple disciplines adopted various epistemologies and why each chose certain practical methodologies as their preferred ones. We also tried to explain why the post-1945 world found this logic constraining and set in motion a series of changes in the academy which had the effect of undermining the distinctions among the disciplines.

The picture that we drew of the history of the social sciences was that of a U-shaped curve. Initially, from 1750-1850, the situation was very confused. There were many, many names being used as the appellations of proto-disciplines, and none or few seemed to command wide support. Then, in the period 1850-1945, this multiplicity of names was reduced to a small standard group clearly distinguished the ones from the others. In our view, there were only six such names that were very widely accepted throughout the scholarly world. But then, in the period from 1945 on, the number of legitimate names of fields of study has been once again expanding and there is every sign that the number will continue to grow. Furthermore, whereas in 1945 there still seemed to be clear demarcations that separated one discipline from another, these distinctions have in the subsequent period been steadily eroded, so that today there is considerable de facto overlap and confusion. In short, we have in a sense returned to the situation of 1750-1850 of a large number of categories which do not provide a useful taxonomy.

But this overlap and confusion is the least of our problems. This process of defining the categories of the social sciences has been occurring within the context of a much larger turmoil that goes beyond the social sciences and implicates the entire world of knowledge. We have been living for 200 years in a structure of the organization of knowledge in which "philosophy" and "science" have been considered distinctive, indeed virtually antagonistic, forms of knowledge. It is salutary to remember that this was not always so. This division between the so-called "two cultures" is also a rather recent social construction, only a bit older than that which divided up the social sciences into a specified list of disciplines. It was in fact virtually unknown anywhere in the world before the middle of the eighteenth century.

The secularization of society, which has been a continuing feature of the development of the modern world-system, expressed itself in the world of knowledge as a two-step process. The first step was the rejection of theology as the exclusive, or even the dominant, mode of knowing. Philosophy replaced theology; that is, humans replaced God as the source of knowledge. In practice, this meant a shift of locus of the authorities who could proclaim the validity of knowledge. In place of priests who had some special access to the word of God, we honored rational men who had some special insight into natural law, or natural laws. This shift was not enough for some persons, who argued that philosophy was merely a variant of theology: both proclaimed knowledge as being ordained by authority, in the one case of priests, in the other of philosophers. These critics insisted on the necessity of evidence drawn from the study of empirical reality. Such evidence, they said, was the basis of another form of knowledge they called "science." By the eighteenth century, these protagonists of "science" were openly rejecting "philosophy" as merely deductive speculation, and proclaiming that their form of knowledge was the only rational form.

On the one hand, this rejection of philosophy seemed to argue a rejection of authorities. It was in that sense "democratic." The scientists seemed to be saying that anyone could establish knowledge, provided he (or she) used the right "methods". And the validity of any knowledge that any scientists asserted could be tested by anyone else, simply by replicating the empirical observations and manipulation of data. Since this method of asserting knowledge seemed to be capable of generating practical inventions as well, it laid claim to being a particularly powerful mode of knowing. It was not long therefore before "science" achieved a dominant place in the hierarchy of knowledge production.

There was one major problem, however, in this "divorce" between philosophy and science. Theology and philosophy had both traditionally asserted that they could know two kinds of things: both what was true and what was good. Empirical science did not feel it had the tools to discern what was good; only what was true. The scientists handled this difficulty with some panache. They simply said they would try only to ascertain what was true and they would leave the search for the good in the hands of the philosophers (and the theologians). They did this knowingly and, to defend themselves, with some disdain. They asserted that it was more important to know what was true. Eventually some would even assert that it was impossible to know what was good, only what was true. This division between the true and the good constituted the underlying logic of the "two cultures." Philosophy (or more broadly, the humanities) was relegated to the search for the good (and the beautiful). Science insisted that it had the monopoly on the search for the true.

There was a second problem about this "divorce." The path of empirical science was in fact less "democratic" than it seemed to claim. There rapidly arose the question of who was entitled to adjudicate between competing scientific claims to truth. The answer that the scientists gave was that only the community of scientists could do this. But since scientific knowledge was inevitably and increasingly specialized, this meant that only subsets of scientists (those in each subspecialty) were deemed part of the group that had a claim to judge the validity of scientific truth. In point of fact, these groups were no larger than the group of philosophers who had previously claimed the ability to judge each other's insights into natural law or laws.

There was a third problem about this "divorce." Most persons were unwilling truly to separate the search for the true and the good. However hard scholars worked to establish a strict segregation of the two activities, it ran against the psychological grain, especially when the object of study was social reality. The desire to reunify the two searches returned clandestinely, in the work of both scientists and philosophers, even while they were busy denying its desirability, or even possibility. But because the reunification was clandestine, it impaired our collective ability to appraise it, to criticize it, and to improve it. All three difficulties were kept in check for 200 years, but they have returned to haunt us in the last third of the twentieth century. The resolution of these difficulties constitutes today our central intellectual task.

There have been two major attacks on the trimodal division of knowledge into the natural sciences, the humanities, and the social sciences. Neither of these attacks has come from within the social sciences. These attacks have come to be called "complexity studies" (in the case of the natural sciences) and "cultural studies" (in the case of the humanities). In reality, starting from quite different standpoints, both of these movements have taken as their target of attack the same object, the dominant mode of natural science since the seventeenth century, that is, that form of science which is that based on Newtonian mechanics.

To be sure, in the early twentieth century Newtonian physics had been challenged by quantum physics, but quantum physics still shared the fundamental premise of Newtonian physics that physical reality was determined and had temporal symmetry, that therefore these processes were linear, and that fluctuations always returned to equilibria. In this view, nature was passive, and scientists could describe its functioning in terms of eternal laws, which could eventually be asserted in the form of simple equations. When we say that science as a mode of knowing became dominant in the nineteenth century, it is this set of premises of which we are speaking. That which could not be fit into this set of premises, for example, entropy (which is the description of necessary transformations in matter over time), was and is interpreted as an example of our scientific ignorance, which could and would eventually be overcome. Entropy was seen as a negative phenomenon, a sort of death of material phenomena.

Since the late nineteenth century, but especially in the last twenty years, a large group of natural scientists has been challenging these premises. They see the future as intrinsically indeterminate. They see equilibria as exceptional, and see material phenomena as moving constantly far from equilibria. They see entropy as leading to bifurcations which bring new (albeit unpredictable) orders out of chaos, and therefore the process is not one of death but of creation. They see auto-organization as the fundamental process of all matter. They resume this in two basic slogans: not temporal symmetry but the arrow of time; not simplicity as the ultimate product of science, but rather the explanation of complexity. It is important to see what complexity studies is and what it is not. It is not a rejection of science as a mode of knowing. It is a rejection of a science based on a nature that is passive, in which all truth is already inscribed in the structures of the universe. What it is rather is the belief that "the possible is 'richer' than the real."[1] It is the assertion that all matter has a history, and it is its sinuous history which

[1] **Ilya Prigogine**, *La fin des certitudes*, Odile Jacob, Paris, 1996, p. 67.

presents material phenomena with the successive alternatives between which each "chooses" throughout its existence. It is not the belief that it is impossible to know, that is, to understand how the real world operates. It is the assertion that this process of understanding is far more complex than science traditionally asserted that it was.

Cultural studies attacked the same determinism and universalism under attack by the scientists of complexity. But for the most part those who put forward these views neglected to distinguish between Newtonian science and the science of complexity, or in many cases to be aware of the latter. Cultural studies attacked universalism primarily on the grounds that the assertions about social reality that were made in its name were not in fact universal. It represented an attack against the views of the dominant strata in the world-system which generalized their realities into universal human realities, and thereby "forgot" whole segments of humanity, not only in the substantive statements but in the very epistemology of their research.

At the same time, cultural studies represented an attack on the traditional mode of humanistic scholarship, which had asserted universal values in the realm of the good and the beautiful (the so-called canons), and analyzed texts internally as incarnating these universal appreciations. Cultural studies insists that texts are social phenomena, created in a certain context, and read or appreciated in a certain context.

Classical physics had sought to eliminate certain "truths" on the grounds that these seeming anomalies merely reflected the fact that we were still ignorant of the underlying universal laws. Classical humanities had sought to eliminate certain appreciations of "the good and beautiful" on the grounds that these seeming divergences of appreciation merely reflected the fact that those who made them had not yet acquired good taste. In objecting to these traditional views in the natural sciences and the humanities, both movements—complexity studies and cultural studies—sought to "open" the field of knowledge to new possibilities that had been closed off by the nineteenth-century divorce between science and philosophy.

Where then does social science fit in this picture? In the nineteenth century, the social sciences, faced with the "two cultures," internalized their struggle as a *Methodenstreit.* There were those who leaned toward the humanities and utilized what was called an idiographic epistemology. They emphasized the particularity of all social phenomena, the limited utility of all generalizations, the need for empathetic understanding. And there were those who leaned towards the natural sciences and utilized what was called a nomothetic epistemology. They emphasized the logical parallel between human processes and all other material processes. They sought to join physics in the search for universal, simple laws that held across time and space. Social science was like someone tied to two horses galloping in opposite directions. Social science had no epistemological stance of its own and was torn apart by the struggle between the two colossi of the natural sciences and the humanities.

Today we find we are in a very different situation. On the one hand, complexity studies is emphasizing the arrow of time, a theme that has always been central to social science. It emphasizes complexity, and admits that human social systems are the most complex of all systems. And it emphasizes creativity in nature, thus extending to all nature what was previously thought to be a unique feature of Homo Sapiens.

Cultural studies is emphasizing the social context within which all texts, all communications, are made, and are received. It is thus utilizing a theme that has always been central to social science. It emphasizes the non-uniformity of social reality and the necessity of appreciating the rationality of the other.

These two movements offer social science an incredible opportunity to overcome its derivative and divided character, and to place the study of social reality within an integrated view of the study of all material reality. Far from being torn apart by horses galloping in opposite directions, I see both complexity studies and cultural studies as moving in the direction of social science. In a sense, what we are seeing is the "social scientization" of all knowledge.

Of course, like all opportunities, we shall only get *fortuna* if we seize it. What is now possible is a rational restructuring of the study of social reality. It can be one that understands that the arrow of time offers the possibility of creation. It can be one that understands that the multiplicity of human patterns of behavior is precisely the field of our research, and that we may approach an understanding of what is possible only when we shed our assumptions about what is universal.

Finally, we are all offered the possibility of reintegrating the knowledge of what is true and what is good. The probabilities of our futures are constructed by us within the framework of the structures that limit us. The good is the same as the true in the long run, for the true is the choice of the optimally rational, substantively rational, alternatives that present themselves to us. The idea that there are "two cultures," *a fortiori* that these two cultures are in contradiction to each other, is a gigantic mystification. The tripartite division of organized knowledge is an obstacle to our fuller understanding of the world. The task before us is to reconstruct our institutions in such a way that we maximize our chances of furthering collective knowledge. This is an enormous task, given the inherent conservatism of institutional authorities and the danger such a reconstruction poses to those who benefit from the inegalitarian distribution of resources and power in the world. But the fact that it is an enormous task does not mean that it is not doable. We have entered a bifurcation in the structures of knowledge, which appears in many ways to be chaotic. Of course, we shall emerge from it with a new order. This order is not determined, but it is determinable. We can only have *fortuna* if we seize it.

AFFILIATION

Immanuel Wallerstein
Fernand Braudel Center
Binghamton University
United States

NOTE

Presentation at "Which Sciences for Tomorrow? Dialogue on the Gulbenkian Report: *Open the Social Sciences*," Stanford University, June 2-3, 1996.

ROBERT MAIER

DISCOURSES OF GLOBALISATION AND WORLD VIEWS

INTRODUCTION

globalisation, as a term and as a concept, is used in a variety of political, economic and social contexts. Not only can we find lots of conferences and books on 'globalisation', there are also several bastard forms of use, such as 'glocalization' [13] or 'globalitarian regimes' [12], which constitutes a good proof of the discursive success of this concept. This success is not based on any precise meaning of the concept; on the contrary, the multiplicity of meanings and contexts of use seem to be characteristic of the discursive success.

globalisation can be used to designate a particular phase of technological or economic development, or to point to the role of financial capital, or to the reorganization of the production process. On the other hand, globalisation seems to be the culprit when trying to understand the decline of Nation States. Others use the concept as a characteristic of the present-day class struggle. These examples are far from exhaustive, but they indicate the fact that various discourses have emerged or are brought up to date based on the concept of globalisation.

These discourses use this concept either as a description of a state of affairs or of an ongoing process or as an explanation of supposedly factual phenomena, such as the decline of the Nation State. In all these cases, these discourses articulate also a more or less explicit world view. While this is true of any discourse whatsoever, it is particularly true of the discourse of globalisation, because these discourses have as central theme the state of affair and the processes of the social world. The notion of world view will be defined in detail later on, at the moment it is sufficient to offer the following short description: by world view is meant the representation of individuals and society (and eventually of the cosmos), the representation of their history, their goals and their motives of action.

In this article I will examine the following two questions: (1) which world views are more or less explicitly promoted and articulated by the various discourses of globalisation? (2) Starting from a typology of world views I will ask the question how the various discourses of globalisation can be understood?

In order to answer these two questions, I will start by presenting a method how discourses can be classified. In a following part I will present a typology of world views. With the help of these two points I will try to answer the two questions asked.

CLASSIFYING DISCOURSES

A discourse is a social form of action by which a text, written as well as spoken, is produced, distributed and consumed in a specific way. As a social form of action, discourse is goal-oriented, but can as any social action also produce un-intended consequences. A good overview of discourse theories is offered by Fairclough [3].

With this definition one can identify a certain number of criteria which permit to classify different discourses. For the purpose of this article it is sufficient to use the following criteria:

(1) the source of the discourse. This can be an individual author who writes as an autonomous person, but that is rather an exception. In general, the author is clearly situated socially and politically, as a member of a political movement or as working for a certain institution or as representing a social movement;
(2) the constructed auditorium, or in other words the intended addressee of the discourse. A discourse can be addressed to all reasonable individuals, or specifically to scientists or to citizens of a certain Nation State;
(3) the form of distribution; a given discourse can be distributed in the specific form of a book, or as more or less explicit presuppositions in a variety of news items in different media;
(4) the main characteristics of the discursive object. Any discourse presents a representation of some part of the world, and this representation as a discursive object has specific characteristics. For globalisation, for example, I distinguished in the introduction different meanings, and these meanings can be used in order to classify discourses of globalisation;
(5) the social identities shaped with the help of the discourse and their relationships. In any discourse appear social actors, such as the social classes, Nation States or politicians in the discourses of globalisation, actors which are related to each other in specific ways. An examination of the actors put on the scene by the different discourses of globalisation permits to distinguish families of such discourses which introduces different actors or at least different types of relationships between them;
(6) the explicit aims formulated by the discourse and the actions which are supposed to be successful for reaching these goals. Discourses of globalisation, for example, formulate quite different aims. On the one

hand, globalisation as a process can be depicted as almost natural, which should not be hindered at all, or on the other as a very undesirable tendency which has to be stopped or which should be controlled.

With the help of these criteria I will suggest in a following section a classification of the main discourses of globalisation. However, it should be noted, that the use of these criteria is not altogether an innocent and neutral affair.

First of all, the use of these criteria implies that in what follows I will not be concerned with a quest of what is true or not, in the various discourses. With these criteria one can only describe the reasons offered and the arguments formulated by the different discourses of globalisation without any possibility to reject one or more of these discourses. For this goal, other instruments will be needed [9].

Another reservation is related more intimately with the subject of this article. My purpose is to examine the world views which are put forward by the various discourses of globalisation. In order to do this I will first establish a classification of the discourses of globalisation and look for the ways by which each of these discourses elaborates at least partly a world view. Such a procedure supposes that the established classification of discourses is not biased with respect to world views. In other words, the criteria used should not entail consequences concerning conceptions of world views in the first place. Because if these criteria have such consequences, the possible results will at least be partly an effect of the method used for establishing them. However, such a guarantee can not be offered. Why? The conception of discourse used here in order to formulate the six criteria cannot be said to be totally independent of any world view. On the contrary, this theory of discourse belongs to a specific world view, and other world views would reject completely such a conception of discourse. This is really a very serious reservation which I mention here, but which cannot be avoided. In the last part of this paper I will come back to this delicate point.

THE FIVE MAIN WORLD VIEWS

"A world view is a coherent collection of concepts and theorems that must allow us to construct a global image of the world, and in this way to understand as many elements of our experience as possible" [16, 17]. This definition of world views formulated by the Worldviews Group should be sufficient for our purpose. In this perspective world views are attempts of individuals or societies to construct a global framework for the answers to the questions concerning their being and becoming and the being and becoming of the world. A world view is a frame of reference in which everything presented to us by our diverse experiences can be placed. It is a symbolic system of representation offering a global picture.

At present, in the modern (or post-modern according to some) world, characterized by pluralism and individualism, the classic world views have not completely disappeared. On the contrary, some of these world views which have been constructed in specific historic periods are reactivated and transformed, and others are rather original creations of the present time.

A world view is a global framework, which offers schematic answers to the following questions [16, 25]: What is the nature of our world? How is it structured and how does it function? Why do we feel the way we feel in this world? How are we to act and to create in this world? How, and in what ways can we influence the world and transform it?

World views as conceptual and symbolic frameworks are not necessarily explicitly formulated, that is why they belong to the historical and cultural context. World views are not exclusively concerned with questions of knowledge, also emotions and values have a place in a world view, as can be seen by the questions formulated in the last paragraph.

At the moment there are only some certainties about certain uncertainties concerning world views, which I will shortly present here. None of the known world views is completely worked out, we dispose at the moment only of partial constructions of world views. As each of the known world views offers a global, symbolic framework, it is possible to integrate the same experience in different world views. However, the different world views exclude each other on the one hand, but on the other hand they need each other for further constructions.

These remarks underline the limitations of the present world views. Given these limitations, the question arises: why should we be concerned with them at all? Indeed, there are many who reject the preoccupation with world views with the argument that it is only a speculative quest for global meaning.

I would like to offer here just one good reason for taking world views seriously. If we reject the idea that there is a simple and single clear answer to the question of how we live in this world and what the future of it can and should be, we can at least understand the importance of the question in relation with the ongoing social transformations of the world at the present time. If we do not want to be subjected to unwanted consequences of these transformations and their meanings, we should at least try to understand the possible meanings and directions of these transformations. This task can only be undertaken by interpreting the effects of these transformations within the various global frameworks at our disposition; in other words, with the help of the existing, partially elaborated world views. But this task also entails that the different world views will have to be explicated and elaborated.

In this paper I will limit myself to use some results of the work of the Worldviews Group and in particularly of Leo Apostel [1]. Apostel, building on the works of Dilthey and Pepper, has distinguished five families of world views. These world views have a certain robustness, which means that even if they have been elaborated in specific historic periods, they are constantly taken up again and re-elaborated in later periods up to the present. In other words, these five world views have a high degree of fitness, adaptability and internal variability.

(1) *Formism*: similarity is the basic metaphor of this world view. Reality is made up by elementary entities and by characteristics which some of the entities have in common. These characteristics can either belong to a separate sphere or be immanent; think of Plato or Aristotle. The elementary concrete entities and the concrete elementary characteristics can constitute more complex entities and characteristics which can be ordered in hierarchies. The two basic experiences of formism are on the one hand the work of producers who manufacture many exemplars according to a basic plan and on the other the existence of natural kinds, such as minerals or plants. The duality between entities and characteristics entails also a duality between forms and materials.

The ethic which belongs to this world view consists of norms which specify static or dynamic forms of equilibrium between persons and groups. Deviations from these norms will cause suffering. Because of the dualist nature of this world view norms have to be realized as ideals. However, it will be an objective ethic. What is of value for a species will depend on the place this species occupies in the hierarchy of reality. Truth will be defined as a correspondence with reality.

Modern versions of this world view use set theory and the relation of equivalence. In other words, modern formulations of such a world hypothesis will conceive reality as constituted by and through sets, relations and groups, with the emphasis on 'by' and 'through'.

(2) A second world view with a long history is *mechanism*. The basic metaphor of this world view is the lever, or later the clockwork or more recently the computer and the electromagnetic field. There are continuous and discontinuous versions of this world view. As the first metaphor shows, the machine is a central figure. A machine is a configuration of discontinuous places in a continuous space-time; typical are the number of components, their relations and the masses and forces involved. Conceived dynamically in time, the components describe movements with

regular relations. There are strictly deterministic versions of the mechanistic world view, but also probabilistic ones have been formulated. The general tendency of a mechanistic world view consists in the unification of all discontinuities in one general field.

Next to the primary qualities, one can find derived, secondary qualities as for example in the conceptions of a mechanistic psychology. Human individuals are conceived as material systems with specific characteristics which are derived from the primary ones. Truth is more problematic in this world view because of this conception of individuals. A judgement will be true if the reactions which this judgement causes are adapted to the facts on which the judgements bears. There is no immediate relation with an ethic, because values or norms are evidently secondary qualities. But there is no evident and satisfactory derivation of secondary qualities from primary ones. However, a pragmatic and purely operational conception of value can be formulated by fitness.

The modern versions of mechanism use the language of automata, such as the quantum computer.

(3) A third world view is *contextualism or process*, with the complex process (historic action or event) as basic metaphor. Acts and events are in principle complex, they are composed of a great number of interrelated activities with continuously changing patterns. Fundamental for this world hypothesis are the passages from chaos to order and from order to chaos and the importance attributed to transformation and novelty as a global quality. The quality of a process will be described by its extension (the number of objects the process will involve), the degree of change it can bring about and the continuity of the process. Important will be the number of partial (sub)processes, their interrelations and their relations with other processes. However, no definitive analysis of a process will be possible. Pragmatic philosophy can stand for this world view.

Persons and also groups are conceived as historical processes. Values are defined in an indirect way, by the degree of complexity, novelty and emergence (in the sense of new qualities which arise in a process) of actions. A rather dynamic version of ethic has to be formulated.

(4) *Organicism* offers a fourth world view, using integration or the organism as a basic metaphor. This world view also refers to processes as the previous one, but in difference with it, this world view inscribes processes and time in a totality. A model thinker of this world view is Hegel, but there are many other versions.

The central categories of this world view are parts, connection, contradiction, totality and synthesis. A value theory will de derived from the idea that 'the ideal is the real thing'. A person (or group) will realize more reality and value if more real internal and external conflicts are lived through and integrated in a higher, however, conserving synthesis.

There are many possible versions because the basic metaphor 'organism' or 'integration' can be conceived in a variety of ways. For example, an organism can be conceived (a) from an embryological perspective, (b) from an evolutionary perspective, (c) as a metabolism, which means as a dynamic, open system, or (d) by its morphology or building plan.

System theory and organization theory are modern variations of this world view.

(5) Finally, a new world view has been formulated more recently, which can be called *chaoism*, with play as a basic metaphor. The world is conceived as having no cause, no aim, no reason, no value and also no plan. In other words, there is no totality but only chaos. Conceiving the world as play in this sense, means that there are no players and that the play is not performed for anybody. The value theory of Nietzsche goes with this world view. Human individuals should be maximally open towards the world, an openness which is needed in order to relativize all norms and values and to be able to sing and dance in a detached way.

This schematic presentation of the five world views leaves many questions open. I have tried here to present not only the general characteristics of these world views, but also some of their variations and modern reformulations and their possible links with ethic. Once more it should be underlined that none of these world hypotheses is fully worked out, and that each one has some strong points but also weak points. It should also be mentioned that there are several attempts to combine or integrate a specific world view with another one; an example of such an integration will be given later.

A CLASSIFICATION OF DISCOURSES OF GLOBALISATION

I will now apply the criteria established in the second section and examine the discourses of globalisation in order to establish a classification. Five families of discourse can be distinguished clearly with the help of these criteria. For each type or family, I will also examine if this discourse uses and elaborates one or more of the five world views.

(1) A first family of discourses of globalisation can be identified as originating from the New Right movements or parties in various European countries. These discourses are addressed to all the citizens of a Nation State who feel threatened by some of the transformations which are considered as being brought into being by globalisation. This discourse is

distributed mainly by speeches and pamphlets. The central themes of this discourse are 'flows' of migration and the threat of a deterioration of working, housing and living conditions in the given Nation State. In other words, globalisation or at least some of its consequences undermine according to this discourse the national identity and culture and the national well-being. Three social identities play a central role: the good, national citizens, the cosmopolists, linked to financial capital who have no consideration for national identity and well-being and the migrants who threaten the working and living conditions of the good citizens. In other words, the good, the bad and the ugly.

The explicit aim of this discourse consists in the defense of national identity and well-being. Barjon [2] offers a good analysis of the French situation.

This discourse considers only very specific aspects of globalisation, such as the flows of migration which are supposed here to be due to globalisation or the consequences of globalisation on the working and living conditions in a given state.

In this discourse we find a very limited and specific elaboration of an organismic world view. A sub-species and its ecosphere are considered, and the threat to the survival of this sub-species by an unwanted mix with other sub-species and by a destruction of its ecosphere. The intrinsic value of this sub-species is presupposed, and the aim is its defense and preservation. In the language of the organismic world view one could say that this discourse is centered on one part and its intrinsic value, with the aim to defend the identity and living conditions of this part. There are conflicts and contradictions, but there is no search for any new synthesis. The totality, in this case the whole of mankind, is conceived in a static way, being composed by a certain number of sub-species with their particular environments.

In short, this family of discourses uses only a very poor, and problematic part of the possibilities offered by the organismic world view. Even if we consider only one possible elaboration of this world hypothesis with the help of biological metaphors, namely an evolutionary perspective, many other questions could be formulated. For example, one might ask if a certain mix would not favor the adaptability or offer new possibilities?

(2) A second family of discourses depicts globalisation as a quasi-natural consequence of technological and/or economic developments. The authors of this discourse are international organizations, such as the IMF (International Monetary Fund) or the WTO (World Trade Organization), many politicians and managers. These authors belong to what is called the neo-liberalist tendency. The discourse is addressed to 'reason-

able' and 'rational' citizens of the world. The discourses are distributed through books [7], by declarations, reports, interviews, etc. globalisation is represented as a natural consequence of technological and/or economic factors. In order to avoid misunderstandings is should be specified that from an economic point of view technological innovations comprise all innovations of the production process. These may be technological innovations in the usual sense of the term, such as the new information technology, but also new ways of organizing production, as for example by a new management of communication.

According to this discourse the new technological possibilities of production, ranging from financing, the distribution of production and its management to the technical procedures involved push the production and consumption of goods and values in a more global order.

In this discourse three main social identities are set up. First of all, the responsible managers of this globalizing venture. In the second place one encounters the 'world citizens' who can in the long run all profit from this global extension. However, the discourse distinguishes between two sub-categories of this social figure: one sub-category of well qualified producers can profit immediately from the new opportunities offered, whereas a second sub-category, generally called the 'uneducated' or 'unqualified' face eventually more or less serious problems in the short run. However, there is no need to deplore them because in the long run they will also profit from the globalizing venture. Finally, a third category is introduced, which is formed by persons and groups who do not understand what is going on. They try to negate the natural course by resisting the changes and opportunities offered by the new technological form of production, by being attached to quite conservative values and forms of life. The relationships between these three social identities are quite peculiar, because the first one is just a specific designation of the first sub-group of the second identity, whereas the third type of identity may correspond with parts of the second sub-group of the second identity.

The explicit aim of this discourse is to point out the pointlessness of throwing sand in the machine. If globalisation is the automatic consequence of new technological possibilities, any resistance or artificial attempt to stop the ongoing tendencies will only have problematic effects for everybody. Or in other words, trying to stop time is either a sign of madness or of total irresponsibility.

At present, this rich family of discourses occupies a rather dominant position, which is evident as well by the social position of its producers as by the powerful means of distribution involved.

This discourse elaborates a world view of mechanism. The production process and its technological innovations occupy a central place. Human individuals are in the first place conceived as productive forces which contribute to production. The production and distribution of goods is conceived as the autonomous motor which functions according to its own laws. The new technological opportunities entail quite naturally a globalizing tendency of unification. Psychological and social characteristics are derived qualities which should not stand in the way or hinder the natural functioning. Being good and truthful means to accept the natural laws, by taking account of them and by not opposing them.

The mechanistic world view, even in its modern formulations, is more or less fully used by the various discourses of this family. This is an important difference with the first family of discourses of globalisation. This is the strength of this discourse, it is within this world view quite consistent. But at the same time, this is also a sign of weakness, because moral, psychological, social and political characteristics are conceived as derived and secondary.

(3) A third family of discourses of globalisation is based on a fundamental distinction: there are on the one hand technological and economic structures with their own laws, and on the other hand social relations which are and can be based on idealistic principles. This type of discourse is produced by intellectuals and politicians who accept on the one hand economic laws, but who consider that a certain quality of social relationships is and should be independent of economy. This discourse is addressed to the responsible citizens of the world, who are supposed to be accountable for the quality of social relations. This discourse is distributed by books, appeals, declarations, new items, etc.

The main theme of this discourse is about the possible dangers of the ongoing forms of globalisation for the quality and conditions of social relations. This discourse does not negate the possible positive effects of globalisation which offer new opportunities for human individuals and groups, nor does it reject globalisation en bloc. However, globalisation considered basically as an economic enterprise may disturb and eventually destroy the acquired qualities of social relations. In the first place, democracy appears as threatened, but good candidates—in the sense of being disturbed—are also the environment, the Nation State, international negotiations, etc.

This discourse introduces the following social identities: on the one hand there are the managers of the global economy and on the other hand are the responsible world citizens. The first ones are blindly concerned with the economic affairs, whereas the second ones are mainly concerned with the preservation of the quality of social relations. There is a possible third form of identity, constituted by persons who are either excluded and marginalized or who are not conscious of the effects of the ongoing waves of globalisation. These groups have not at all a fixed membership, members of the first and of the third group can join forces with the second, responsible one. This is also the explicit aim of this discourse which is directed against the unnecessary and destructive influences of globalisation by trying to set up political forms of management and control of the problematic consequences of globalisation.

A typical example of this discourse can be found in the work of the group of Lissabon [5]; in their book they suggest that a certain number of new international contracts concerning democracy and the environment have to be concluded in order to preserve the quality of life, which is seriously threatened by the consequences of globalisation. Other authors are Martin and Schumann [10].

In this family of discourses one can find many variations, with centers of attention ranging from the problems of the Nation State in the age of globalisation, to the quality of democracy or of international relations, to the environment or human rights. In all these cases, however, a solution can be found by political decisions. More or less implicitly, this discourse accuses the managers of globalisation to have sold out on democracy, human rights, etc.

This discourse is formulated with the help of an organismic world view. Indeed, according to this discourse the totality is threatened by the effects of one part, namely the economy with its globalizing tendencies. Because of the autonomization of the economic sphere the totality can break apart. This threat can only be contained by the political sphere, which must affirm its predominance by setting up democratic forms of control, management and regulation at the level of the Nation State and on an international level. In other words, the conflicts and contradictions between economy and politics have to be transcended by a new synthesis which recognizes the relative autonomy of the economy, while regulating it by new political forms of control.

A rather problematic dualism is involved in this discourse, with the economy and its tendencies on the one hand, and politics as an expression of the general will on the other.

(4) A fourth family of discourses of globalisation is formulated by some post-marxists, sometimes called regulation theorists, such as Lipietz [8], Went [15] and Hirsch [6]. The authors are academics or critical intellectuals. They address the general public, but the fact that this discourse is only distributed by books and articles in scholarly journals imposes a serious limitation on the real auditorium.

This discourse attempts to integrate a structural approach with a genetic, historical one. According to this discourse, there are socio-historical periods, and in each of these periods one can find a specific form of economic activities and of social-political forms. In other words, economy and politics are not conceived as independent entities of a different nature. They are interdependent, and in each period there will be a new structure which specifies the boundaries of the economic and of the political domain as well as their interdependencies. This structure is called a regime of regulation. In our century, there was a dominant regime of regulation, called 'fordism' up to the seventies. For several reasons this regime could not function any more, and in the last twenty years we are according to this discourse in a phase of transition characterized by the emergence of new forms of production and above all by struggles about the new definition of what belongs to the economic domain and what to the social and political one. globalisation designates these transformations. These movements and struggles have not yet produced a new regime of regulation.

In this discourse one finds many social identities, related to the various events and actions which are part of the ongoing, very much disputed, construction of a new regime of regulation. The aim of this discourse is mainly directed at a clarification of what is going on, in order to understand and to orient practical interventions.

This discourse attempts to integrate two world views, formism and contextualism. Indeed, this discourse introduces general structures, called regimes of regulation, which are supposed to govern social and economic activities for a certain period of time. At the same time, any of these structures, by its functioning itself can be destroyed, and be replaced by a new one through a multiplicity of processes of new constructions during a phase of transition. Such an integration of world view, how interesting and challenging it may be, raises many questions. I will limit myself to one problem which is quite central for such an integration. There is a fundamental tension between the reality of structures on the one hand, and the reality of the processes of construction on the other. At moments, when a regime of regulation is well established, the reality of structures is predominant, and in phases of crisis and transitions, the processes of construction will be dominant. This tension can be seen in the discourses of this family which either pretend that fordism is a regime of the past or that it is still partly in existence, or that globalisation as a multiplicity of processes is already replacing it by parts of a new regime of regulation or only preparing a new regime.

(5) A fifth type of discourse on globalisation attempts to offer an original interpretation of globalisation. globalisation as a general tendency or as

a basic process is rejected as a substantialist illusion. In fact, what we encounter is an ever increasing mix of traits and aspects of activities from all over the world. This discourse is produced by some intellectuals [11] and cultural agents. It is addressed in principle to the general public, but the actual auditorium is mainly constituted by young people. Its distribution occurs by articles and declarations in the fora of the cultural avant-garde.

The main theme of this discourse are the ever-increasing and ever-changing forms of mix of elements of practice (information, clothing, food, articles, music, activities, etc.) from all over the world. According to this discourse there are not only more intercultural encounters, but there is a general tendency to break down global practices in bits and pieces and to recombine them in various ways. Through these fortuitous encounters and recombinations new, rather unstable and fragmented identities [4] are constituted, which can be called hybrids. This discourse rejects the substantialist illusions of other discourses of globalisation and underlines the value of the playful, open combinations of hitherto separated blocks of practices.

This discourse is in an evident way related to the world view of chaoism, with its unplanned, playful activities which have no meaning as such.

In this analysis only the central references to world views in the various discourses of globalisation have been examined. In these discourses one can also find bits and pieces of other world views, for example most of these discourses introduce some elements of chance.

Two of the mentioned families refer to an organismic world view. In fact, one could qualify the first family of discourses by the New Right movements as a special and limited case of the third family. However, according to the criteria used, there are also important differences concerning the production and distribution of the discourse, and also of the identities which are constructed.

Are there other discourses of globalisation which do not belong to one of the five families I have distinguished? This may be the case; for example, I have neglected to consider the mystical discourses (New Age, integrism, new religions), which affirm a variety of globalisation.

Furthermore, within each of the five families of discourses one can find many variations, and also some attempts to combine the central world view characteristic for each family with aspects of other world views. For example, within the discourses based on a mechanistic world view there are some attempts to present the functioning of globalisation as a universal, dominant structure. In this way, a combination of a mechanistic world view is realized with formism.

The most important result of this analysis shows that the various discourses of globalisation elaborate indeed quite diverse world views, using the possibilities of each one.

FROM WORLD VIEW TO DISCOURSES

The preoccupation with world views offers possibilities, which I will shortly point out in this last part. Examining discourses of globalisation and their relation with world views has been undertaken by starting with a classification of five families of discourses, and for each discourse the question has been asked what kind of world view is elaborated by this discourse. But it is also possible to turn the question around, and start with the different world views. In this case we can analyze each of the different discourses from the perspective of the various world views. As each world view as a global framework gives indications about what is truth and how to consider values, we could undertake the task to subject the various discourses of globalisation to the test of the criteria of truth and value of the various world views. For example, the first world view defines truth as a correspondence with reality. This criterion could be used in order to examine if the different discourses take into account the known facts and phenomena of globalisation or not. In other words, by applying the test of truth of the first world view, formism, we would analyze the discourses of globalisation as discourses *on* globalisation. A similar analysis could be realized with the help of the criteria of truth of the other world views, and the same procedure could be used for analyzing the value perspectives of the various discourses. Here is not the place to do this, but I hope to make it clear that such an opportunity is directly offered when working with world views.

Here I will only consider one other point. When presenting the criteria for classifying world views, I admitted that the suggested criteria are not innocent because they are formulated with the help of a theory of discourse which intimately related with a specific world view. It is now easy to see that the adopted theory of discourse, which defines discourse as a situated social action belongs to the world view of contextualism or process. This world view has no correspondence theory of truth which entails serious consequences which have already been mentioned before. Truth and value are defined in this world view in an indirect way, by the degrees of complexity and novelty which discourses as situated social actions can produce. As we have seen, the different discourses operate with varying degrees of success for different auditoria. The question whether they also stimulate new and more complex actions with emergent qualities, I raise here but I leave at the moment the answer to the reader.

AFFILIATION

Robert Maier
Faculty of Social Sciences
University of Utrecht
The Netherlands

REFERENCES

[1] **Apostel, L.**, "Wereldbeelden, Zingevingssystemen en ethische stelsels. Schets voor een typologie", in: **Apostel, L.**, *Wereldbeelden en ethische stelsels* (in press).

[2] **Barjon, C.**, "Enquête sur les réseaux Le Pen", *Nouvel Observateur*, **1669**, 30/10/1996.

[3] **Fairclough, N.**, *Discourse and social change*, Cambridge, Polity Press, 1992.

[4] **Gergen, K.**, *The saturated self. Dilemmas of identity in contemporary life*, New York, Basic Books, 1990.

[5] **Groep van Lissabon, R. Petrella**, *Grenzen aan de concurrentie*, Brussel: VUB Press, 1994.

[6] **Hirsch, J.**, *Der nationale Wettbewerbsstaat. Staat, Demokratie und Politik im globalen Kapitalismus*, Berlin, Edition ID-Archiv, 1995.

[7] **Krugman, P.**, *De Borreltafeleconomie. Drogredenen over Globalisering*, Utrecht, Het Spectrum, 1996. [First published in english in 1996, title: Pop Internationalism, MIT Press]

[8] **Lipietz, A.**, *La société en sablier. Le partage du travail contre la déchirure sociale*, Paris, Editions La Découverte, 1996.

[9] **Maier, R.**, "Forms of identity and argumentation." *Journal for the theory of social behaviour*, **26**, 1, 1996, pp. 35-50.

[10] **Martin, H.-P. & Schumann, H.**, *Die Globalisirungsfalle. Der Angriff auf Demokratie und Wohlstand*, Hamburg, Rohwohlt, 1996.

[11] **Nederveen Pieterse, J.**, "globalisation as hybridisation" in: **Featherstone, M.. Lash, S. and Robertson, R.** (eds.), *Global modernities*, London, Sage, 1995, pp. 45-68.

[12] **Ramonet, I.**, "Régimes globalitaires", *Monde diplomatique*, janvier 1997.

[13] **Robertson, R.**, "Glocalization: time-space and homogeneity-heterogeneity", in: **Featherstone, M.. Lash, S. and Robertson, R.** (eds.), *Global modernities*, London, Sage, 1995, pp. 25-44.

[14] **Walry, J.**, *(Moderne) Glocalisatie en (postmoderne) Vermenging als maatschappelijke fenomenen* (in press).

[15] **Went, R.**, *Grenzen aan de globalisering?*, Amsterdam, Het Spinhuis, 1996.

[16] **Worldviews group**, *Perspectives on the world. An interdisciplinary reflection*, Brussel, VUB Press, 1995.

DIEDERIK AERTS

THE GAME OF THE BIOMOUSA : A VIEW OF DISCOVERY AND CREATION

1. INTRODUCTION

No branch of science is explicitly devoted to studying world views and there are probably several reasons for this. First and foremost there is so little consensus within the same discipline, that very different and generally incompatible fragments of world views exist. On the other hand, there is a certain diffidence, perhaps even a taboo, about putting these fragments of world views, often the subject of enthusiastic discussions over coffee at scientific gatherings, into the form of a text. This diffidence is very marked amongst physicists because they know from experience how quickly, and for various reasons, such an attempt is dismissed as "unscientific" and "speculative". The world view fragment I would like to present here from the viewpoint of quantum mechanics should therefore be looked at in this light. It contains a number of aspects that are speculative and others that are directly based on recent experimental and theoretical data in quantum mechanics. Some parts are almost exclusively the result of the author's own personal beliefs.

2. QUANTUM MECHANICS AND THE SPACE VIEW

Within the limited scope of this article it is not possible to discuss all aspects of the problems caused by quantum mechanics. We therefore refer the reader for a more detailed analysis to [1,2,3,4], were more explicit details of many aspects are contained as well as further references to other articles. Without fear of exaggeration, however, we can say that quantum mechanics differs from all previous physical theories in one very fundamental respect.

Until quantum mechanics was developed, a physical theory could, without causing too many problems, be regarded as a description of a part of reality that we can imagine. By this we mean the following: if we describe a classical physical entity using a classical theory, we can always "imagine" what happens to that physical entity and the description refers to what happens. For example, if we are describing a particle that is moving through space, we can "imagine" how this particle moves through space and classical mechanics describes what we have imagined.

This no longer seems to be possible in the case of a quantum entity. Quantum mechanics gives us a number of rules that provide us with predictions about the chances of detecting a quantum entity in space; however, we are not able to imagine an object that is moving through space that is also compatible with the chances of detection offered by quantum mechanics.

Recently a lot of experimental and theoretical data has been gathered, which can be used to put forward a new view of the behaviour of a quantum entity. The aim of this article is to examine these data and to attempt to put forward a global view of reality that takes them into consideration. We shall call this global view the creation-discovery view. We shall be forced to abandon an old and profound preconception about the nature of reality if we adopt the creation-discovery view. This old preconception, which hinders our understanding of the micro-world, consists in believing that reality exists within space; we shall therefore refer to it as the space preconception.

Classical physics is also based on this intuition and there are various ways of describing the place occupied by a material entity in three-dimensional space. In the case of a rigid body, the position of the mass centre can be considered and the entity can then be described in a relative coordinate system with its origin in this mass centre. In the case of a liquid or gas, continuum mechanics is used, with the liquid or gas being described by means of a collection of points, present in that part of space where the mass density of the liquid or gas is different from zero. A wave, though often spread out, also has a place in our space. Whatever description of whatever entity may be considered in classical mechanics, the entity always has a place.

It does seem to be pre-scientifically a priori to assume that all material objects, both macroscopic and microscopic, are present at any given moment somewhere within our three-dimensional Euclidean space. Recent experimental and theoretical data have shown that it is very plausible to assume that quantum entities can find themselves in states, in which they are not present in space. According to the creation-discovery view, the view we would like to put forward here, space is no longer regarded as an all-encompassing setting. According to this view, detection of a quantum entity is not an "observation", but rather a "process", in which the detection apparatus "sucks" or "pulls" the quantum entity into space. This conflicts with our intuition since in our everyday reality every material entity has a place at any given moment.

In the creation-discovery view, we assume that the detection experiment contains an element of creation that partly creates the place "itself". This means that before the experiment, the quantum entity did not necessarily have a place and that the place is created by the experiment itself. An analogous process can be seen when determining the linear momentum (mass times velocity) of the quantum entity. The quantum entity has no particular momentum before the experiment that results in its momentum being created. Hence within this view, we no longer see space as an all-encompassing stage on which the whole drama of reality is enacted, but rather as a structure that we, as humans, have experienced, relying on our everyday experiences with the material macroscopic entities around us. We differentiate between the following two characteristics: 1. Every entity can be detected in space; space is therefore one of the structures through which we, as humans, encounter and create reality. 2. Every entity is present in space; space is therefore the stage on which everything real is enacted. The first characteristic is also valid for quantum entities, the second is not. In this way we adopt a new reality "statute" for space. Space as an "intermediate" meeting structure and not as an "all-encompassing" stage. Things make their place rather than simply have one. We can still find evidence in our everyday language of the change in meaning that the concept of space has undergone. We still say that space is the stage on which an event takes place. Events, since we still do not regard them as entities, can apparently still "find" their place in space, which means that they are not necessarily present in space before they have "found" that place.

To show why research in quantum mechanics has led us to the creation-discovery view, we shall briefly outline the latest findings in this discipline.

3. IS SCHRODINGER'S CAT DEAD OR ALIVE, OR NEITHER?

If we want to use quantum mechanics as a universal theory and also use it to describe the entity comprising the measuring apparatus (a macroscopic entity) and the quantum entity, we find some very peculiar predictions. Schrödinger studied this problem in detail and we would therefore like to consider it from his cat's point of view [5]. Schrödinger devised the following mental experiment. Consider a room in which there is a radioactive source and a detector that can detect radioactive particles. There is also a glass bottle containing poison and a live cat in the same room. The detector is turned on for a period of time, during which it has a 1 in 2 chance of detecting a radioactive particle emitted by the source. If the detector detects a particle, a mechanism is activated that breaks the glass bottle, thereby releasing the poison and killing the cat. If the detector does not detect any particles, nothing happens and the cat lives. We can only learn the outcome of the experiment by opening the door of the room to see what has happened. If we make a quantum description, within orthodox quantum formalism, for the whole system (including the cat), the state of the cat (which we shall call p_{cat}) remains a "superposition state" of the two states "the cat is dead" (which we shall call p_{dead}) and "the cat is alive" (which we shall call p_{alive}) until the very last moment, i.e., when we open the door. To recapitulate,

$$p_{cat} = \frac{1}{\sqrt{2}}(p_{dead} + p_{alive}) \tag{1}$$

and this superposition state only ceases when we open the door to see what is happening. A considerable part of the basic problem of quantum formalism can be demonstrated by means of this example; we shall therefore examine it in detail. If we interpret the state, as it is described by the wave function of orthodox quantum mechanics, as a mathematical object that simply and solely describes our knowledge about the physical system, there is no problem with Schrödinger's cat. In fact, within this "perception of knowledge view" we can assume that the cat was already dead or was still alive before we opened the door, and that the quantum mechanical change in state only describes our cognisance of this fact. This "perception of knowledge view" also disposes of another problem: according to quantum formalism, while the door is being opened the superposition state $p_{cat} = \frac{1}{\sqrt{2}}(p_{dead} + p_{alive})$ "suddenly" changes into one of the component states p_{dead} or p_{alive}. This sudden change in state is known in quantum mechanical jargon as "the collapse of the wave function" and there is a natural explanation for it within the "perception of knowledge view". If the quantum wave function does in fact describe our knowledge of the situation, then this knowledge suddenly changes, as does this wave function, when we receive new information, such as when the door of the room is opened.

The "perception of knowledge view" therefore assumes that the wave function of quantum mechanics does not describe reality, which is independent of our knowledge of it, but rather it represents the cognition we have of this reality. It follows from this that, if the "perception of knowledge view" is right, an underlying reality must exist, which is not described by the wave function of quantum mechanics. In the example of the cat, this underlying theory provides a description of the cat's state, dead or alive, irrespective of whether we open the door to find this out. The "perception of knowledge view" therefore immediately gives rise to the hypothesis of the existence of a "hidden variable theory", which describes this underlying reality. And it is here that the problem with the "perception of knowledge view" arises. It can be demonstrated that the probability model of a theory, in which a "lack of knowledge" about an underlying reality is what causes the probability (called a hidden variable theory), always satisfies Kolmogorov's axioms. Kolmogorov's axioms (advanced by Kolmogorov in 1933) are those of classical probability theory, as already specified by Simon Laplace in the previous century. The quantum probability model does not satisfy Kolmogorov's axioms. John von Neumann was the first to prove a no-go theorem for the hidden variable theories [6]. The proof of the impossibility of reproducing quantum probabilities using a hidden variable theory was gradually perfected later [7]. Since the probability model of quantum mechanics does not satisfy these axioms, this theorem shows that the "perception of knowledge view" is not right. Recent experiments (concerning Bell inequalities) have confirmed that the "perception of knowledge view" is wrong.

It is now virtually a foregone conclusion that the wave function of quantum mechanics does not describe our knowledge of the system, but represents the actual state of the system, irrespective of whether we know this or not. However, if this is the case, then Schrödinger's cat creates a serious problem. Could it be that, before we open the door, the cat is in a superposition state, neither dead nor alive, and that this state is transformed into a state of being dead or alive by the door being opened? It does seem impossible to us that reality would react to our observations in this way. A reality where a state comes into existence because we take cognisance of it contradicts our everyday view of reality in so many respects that we can scarcely take the idea seriously. Nevertheless it seems to be an immutable consequence of orthodox quantum mechanics, applied to reality as a whole. Recently, with the emergence of the new formalisms, the fundamental concepts of which are much closer to reality, a completely different light has been shed on this problem.

4. NEW FORMALISMS AND FALSE PARADOXES

The paradox of Schrödinger's cat and many similar paradoxes of quantum mechanics are partly the reason why it is impossible to put forward a view of reality that is compatible with quantum mechanics. In fact, if the cat only lives or dies when we open the door, then only a subjective world view seems possible. There is a fundamental confusion between discovery and creation involved here. According to our everyday perception of reality, we believe that the cat is already dead or alive before the door is opened, and that finding a dead or a living cat is only a "discovery" of a reality that already existed. The application of quantum mechanics to this situation forces us to interpret this discovery as a "creation". The theoretical advances made in research in quantum mechanics, and more specifically the development of new quantum formalisms, have resolved paradoxical situations such as that of Schrödinger's cat. The conclusion is that orthodox quantum mechanics is not a universal theory, but rather a formalism that is only valid under limited circumstances. Since these findings lie partly at the origin of the development of the creation-discovery view, they will be discussed briefly.

These new formalisms are very general. A very general formulation of the new formalisms is that of Charles Randall and David Foulis [8] and a more physical formulation is that of Piron and Aerts [9, 10, 11]. An entity S is described by means of the collection of its states. A state describes the reality of the entity S. No specific mathematical structure is imposed a priori on the collection of states, as is the case in quantum mechanics (a Hilbert space) and in classical mechanics (a phase space). Furthermore, it is assumed that experiments are carried out on the entity S and the collection of relevant experiments is explicitly examined. Once again no mathematical structure is fixed a priori for the collection of these experiments, as is the case in quantum mechanics and classical mechanics. It is merely assumed that when an entity S is in a particular state and an experiment is carried out, a result is achieved with a certain probability. The original state is hereby changed into a new state. In this way the measurement process can be described in general terms. If no measurement is made, the entity S is still in a particular state, which then changes as time goes on. This dynamic change is described by the Schrödinger equation in the case of quantum mechanics, and by Newton's equations in the case of classical mechanics.

In this new general description it is very possible, even natural, to discern special experiments. Hence the concept of "classical experiment" is introduced: an experiment devised in such a way that there is always a pre-determined result for every state of the entity S. A classical experiment is therefore an experiment where the result is fixed with certainty, even before the experiment is carried out. Generally speaking, the collection of relevant experiments will include some classical experiments and some non-classical experiments. It is possible to prove a theorem in which the classical part of the description of an entity is kept separate [12, 13]. The collection of all states can then be described as the union of a collection of classical mixtures, where every classical mixture still contains a collection of microstates that are non-classical. If quantum mechanical axioms are formulated based on this general situation, it can be demonstrated that the collection of states within one classical mixture can be represented by a Hilbert space. The collection of all of the states of the entity is then described by means of an infinite collection of Hilbert spaces, one for each classical mixture. Orthodox quantum mechanics emerges here as a limiting case, in which not a single classical measurement exists, and the representation then gives rise to one Hilbert space. Classical mechanics is the other limiting case, in which only classical measurements exist, and the representation then gives rise to a phase space description. The general situation of an arbitrary physical entity is neither pure quantum nor pure classical and can only be described by means of a collection of various Hilbert spaces. If the measurement process is viewed within this general framework, the problem of Schrödinger's cat disappears. Opening the door is a classical measurement that does not change the state of the cat in any way, and it can be described in this way within this general formalism. The quantum collapse occurs when the radioactive particle is detected by the detector, and this process is non-classical, even in the description within the general formalism.

It's not only the paradox of Schrödinger's cat that is resolved within this general formalism. It is also possible to regard quantum mechanics and classical mechanics as two special cases of a more general theory. This general theory is "quantum-like" but causes no paradoxes for the measurement process because the measuring apparatus is described within the same formalism as a classical entity and the entity to be measured as a quantum entity. In conclusion we can state that the measurement paradox is due to the structural limitations of the orthodox quantum formalism.

5. THE EINSTEIN-PODOLSKY-ROSEN PARADOX AND THE QUANTUM-CLASSICAL RELATIONSHIP

The perpetual existence of the superposition states, which caused paradox of Schrödinger's cat (the superposition state of the living and the dead cat), was used by Einstein, Podolsky and Rosen to formulate a much more subtle paradoxical situation. Einstein, Podolsky and Rosen considered the situation of two separated entities S_1 and S_2 and the composite entity S composed of these two separated entities. They demonstrated that it is always possible to put the composite entity S into a state such that measurement of one of the component entities determines the state of the other component entity. In a situation involving separated entities, this is a prediction of quantum mechanics that conflicts with the "separated" concept itself. In fact, in the case of separated entities, the state of one of the entities is not determined by what is done with the other entity, a fact confirmed by the experiments that we can carry out on separated entities.

Once again if we look at this situation from the point of view of the new formalisms, the paradox can be explained. If we consider the situation of two separated entities, it is possible to prove that the entity S, which is composed of the two separated entities S_1 and S_2, never satisfies the axioms of orthodox quantum mechanics, not even if classical experiments are conducted, as was the case in the measurement paradox [10, 11, 14]. There are two axioms in orthodox quantum mechanics (known as "weak modularity" and "the covering law" in the jargon), which are never satisfied for the situation involving an entity S composed of two separated quantum entities S_1 and S_2. The shortcoming of orthodox quantum mechanics identified here goes much deeper structurally than the one discussed earlier in connection with the measurement problem. In the latter case it is possible to put forward a solution where one Hilbert space in orthodox quantum mechanics is replaced by a collection of Hilbert spaces. This solution is manageable within the framework of the Hilbert space formalism. The inability of orthodox quantum mechanics to describe separated entities lies in the vector space structure of the Hilbert space itself. The two "bad axioms" are the axioms that give rise to the vector space structure of the Hilbert space and if we remove these axioms in order to describe separated entities, we have to construct a completely new mathematical structure for the state space. This, however, is what recent findings tell us to do and it is the only way to free quantum formalism from EPR-type paradoxes in connection with the description of separated entities. Work is under way on this within the new formalisms.

It is a serious mathematical step to abandon the vector space structure for the collection of states but recent new findings have confirmed the necessity of such a step. The possibility of examining both classical entities and quantum entities from the viewpoint of the general formalism resolves the measurement paradoxes. However, the possibilities as regards description remain polarised between classical and quantum or generally a mixture of both. Very recently we began to explore intermediate regions (between classical and quantum; the so-called mesoscopic region) from the viewpoint of this general formalism. Once again the same two axioms seem to make a description of intermediate regions impossible [15, 16]. If we formulate a theory without these two axioms, we can describe quantum, classical, a mixture of both, and also intermediate mesoscopic entities, which are neither quantum nor classical. This approach would enable us to describe a continuous transition from quantum to classical.

In conclusion we can state that orthodox quantum formalism has several fundamental shortcomings, which have now been identified. By resolving these paradoxes, quantum mechanics (more specifically the "quantum-like" generalised formalisms) is freed from the subjective predictions that were an immutable part of the orthodox theory.

6. THE NEW EXPERIMENTS

It is evident to us from experiments that Schrödinger's cat does not live or die because we open the door and that the state of one of the two separated entities is not determined by what is done with the other entity. We have been able to explain these paradoxical predictions of orthodox quantum mechanics as being the result of a structural shortcoming of the mathematical formalism. However, it is not at all evident from experiments what these special quantum effects, connected with these superposition states, where they do exist, actually mean. Some physicists have gone so far as to insinuate that perhaps certain superposition states in question do not exist at all and are just mathematical artefacts of the theory. Recent experiments have shown, however, that it is actually possible to prepare quantum entities in these superposition states. We shall briefly mention a few aspects of these experimental results.

In experiments with very low-energy neutron beams, Helmut Rauch and his group succeeded in putting one neutron within a silicon crystal into a superposition state of two states located far apart [17, 18]. In the experiment, the silicon crystal has a diameter of over five centimetres and the two component states are states of a neutron that is located within cubes A and B with edges measuring one millionth of a centimetre. Location A of one component state, which we shall call p_A, is over three centimetres away from location B of the other component state, p_B. The neutron was prepared by Rauch in a superposition state p_{sup} of these two component states, in other words $p_{sup} = \frac{1}{\sqrt{2}}(p_A + p_B)$. If the neutron is in state p_{sup} and detection is initiated in one of the two areas A or B, there is a 1 in 2 chance of detecting the neutron in area A and a 1 in 2 chance of detecting it in area B. Rauch and his group succeeded in verifying that the neutron was in a superposition state p_{sup} because they conducted additional experiments on the neutron in this state. One of the most fascinating of these additional experiments involved rotating the spin of the neutron. Using a magnetic field located in area A, Rauch rotated the spin of the neutron in the superposition state p_{sup}. The result of a precession over an angle of n degrees through the magnetic field in area A on the component state p_A results in an actual rotation of $\frac{n}{2}$ degrees, as predicted by means of the superposition principle. A simultaneous rotation over n degrees of the other component state p_B, using a magnetic field in area B, also results in the spin of the neutron in the superposition state rotating $\frac{n}{2}$ degrees. Rauch conducted all kinds of additional experiments, which all correspond with the quantum description of one neutron in the superposition state of the two component states located far apart [19]. These experiments, and many others, prove the existence of the superposition states in question, and the question we can now ask ourselves is whether we can draw conclusions from these experiments concerning the physical meaning of the superposition state. Before answering this question, we would like to discuss another problem of quantum mechanics that is closely linked with the possibility of suggesting a physical meaning for the superposition states.

7. THE ORIGIN OF QUANTUM PROBABILITIES

We have already mentioned that the "perception of knowledge view" is not right and we have also explained how it offered a natural solution for the situation of Schrödinger's cat. The "perception of knowledge view" did more than provide a solution for Schrödinger's cat; it also presented a simple "classical" explanation for the existence of quantum probabilities. In fact let us argue once more from the point of view of the "perception of knowledge view", where the underlying reality is not described by the wave function, but by a hidden variable theory. In this situation quantum probability is the natural result of our lack of knowledge about this underlying reality. This is also how classical probability is explained, so it should come as no surprise to us that the aforementioned theorem exists, in which theorem it is proved that every hidden variable theory gives rise to a probability model that satisfies Kolmogorov's axioms. Kolmogorov did, after all, formulate his axioms for the classical probability theory. Since the probability model of quantum mechanics does not satisfy Kolmogorov's axioms, with the result that the "perception of knowledge view" cannot be sustained, we have to look for a new non-classical explanation for the origin of quantum probability. Quantum probability is not the result of our lack of knowledge about a deeper reality, as is the case with classical probability.

It is sometimes suggested that these quantum probabilities are intrinsically part of nature, and we then talk about "ontological probabilities". No-one seems to be able to "understand" this kind of probability, however; hence it has never gone beyond a vague abstract concept of "ontological probabilities". There is a third possibility, though, and we would like to illustrate this by means of an example.

Suppose that we are considering the following experiment: "We take a walnut from a basket and we crack the walnut in order to eat it." Let us elaborate on the way in which we crack the walnut. We do not use nutcrackers, but simply take it between both hands and squeeze as hard as we can and see what happens. Anyone who has ever cracked walnuts in this way knows that various things can happen. The first occurrence we want to identify is when the walnut turns out to be mouldy. (1) If a cracked walnut is mouldy, we do not eat it.

Let us now suppose that there are N walnuts in the basket. This means that for a given nut, which we shall refer to as H_k, there are two possible outcomes of our experiment, which we shall refer to as E_1: we crack the nut and eat it, and E_2: we crack the nut and do not eat it. Suppose that of the N nuts in the basket, M are mouldy. The probability that our experiment will result in E_1 for a nut H_k is $\frac{N-M}{N}$, and the probability that our experiment will result in E_2 for the nut H_k is $\frac{M}{N}$. This probability is the result of our lack of knowledge about the complete reality of the nut. In fact, before we start cracking the nut H_k, it is either mouldy or not. If we could gain this knowledge without having to crack the nut, we could eliminate the probability arising from this lack of knowledge by only considering the nuts that are not mouldy. Classical Kolmogorovian probability theory is based on this assumption about the nature of the probability that exists.

Anyone who has experience of cracking walnuts knows that other things can happen too. Sometimes we destroy the nut by breaking it, so that it becomes mixed up with the broken shell. If that happens, we generally make a quick assessment of how serious the situation is and decide whether it is worthwhile separating the nut from the shell. If it is not worthwhile we do not eat the nut. Hence there are a further two possible outcomes of our experiment: E_3, which corresponds to a "badly cracked nut", in which case we do not eat the nut, and E_4, which corresponds to a "well-cracked nut", in which case we do eat the nut. We can again state that for a given nut H_k, the two outcomes are possible and each outcome will occur with a certain degree of probability. We sense right away, however, that this type of probability is different to the previous type because it depends on the way in which the nut is cracked. Unlike the previous case, where M walnuts are mouldy and $N - M$ walnuts are not mouldy, we cannot divide the nuts in the basket "beforehand" into those that will be "well cracked" and those that will be "badly cracked". This kind of division does not exist because it is created by the cracking experiment itself. We have here a good example of how part of reality is created by the measurement itself, namely the cracking of the walnuts.

The most interesting aspect is that the mathematical structure of the probability model needed to describe the probabilities that ensue from cracking the walnuts well or badly is different from the mathematical structure of the probability model needed to describe the probabilities

that ensue from the walnuts being mouldy or not mouldy. More particularly:
-The probability structure that describes the indeterminism that is the result of a lack of knowledge about a more complete reality of the occurrence in question is a classical Kolmogorovian probability model (this situation fits within the "perception of knowledge view").
-The probability structure that describes the indeterminism that is the result of the fact that while the measurement is being carried out a new part of reality is created, which did not exist before the measurement was carried out, is a quantum-like probability model.

We cannot demonstrate these two statements within the scope of this article but we can refer the reader to the articles in which these statements are proved and illustrated by means of examples [1, 20, 21, 22].

It can also be proved that every quantum mechanical entity can be obtained by means of a model, where the cause of the quantum probability is a lack of knowledge about the interaction of the measuring apparatus with the quantum entity during the measurement experiment, in the course of which a new part of reality is created that did not exist before the measurement was carried out [11, 21, 22]. This is the explanation for quantum probability we would like to advance.

8. DISCOVERY AND CREATION: THE ROLE OF SPACE

Let us assume that we have been able to remove all of the mouldy nuts from the basket, leaving only nuts that are not mouldy. In the jargon of physics, we shall say that each individual nut is in a pure state, with regard to the property of being mouldy or not. In the original situation, when the mouldy nuts were still present in the basket, an individual nut was in a mixed state of mouldy and not mouldy, with weights $\frac{M}{N}$ and $\frac{N-M}{N}$. In the situation under consideration, we have a basket of walnuts that are not mouldy and with reference to this we would like to introduce the concept of "potential". With regard to being mouldy or not mouldy, we could claim for each walnut that it was mouldy or not mouldy before the experiment. With regard to being "well cracked" or "badly cracked", we cannot describe the walnut as such before cracking is measured. What we can claim is that each walnut is potentially well cracked (and is then eaten) or potentially badly cracked (and is not eaten).

No-one has any difficulty understanding the example of the walnut. Our proposition is that we should try to understand quantum probability in the same way. The only difference is that the measurements in quantum mechanics, where the second type of probability is introduced (due to the fact that a new part of reality is created during the measurement), are measurements for which such a creation is difficult to imagine. For example, the detection of a quantum entity is just such a measurement: whereas we would intuitively like to consider detection as the "determination of position", a position that already existed before we began the measurement to determine it, we have to learn to accept that detection of a quantum entity contains an element of creation of the position of this entity during the process of detection. Walnuts are potentially "well cracked" or "badly cracked" and quantum entities are potentially within or outside a particular area of space. The experiment that consists in finding the quantum entity within this area of space or not finding it within this area, takes place after the measuring apparatus needed to detect this has been set up in the laboratory, and the interaction of the quantum entity with this measuring apparatus has begun. Before that the quantum entity is potentially present and potentially not present within this area of space.

Note that this explanation for quantum measurements forces us to look at the concept of "space" in a new way. If a quantum entity, in a superposition state, is only potentially present within an area of space, we can no longer regard space as the stage for reality as a whole. Rather space is a structure that has developed along with the classical relationships between macroscopic physical entities. These macroscopic physical entities are always present within space because space is simply a structure in which they are always present, but this is not the case for quantum entities. In a normal state a quantum entity is not present within space; it can only be pulled in by a detection experiment. This process of pulling into space is associated with the second type of probability (as with the cracking of the walnuts) because the place of the quantum entity is partly created during the process of detection. The neutron in the Rauch experiment is not within space. It can be detected in two different areas of space, A and B, but the fact that this always occurs in these two areas is due to the fact that a detection experiment pulls the neutron into one of these two areas.

The experiments concerning the Einstein-Podolsky-Rosen paradox can also be easily understood within this discovery-creation view. Involved here are two quantum entities S_1 and S_2, which can be prepared in a superposition state p_{sup} of two component states p_{12} and p_{21}. The component state p_{12} is a state where entity S_1 is present within an area of space A_1 and the other entity S_2 is present within an area of space A_2, while component state p_{21} is a state where entity S_2 is present within area A_1 and entity S_1 is present within area A_2. These two areas A_1 and A_2 are located far apart (12 metres in the case of the photon experiments) and measurements are carried out within areas A_1 and A_2. These measurements produce results that would seem very contradictory if we were to interpret the situation from the viewpoint of a "perception of knowledge view" or "hidden variable theory", where we would assume that the two entities S_1 and S_2 are already present within areas A_1 and A_2 before the measurement is carried out, and we only have a lack of knowledge about exactly where the entities are. With the creation-discovery view, where we assume that the detection measurement involves an essential creation of place, and hence that before the measurement is carried out the two entities S_1 and S_2 are not already present within areas A_1 and A_2, there is no problem interpreting the measurement results. The violation of Bell inequalities is even a natural phenomenon in this creation-discovery view, which can be perfectly imitated by means of macroscopic physical entities [1, 22].

9. THE CREATION PROCESS: THE BIOMOUSA

We would like to try to extend the findings concerning the non-spatial character of quantum entities, and the way in which this is explained in the creation-discovery view, to reality as a whole. Hence we can distinguish different layers of reality: pre-material layers, the material layer, the biological layer, the social layer and the cultural layer. Note, however, that this is a greatly simplified representation. The localisation process of quantum entities is the bridge between the pre-material and the material layer of reality, and that is the way in which the material entities are rooted in the pre-material layer. Similarly, every layer is rooted in the previous layer (biological in material, social and cultural in biological) and a similar structuring process describes the "existence" of the entities over the layers. In the creation-discovery view, the entities in the cultural layer (languages, communications systems, works of art, theories, and so on) are not simply regarded as human creations, but also as new entities for a nascent reality. Things have always happened in this way: what we now call matter was once a vague and rather insubstantial structure. When neutrons, protons and electrons were busy deciding whether to organise themselves into atoms, atoms were world views. When atoms organised themselves into macroscopic matter, this macroscopic matter was a world view. When cells organised themselves into plants, animals and humans, these entities were world views. World views are precursors of reality.

This classification of reality into different layers contains an explicit idealisation because the different layers are not really separated. The material layer is the most important one for our present way of life. It is made up of the organisation of atoms. This organisation is so complex that the atoms, as individual entities, no longer fit into the most fundamental aspects of this reality. According to quantum mechanics, the atoms, and even the initial structures in which atoms started developing material reality, are not in space in most of the cases. The biological layer is the layer of living matter. It is not a fundamental new layer, but rather a choice in favour of the power of perpetuation of macromolecules, which have organised themselves into self-replicating organisms, with the *DNA* molecule as the basic module. The social layer is the layer where living organisms interact with one another and try out new perpetuating entities: the hunt as the entity that provided food for everyone, the table as the entity where people eat together, and the house as the entity where people live together. The cultural or intellectual layer is the latest and most fragile achievement. In this layer world views provide fresh impetus for developing greater perpetuating entities. Cultural products and the creations of the human mind exist in this layer and are seeking the space, the world view, in which they belong. Individuals who travel between two cultures will not necessarily be present in the reality of a culture, just as quantum entities that travel between two macroscopic entities are not necessarily present in the reality of these macroscopic entities, a reality that we situate within space.

Long ago only pre-material quantum entities existed in a pre-material layer. They organised themselves into matter and space as a meeting place for this matter was created. The same creation process that began in the pre-material layer is now fully under way in the cultural and intellectual layer, and new small phases constantly appear. We shall call this creation process "the Biomousa" ("Biomousa" or "muse of life").

10. ENTROPY AND CREATION: BOLTZMANN AND STATISTICAL MECHANICS

According to the second law of thermodynamics, the entropy of a closed system cannot decrease. It is often maintained that entropy is a measure of the disorder of a system and if this is so, the second law of thermodynamics seems to be at odds with the idea of a creation process that produces entities with a great power of perpetuation. We would like to examine how this situation arises within the creation-discovery view. First let us clarify the concept of entropy. There appear to be two layers of reality where entropy plays a fundamental part and we shall see that this is no coincidence.

Originally entropy was introduced in the material layer of reality. Ludwig Boltzmann was fascinated by the idea that matter is made up of a vast number of atoms. At the end of the nineteenth century, when Ludwig Boltzmann developed his theory, atomic theory had not yet been proved and was far from being generally accepted. Boltzmann believed in the existence of atoms but many important physicists did not share his view. In his search Boltzmann tried to understand some important, and at the time very topical, problems, using this theory as his starting point. The industrial revolution in the nineteenth century had created tremendous interest in the steam engine, more specifically physicists and engineers were working intensively on the problem of converting heat into mechanical work. It was known that mechanical energy could easily be converted into heat, for example by friction, but the reverse process did not seem so simple. Cold water can be mixed with hot water to obtain lukewarm water, but what about the other way round? What were the laws of nature that stood in the way?

The introduction of the concept of entropy shed some light on the subject. Let us consider an example: a drop of hot water has a certain entropy, which we shall call $S(hot)$, and a drop of cold water also has a certain entropy, which we shall call $S(cold)$. If we consider two drops of water, one hot and one cold, then these two drops of water have an entropy $S(hot + cold)$, which is the sum of the original entropies, $S(hot + cold) = S(hot) + S(cold)$. If we mix the drop of hot water and the drop of cold water, thereby obtaining two drops of lukewarm water, then the entropy of these two drops of lukewarm water, $S(lukewarm)$, is much higher than the sum $S(cold) + S(hot)$. By mixing hot and cold water we have increased the entropy, and this is an irreversible process. This principle is expressed in the second law of thermodynamics: in every physical process for a closed system entropy remains constant or increases. This principle explains why heat can only partially be converted into mechanical work; the concept of entropy was therefore vital for what Boltzmann was trying to understand.

Let us consider for a moment what a drop of water is according to atomic theory, in which Boltzmann firmly believed. The molecules contained in a drop of cold water can occur in many different configurations. They dance about and vibrate and their configuration changes constantly. All of these configurations would look different if we could observe them at microscopic level, but with the naked eye they all appear alike, i.e., like a drop of cold water. When we speak of a drop of cold water, then we are referring to an entity that has many different states at microscopic level, without this changing its macroscopic aspect in any way. Boltzmann's discovery was that entropy is a measure of this indeterminacy at microscopic level.

The entropy of a drop of water is the logarithm of the number of "microscopic" states that macroscopically give rise to an identical drop of water, multiplied by a constant k, which is known as Boltzmann's constant. Can we understand this effect of increasing the possible microscopic states of a mixture? Let us try with the help of an example. Let us assume that we have some red balls and some yellow balls that we can put into compartments. To make the experiment more specific, let us consider a case where we have three different compartments in which we can place the red balls. Only one ball can fit into each compartment and this can result in $2^3 = 8$ different configurations. Let us list the possibilities: $(\, , \, , \,), (\, , \, , r), (\, , r, \,), (r, \, , \,), (\, , r, r), (r, \, , r), (r, r, \,), (r, r, r)$, where $(\, , r, \,)$ means "first compartment empty", "second compartment occupied", and "third compartment empty". In general, if n compartments are available, this gives us 2^n different configurations. Where $n = 2$, this gives $2^2 = 4$ configurations; where $n = 3$, this gives $2^3 = 8$ configurations, which are listed above; and where $n = 4$, we have $2^4 = 16$ different configurations. It can be seen that if we allow n to increase, we quickly reach a very large number of configurations. For example, $n = 25$ gives us $2^{25} = 33,554,432$ configurations and $n = 100$ gives us $2^{100} = 126,750,600,000,000,000,000,000,000,000$ configurations, which is a huge number. Let us now assume that we also have yellow balls that can be put into compartments too. For three compartments, this gives eight different configurations once again: $(\, , \, , \,), (\, , \, , y), (\, , y, \,), (y, \, , \,), (\, , y, y), (y, \, , y), (y, y, \,), (y, y, y)$. Now let us look at the two entities together, in other words on one side red balls and on the other yellow balls that can be put into compartments. In the case of two times three compartments, the number of configurations possible is $2^6 = 64$. We shall not list them all but instead we shall give an example: $(\, , y, \, , r, \, , r)$, a configuration that is the combination of $(\, , y, \,)$ and $(r, \, , r)$. In this case the yellow balls were not mixed with the red. If we now consider mixtures, then $(r, \, , y, \, , y, r)$ is also a possible configuration. How many mixture configurations of this kind are there? Each of the six compartments can either be occupied or not occupied by a red ball or a yellow ball. This gives $3^6 = 729$ configurations, which is much more than the 64 non-mixed configurations. This difference between mixed and non-mixed configurations becomes much greater as the number of compartments increases.

Why do drops not separate out? Boltzmann's line of reasoning went as follows: given that no preferred microstates exist and that the chance that a certain mixture will move towards a particular microstate owing to arbitrary external influences is therefore the same for every microstate, very improbable states will virtually never occur. Let us look at our example again. If every microstate has the same chance of being realised, then this corresponds in our case to 1 chance in 729. The chance of changing into a non-mixed configuration is then 64 in 729 = 0.087, less than one chance in ten. In our example this still amounts to a good chance and in fact in the case of red and yellow balls that are divided between 6 compartments, there will be a fairly probable chance of finding a non-mixed configuration, i.e., only yellow balls in the three left-hand compartments and only red balls in the three right-hand compartments. Let us now try to make our example a bit more realistic in comparison with the actual situation of mixtures of matter. For n compartments, the chance of finding a non-mixed configuration (only yellow balls in the $n/2$ left-hand compartments and only red balls in the $n/2$ right-hand compartments) is $\frac{2^n}{3^n} = (2/3)^n$. This chance becomes very small as n increases because the limit as n approaches infinity of this variable is nil.

The number of molecules in a sample of matter may be estimated at about 10^{24} and the number of microstates of this enormous accumulation of molecules, which give rise to the same macrostate, is very large. If we are describing the sample of matter by means of classical statistical mechanics, as was the case in Boltzmann's day, we have to replace the concept of "number of microstates" with "volume in phase space". The scope of this article does not allow us to put forward an exact calculation because we would have to introduce the concept of phase space. Nevertheless, we would like to further discuss a highly simplified (and strictly speaking mistaken) view, where we would simply present the sample of matter as an accumulation of individual molecules and only the number of these molecules would be used in the calculation of the entropy. We therefore assume that the sample of matter consists of 10^{24} spaces, each one of which can be filled by a molecule. To find out what the chance is of a mixed situation changing into a non-mixed situation, we have to calculate (2/3) to the power 10^{24}. The chance is something like 1 divided by 10 to the power 10^{23}, which is an extremely small chance. Let us develop this example further in order to get an idea of how small this chance really is. Assume that every nanosecond (1/109 seconds) a new configuration of 10^{24} compartments is filled with red and yellow balls, which more or less corresponds to the frequency with which these kinds of changes in configuration could take place for real molecules. For each change in configuration the chance of a non-mixed configuration is 1 divided by 10 to the power 10^{23}. Using probability theory, we can then calculate how many changes in configuration have to take place for there to be more than a 1 in 2 chance of finding a non-mixed configuration at least once. This produces a figure of 0.7 times 10 to the power 10 to the power 23. If we assume that a change in configuration occurs every nanosecond, then we have to wait 0.7 times 10 to the power 10 to the power 23 nanoseconds in order to have more than a 1 in 2 chance of finding a non-mixed configuration at least once. The age of the universe is estimated to be 10 billion years, which is 10 to the power 26 nanoseconds. We would therefore have to wait 10 to the power 10 to the power 22 lifetimes of the universe. This is unimaginable and actually boils down to the fact that the realisation of a non-mixed configuration is so improbable that it will never happen, not even at the level of the age of the universe. The chance of changing from a mixed situation to a non-mixed situation is therefore unimaginably small. The same holds for a mixture of two drops of water, one cold and one hot. The configuration of two drops of water divided up into a cold drop and a hot drop exists, but is so improbable that a spontaneous change to such a state never occurs.

Boltzmann's reasoning is the subject of great debate because it is an attempt to explain the irreversibility of certain macroscopic processes starting from reversible microscopic processes. Although this is a very important issue and debate is still raging, we shall see that the problem of irreversibility manifests itself in a much more crucial way in the creation-discovery view. We would just like to note that Boltzmann's reasoning perhaps does not prove irreversibility, but it does enable us to understand why there is a spontaneous conservation of or increase in entropy before a change occurs in a closed system. The fact that structures we know "go to ruin", attacked by "the ravages of time", is an expression of this reasoning. The pyramids of Cheops are silting up because the configuration of stone crystals, built over three thousand years ago by the Egyptians, has a lower entropy than normal desert rock. Mountain ranges are eroded by rain because a sharp mountain peak has a lower entropy than a flat wasteland of rocks and mud. When you go for a walk along the beach, you will never see the sea and wind form a sandcastle as if by magic because a sandcastle has a much lower entropy than a pile of sand. Sugar never spontaneously separates from the coffee in a cup of sweet coffee because the coffee and sugar mixture has a much higher entropy than coffee and sugar separately. If we watch a film in reverse and see how a cigarette "unsmokes" itself, and how our best friend dives out of the swimming pool onto the diving board, and how eggs jump out of the pan and back into their shell, then we are flouting the second law of thermodynamics.

All entities of the construction process that are fixed in a particular layer of reality are struggling against this second law of thermodynamics. A table has a much lower entropy than a random collection of pieces of wood and iron, and that is the reason why tables do not come into being spontaneously. A hunt, as a social entity, has a much lower entropy than the random actions of a group of people, and that is why a hunt does not arise spontaneously. A book has a much lower entropy than a random collection of sheets of paper, and in turn a sheet of paper has a much lower entropy than a random sample of wood pulp, and that is why no books come into being spontaneously. Living beings (plants, animal and humans) are entities that offer "resistance" to the second law of thermodynamics. The way in which molecules are organised within living matter corresponds to a local decrease in entropy. What is the driving force behind this struggle against the second law of thermodynamics? Do we understand this second law enough to be able to formulate a response to this question? To clarify this point, we would now like to study the other layer of reality where the concept of entropy has had success.

11. SHANNON AND INFORMATION THEORY

As already mentioned, entropy is introduced in the material layer of reality. In 1948 Claude Shannon published a article in which he laid the foundation of information theory [23]. Like Boltzmann in his day, Shannon was interested in an important and topical problem: how can information be efficiently transmitted? In his article he introduces the concept of information content of a message. The length of a message is certainly related to the information content of the message, but it is also evident that messages of different lengths can have the same information content, for example, the same message in different languages. The reason for this is that there is often redundancy, which makes the message much longer than its information content, so that the information content actually depends on what are considered to be permitted messages of a given length. If all the permitted messages of a given length are known and are numbered, there is no redundancy in this numeration and the size of the code number can be regarded as a measure of the actual information content. Hence Shannon defines the information content of a given message as the logarithm of the number of permitted messages. The information content of a message is usually given in bits. A bit is a binary digit. The idea is that the message is translated into a binary alphabet with two symbols 0 and 1 and its length is then measured. If the text I am writing now has an information content of $10,000$ bits, that means that if I were to translate the same text into machine language, using only 0 and 1, I would need a string of $10,000$ characters to set it down, and this string would be chosen from a possible $2^{10,000}$ of these strings of $10,000$ characters.

Can we identify the second law of thermodynamics in this cultural layer of reality? The amount of information in a text is entered by

Shannon and corresponds to the entropy of a material substance. The microstates for the material layer are determined by the states of the molecules that make up the sample of matter in question, and the entropy is the logarithm of the number of microstates. In the case of the cultural layer, the entropy of a text is determined by the amount of information needed to store this text.

What would be analogous in the cultural layer with mixing matter in the material layer? We propose the following: suppose that two people want to write a text together, in other words a joint text, and to make the analogy clear, suppose that one of the two people writes in red ink and the other in yellow ink. A non-mixed text is one in which one person has written the first half and the other person the second half. Suppose that we digitise the text and that there are n spaces available. In a randomly mixed text, every space can be empty, or can contain either a red character or a yellow character. This gives 3^n different possible configurations. In 2^n of these configurations, the first half of the text is written by one person and the second half by the other person. Where a text of n bits is written jointly, the chance of a non-mixed text being spontaneously created in this way is $(2/3)^n$, the same chance of finding a non-mixed configuration in our earlier example of the red and yellow balls that can be put into n compartments. This chance again becomes very small when n is large. Let us look at an example. The text that is being created here takes up $10,000$ bits in memory. If this text were to be written by two people, the chance of a non-mixed text is $(2/3)^{10,000} = (1/10)^{18,031}$ (a decimal point followed by $18,030$ noughts followed by a 1). We can also express this differently: if a text of $10,000$ bits is written and stored by two people, so that the red letters of the first person are still distinguishable from the yellow letters of the second person, this will need a memory of $10,000$ bits multiplied by $log3/log2 = 1.58$, which is equal to $15,800$ bits. In general we can say that a text of n bits jointly written by k people will need a memory of $nx(log(k + 1)/log2)$ bits. For a group of 10 people jointly writing the text, this gives $3.45n$ bits. This example helps us understand what the analogy is with the second law of thermodynamics. Texts are seldom jointly written in such a way that all possible configurations are equally probable. In general, tasks are divided and different sections are written by separate people. Our example, although it clarifies the situation, is therefore a bit unrealistic. Conversations take place in a much more mixed way. The chance that a conversation between several people will deteriorate into a succession of monologues by separate individuals is actually very small. We can generalise this to collaboration on any cultural product and conclude that it is very improbable that such a collaboration would deteriorate into individual actions by the various participants.

Here we encounter the deeper meaning of the second law of thermodynamics. With regard to this we would now like to illustrate the fecundity of our layers perception and our creation-discovery view. In the cultural layer of reality it is abundantly clear that the second law of thermodynamics, as we have illustrated it, is a marked idealisation of a much more profound principle, which is well-known to philosophers, and which we may express as follows: "The whole is greater than the sum of its parts".

12. CREATION AND PERPETUATION IN DIFFERENT LAYERS

Every microstate of an entity S, composed of two non-mixed entities S_1 and S_2, is a product state, determined by a microstate of S_1 and a microstate of S_2. This is the reason why the entropy of this entity S, composed of two non-mixed entities S_1 and S_2, is given by the sum of the entropy of S_1 and the entropy of S_2. The entropy of a mixture S of two entities S_1 and S_2 is higher than the sum of the entropy of S_1 and the entropy of S_2 on account of the fact that the mixture S has microstates, those which describe a mixture and which are not a product of microstates of S_1 and S_2. The mixture S is actually a new entity that does not simply consist of S_1 and S_2, and this explains the fact that S has states that are not product states of states of S_1 and S_2. The increase in entropy for a mixture is a result of the increase in microstates for this mixture.

When two quantum entities come together new states of non-separated entities are created. These new states are represented by the functions that cannot be reduced to waves in three-dimensional space; these are the non-local states. This phenomenon is very well described and predicted by the quantum formalism and has now been confirmed by experiment. In this case, however, we shall not discuss a mixture. With a quantum entity in a non-product state, the subentities lose their individuality. The whole cannot be regarded as the combination of its parts and this fact underlies the many paradoxical quantum effects. This phenomenon is also found in other layers: a jointly created cultural product is not a collection of small basic cultural products. The fact that entropy as a concept can be successfully used in two layers is no coincidence and is associated with the special phase in which the creation process of the biomousa takes place in these two layers. This is the phase we call the "building phase" in the next section of this article. Broad interpretations of the nature of reality, which are often derived from the second law of thermodynamics, are therefore highly over-simplified. For example, the interpretation that the phenomenon of mixtures not spontaneously separating would predestine the whole of reality to evolve towards greater disorder and less order is a local conclusion that is also related to the special phase in which this second law manifests itself, and hence does not constitute a truth in other layers of reality. We should mention that the popular interpretation of entropy as a measure of disorder needs to be refined a great deal. Entropy is a measure of the number of microstates that give rise to the same macrostate and hence it is a measure of the information content of this macrostate. The fact that a situation where there are more microstates corresponds to a situation where there is more disorder is related to the specific phase in which the creation process takes place, and to the various layers that are being compared, and is not a general fact. If, for example, we consider the process of spontaneous crystallisation of a supercooled substance, entropy increases if the process takes place under adiabatic conditions. Using ideas about order and disorder in everyday life, it is difficult to maintain that the crystal is more disorderly than the supercooled substance. The interpretation of entropy as a measure of disorder can be salvaged in this example by assuming that the order we want to consider is made up of two parts: one of a configuration nature and the other of a thermal nature. The configuration-type order produced by crystallisation of the substance is lost because of an increase in disorder, which is caused by the potential energy released (latent heat) being spread over the vibration modes of the crystal. In this way "disorder" has to be related to both the distribution of the particles in space and the distribution of the energy over the energy levels. This is a very sophisticated way of looking at matter, however, which also depends on the theory that is being used. Another situation where it is clear that disorder and entropy are two different concepts is in the analysis of the behaviour of gravitational systems. Here too we find spontaneous spatial structuring, which means increased spatial organisation, even in the case of a closed gravitational system. Only when the process is looked at in phase space, thus introducing a completely abstract notion of order, can entropy still be regarded as a measure of disorder [24].

If we acknowledge that sandcastles are not spontaneously created, we must also acknowledge that people make sandcastles. How should we interpret this? We can now clarify this point. The calculation of the entropy of a sandcastle primarily has to do with how we interpret the sandcastle and what we mean by "spontaneously". We are surprised at the creation of the sandcastle only if we look for the creativity required to create it in the material layer. If by "spontaneously" we mean "by using creativity in the material layer", then the sandcastle is formed simply as a result of the movement of grains of sand, as small material entities, making use of the random forces that prevail between these grains of sand. If we interpret "spontaneously" as "what happens on a beach where people are sunbathing", it is much more probable that a sandcastle will be created. People are therefore needed to make a sandcastle. But now we have only shifted the question because why are these people there? In order to fit this question into the creation-discovery view, we would first like to examine in more detail the analogy between the two layers in which we identified the entropy law.

In the material layer of reality we would like to characterise three organisation levels of atoms and molecules or ions more generally. A crystal (a solid substance) is an explicitly organised form. All of the particles remain in their "place" in the same state and can only exert influence and move locally. This produces a very coherent but rigid form of organisation, which, however, has and requires minimum energy. Various types of structure are possible in this organisational form. A liquid is a more implicitly organised form. Forces hold the particles together but each particle can still move individually to a certain extent. There is no coherence between the particles, only attraction. A gas is an accumulation of virtually free particles that only hit upon each other. The biomousa, however, has not chosen any of these three options. It was not interested in the complete dissociation and excessive vagueness of a gas, or in the wetness and perfect malleability of a liquid, and certainly not in the still, fixed rigidity of a crystal. What kind of backbone does the biomousa have? This question will help us unveil the mystery. There is a fourth way of organising matter. Let us elaborate on this.

The molecule can be regarded as the seed of the crystal. Starting from this molecule there seem to be various ways of building larger structures. We have already discussed one of these ways in detail: that of the crystal. This comprises a constantly recurring pattern of basic structures. Once the periodicity has been established, there is no limit to the size of the crystal. This crystalline mode is averse to any form of creativity. The crystal is like wallpaper with a pattern that is repeated in all directions. Wallpaper is not regarded as an example of creativity in our cultural layer and the biomousa also took the same view in prehistoric times. Although it was fascinated for a time by the creation of the basic molecule, the way of the crystal was not its way. The fourth way is choosing to make creative use of molecules. Complex organic macromolecules are works of art, made by the biomousa, in which every atom and every group of atoms plays a unique and individual part, quite unlike the part played by another group, as in the case of the crystal. The fourth way leads to living matter. It is in fact these molecules that form the basis of the material of living matter. And this is the way that the biomousa has chosen, seeking the power of perpetuation through creativity. The material forms of living beings, the single-cell organisms, plants, animals and humans, are the creations of the biomousa in the material layer of reality. This way immediately takes us away from reductionist territory. A macromolecule is not a collection of interacting atoms, just as an atom is not a collection of interacting pre-material particles. A macromolecule is a structure, a construction: it is greater than the sum of its parts. It has new states, which are not product states and which cannot therefore be regarded as a configuration of atoms.

Now that we have analysed the various forms of organisation within the material layer, let us return to the cultural layer. The basic entities of the most recent layer of reality are the "cultural products" in the broadest sense of the term. An entity is situated in the cultural layer because a meaning is associated with this entity; the possibility therefore exists of "covering" this entity with "symbols" (by introducing symbols in the cultural layer of reality). Symbols originated from sounds and gestures, probably mainly intended as a pure means of communication between humans and animals originally. They gradually became more and more complex and have created their own entities, which are all cultural products. Old material and social entities, such as the table and the hunt and the house, were vested with "meaning", but the "force field" of the meaning has also created really new cultural entities: texts, theories, works of art, cultures, and so on. In this sense we could regard "meaning" as the principal "force field" of this cultural layer of reality. All cultural products are immersed in this force field but entities in the previous layers of reality do not feel this force [25].

Let us pick up the trail of the biomousa once again. We want to think about texts as cultural products and we compare them to material entities. They are made up of words, which we can compare to molecules. The atoms are the letters and letters joined up to make a word are molecules. Words can join up to form sentences, equivalent to chemical bonding to form larger molecules. We can construct texts made up of the same sentence or a collection of sentences repeated over and over. This is a crystal text. This is a crystal text. This is a crystal text. This is a crystal text. This is a crystal text. This is a crystal text. This is a crystal text. This is a crystal text. And so on. These kinds of texts are crystals. If through some process or other, for example an error in the software of our word processing program, the words or letters of these sentences got mixed up, then we get texts that are liquids and gases. Thsi la s cr is yext tat. i le s cr axt ysts i taTh. We would not normally identify these kinds of constructions as being texts. Why not? No "meaning" can be established or passed on because all of the combinations of letters do not result in words, and all of the combinations of words do not result in sentences, and all of the combinations of sentences do not result in texts. Only those combinations of letters that are arranged in accordance with the "force field" of the "meaning" form words, and only those combinations of words that are arranged in accordance with the force field of the meaning form sentences and texts. Once again, meaningful sentences are not a collection of words. They are constructions and structures, which contain new states, which cannot be reduced to the configuration of the words. Shannon's entropy theory does not explore the layer of meaningful sentences.

The same argument applies to material entities. Only those combinations of atoms that are arranged in accordance with quantum mechanical forces result in molecules, larger molecules, and crystals or living matter. And it is only these forms of organisation that support the quantum mechanical force field, and communicate through it, by means of photons. Liquids and gases do not take part in this game, and if we cannot allow random combinations of letters and words to be regarded as sentences or texts, then we should also conclude that liquids and gases cannot be regarded as real matter. They do not take part in the original creativity game. Liquids and gases have only acquired a material aspect by means of the large number of molecules that we find in them at our macroscopic level because disorder is reduced by this large number of molecules and they have become usable "matter" again for the creation of living matter. Crystals and living matter are real matter. Crystals correspond to texts made up of recurring sentences, whose meaning is no greater than the meaning of just one of these sentences. Meaning remains at a local level and cannot really expand. The way of the biomousa, used by the macromolecules, allows the force field of meaning to expand and create living matter. This living matter corresponds to our texts, which we regard as real and valuable cultural products.

13. THE DIFFERENT TONGUES OF THE BIOMOUSA

We would now like to try to identify the different phases of the journey of the biomousa and we are aware that only a broad outline is possible. The contents of this section therefore have to be regarded as an attempt, using broad metaphors, to speculate in very general terms about the nature of the biomousa.

Suppose that we start with the existence of a collection of basic entities. In the material layer these are the atoms, and for part of the cultural layer they are the letters and words that make up the language.

The letters and words of the language are of course only the basic entities for a particular part of the cultural layer, while atoms are the basic entities for all entities in the material layer. For the sake of simplicity, we shall confine ourselves to that part of the cultural layer that relies on language.

One particular phase, which we shall call the *building phase*, is chiefly a construction based on combinations of the basic entities in particular structures. In general there are many possibilities but some will be found to be "better" than others. The building phase in the material layer consists in constructing large quantities of different kinds of matter, all made up of the fundamental building blocks, i.e., molecules. The building phase of the language part of the cultural layer consists in constructing large quantities of different texts, all made up of the same basic entities, i.e., words. This building phase is explosive and partly destroys the old structures and the old order. The less systematically structures are created, the more these structures obey a kind of second law of thermodynamics. In fact, if molecules are simply thrown together any old how, there is little chance of living matter being produced. If words are simply written down in any old order, there is very little chance that a meaningful text will emerge. We would like to mention that in reality this process does not take place in the simplified manner we describe here. Meaning grows when texts are present. Space and spatial forces grow when mixtures of elementary particles are present. Disorder increases, while the entropy view of the mixtures and the increase in disorder provide a good description of this phase.

The second phase, which we shall call the *development phase*, consists in creating functionally oriented aspects, which have more to do with mutual interaction and interaction with the outside world. Some combinations of basic elements are clearly able to interact with their environment more and better than others. Our ancestors searched for caves to live in and, although a cave as a pure material object is highly improbable, its perpetuation is ensured in the form of a house. The cave "clicked" in some way or other with the needs of our ancestors. The model of the cave lead to them constructing house-like buildings themselves. In the cultural layer it is likewise communication that will select the texts as being significant. The development phase is still an explosive phase, like the building phase, but it introduces the element of selection, which tempers the constant tendency towards an increase in disorder. In fact, some texts will be preferred to others and in this new classification, the most disorderly texts will have less chance of survival.

We shall call the next phase the *structuring phase*. The new area is structured. Newly created entities are given a place. This phase corresponds to arranging and structuring particular texts and communications. Cultural elements are assigned a place and the concept of space is introduced. Two very important aspects of the previous phases, creation, which is so essential for exploring all possibilities, and the optimisation of perpetuation and of perpetuation techniques, which often involve copying, or a creative form equivalent to copying, are partly split off and the concept of motion in structured space emerges. Entities can move in space without losing their individuality. Moving is copying oneself in a "stable" way and is reversible over time. This structuring phase is well under way in the cultural layer of our reality. People are constantly engaged in assigning cultural products their place and in defining the space in which they can move. If we return for a moment to the material layer, we can see that the structuring phase is already complete here. All material entities are well arranged in space and can move within it and interact with one another. Space as a stage has become separated from time and has produced an illusion of "reversibility". Our way of thinking about the universe is determined by space to such an extent that we also see time evolving within this space. But this is wrong of course. We can now understand better why pre-material entities are not present in this space. Space is the stable structure, seeking equilibrium between perpetuation and its need for structure, between creativity and its need for exploration. Creation then partly converted itself into "motion". Just as the matter of the ink used to write the letters of a text on a sheet of paper is not present in the cultural space that is now forming in our cultural layer, so too are quantum entities not present in space.

We shall call the fourth phase the *potentiality phase.* This phase clearly starts building a bridge to another layer of reality. The structuring phase introduced a lot of new and perpetuable structures. These structures have a very low entropy compared to the basic elements but they are highly perpetuable and hence they exist. The benefit of modules is discovered, especially in connection with perpetuation. Modules are larger elements than the original fundamental building blocks, which can be used uniformly and which can easily be copied and hence increase perpetuation. All our modern electronic equipment is based on this modular principle. Mathematical theorems are also modules, however, as are large biochemical molecules and genes. Modules are the first attempts to choose new fundamental building blocks and thereby forget the old ones and incorporate them into an automatic reproductive process that has enormous power of perpetuation and, at the same time, a stable structure. That is why the creation of modules is the first step towards a new layer of reality. In general, however, a new layer is not just created immediately since this is not the purpose but rather the ultimate result of the choice of modules. Why do we call this phase the potentiality phase? With the introduction of modules, the interaction with the space extensively created in the previous phase implodes again. A module possesses a special property in that it contains a lot of potential reality in itself, reality that only becomes existent when the module takes up its place. *DNA* is a good example of an unbelievably complex module because it has the potential to allow a whole living being to grow, if it ends up in the right place. It is an implosion of external reality because this living being has the potential to remember and pass on information and hence support the building blocks of the new cultural layer. Plants, animals and humans have developed a very complex game, which we call sex, to ensure that *DNA* always ends up in the right place. Why do we not regard *DNA* as the basic element of a first new layer of reality? We could do this, though how we define these layers is of course arbitrary to some extent. Nevertheless we choose not to call this a new layer because it lacks one essential characteristic of a separate layer. Genes have not explored the universe. They have not seen the stars! They have not expanded their space to infinity. However, the stored potentiality, i.e., the plants, animals and humans in material forms, has set to work to reach the stars. To this end, man, and man alone, had to take a new step. He started to digitise his experiences, he introduced concepts or modules of experiences. He began to ensure the perpetuation of these concepts by inventing writing among other things. And his knowledge, which is the power behind this new cultural layer of reality, began to spread out over the universe. We shall call the next phase the *bridging phase*. New basic elements are introduced in this phase. These basic elements are separate from the previous basic elements yet are grafted onto them. Initially these basic elements are still modules but then steps are taken to digitise all of the important entities. The new layer of reality now becomes a quantised layer. And we have returned to a building phase.

We can now understand what automation means. The introduction of modules to increase perpetuation implies the introduction of automation. Prototypes are in fact forms of modules too. Real creativity swims ahead of this automation, as it were, like foam on the crest of a wave, and takes place on the edge of the new layer forming. We also believe that every layer goes in search of its past and wants to grasp it again in its entirety. The previous layers did this because, as we know, the four fundamental forces in nature (Gravitation, electromagnetism, weak interactions and strong interactions) bind all matter and energy in the universe. If we consider the modern quantised view of interaction between entities by means of a force, this interaction is nothing more than the interaction of a force particle: in the case of gravity, the graviton, in the case of electromagnetism, the photon, in the case of weak interactions, the boson, and in the case of strong interactions, the gluon. And in the case of the cognitive force field in which we live, the word.

The highly speculative view we are putting forward here identifies man as the vehicle for the most recent force of nature, the cognitive force, and as the foundation of the new layer of reality, the cultural layer.

AFFILIATION

Diederik Aerts
Center Leo Apostel
Brussels Free University
Belgium

REFERENCES

[1] **Aerts, D.**, "An attempt to imagine parts of the reality of the micro-world", in: *Problems in Quantum Physics; Gdansk '89*, World Scientific Publishing Company, Singapore, **3**, 1990.

[2] **Aerts, D.**, "Construction of reality and its influence on the understanding of quantum structures", *Int. J. Theor. Phys.*, **31**, 1992, p. 1815.

[3] **Aerts, D.**, "Quantum structures: an attempt to explain their appearance in nature", *Int. J. Theor. Phys.*, **34**, 1995, p. 1165.

[4] **Aerts, D.**, "The Entity and Modern Physics", to be published in: **Castellani, E.** (ed.), *Identity and Individuality of Physical Objects*, Princeton University Press, 1998.

[5] **Schrödinger, E.**, "Die gegenwärtige Situation in der Quantenmechanik", *Naturwissenschaften*, **23**, 1935, pp. 807, 823 and 844.

[6] **von Neumann, J.**, "Grundlehren", *Math. Wiss.*, **XXXVIII**, 1952.

[7] **Kochen, S. and Specker, E.P.**, *Journal of Mathematical Mechanics*, **17**, 1967, p. 59.

[8] **Randall, C.H. and Foulis, D.G.**, "The operational approach to quantum mechanics", in: **Hooker, C.A.** (ed.), *Physical Theories as Logico-operational Structures*, Reidel, 1979.

[9] **Piron, C.**, *Foundations of Quantum Physics*, W.A. Benjamin, Inc. London, Amsterdam.

[10] **Aerts, D.**, "Description of Many Separated Entities without the paradoxes encountered in Quantum Mechanics", *Foundations of Physics*, **12**, 1982, p. 1131.

[11] **Aerts, D.**, "Quantum structures, separated physical entities and probability", *Foundations of Physics*, **24**, 1994, p. 1227.

[12] **Aerts, D.**, *The One and the Many*, Doctoral Dissertation, University of Brussels, Brussels.

[13] **Aerts, D.**, "Classical Theories and non-Classical Theories as special cases of a more general Theory", *Journal of Mathematical Physics*, **24**, 1983, p. 2441.

[14] **Aerts, D.**, "The description of separated systems and quantum mechanics and a possible explanation for the probabilities of quantum mechanics", in: **van der Merwe, A. et al.** (eds.), *Micro-physical Reality and Quantum Formalism*, Kluwer Academic Publishers, 1988.

[15] **Aerts, D., Durt, T. and Van Bogaert, B.**, "Quantum Indeterminism, the Classical Limit and Non-Locality", In: *The Proceedings of the Symposium of the Foundations of Modern Physics*, Helsinki, World Scientific Publishing Company, 1993.

[16] **Aerts, D. and Durt, T.**, "Quantum. Classical and Intermediate, an illustrative example", *Found. Phys.*, **24**, 1994, p. 1353.

[17] **Rauch, H.**, "Neutron interferometric tests of quantum mechanics", *Helvetica Physica Acta*, **61**, 1988, p. 589.

[18] **Aerts, D. and Reignier, J.**, "On the problem of non-locality in quantum mechanics", *Helvetica Physica Acta*, **64**, 1991, p. 527.

[19] The reader can find a detailed description of this experiment and further references in **Aerts, D.**, "The construction of reality and its influence on the understanding of quantum structures", *Int. J. Theor. Phys.*, **31**, 1992, p. 1815.

[20] **Aerts, D.**, "A possible explanation for the probabilities of quantum mechanics", *J. Math. Phys.*, **27**, 1986, p. 203.

[21] **Aerts, D.**, "The origin of the non-classical character of the quantum probability model", in: **Blanquiere et al.** (eds.), *Information, Complexity and Control in Quantum Physics*, Springer-Verlag, 1987.

[22] **Aerts, D.**, "A mechanistic classical laboratory situation that violates Bell inequalities exactly in the same way as the violation by the EPR experiments", *Helvetica Physica Acta*, **64**, 1990, p. 1.

[23] **Shannon, C.**, "A Mathematical Theory of Communication", *Bell System Tech. J.*, **27**, 1948, p. 379.

[24] **Severne, G.**, "Irreversibility in the Large", in: *Frontiers of Physics Lectures*, Physical Research Laboratory, Ahmebadad, 380 009, India.

[25] In "Sinn als Grundbegriff der Soziologie", published in: *Theorie der Gesellschaft oder Sozialtechnologie*, Suhrkamp Verlag, Frankfurt am Main, Niklas Luhmann advances the idea of meaning as a "force" in the social layer.

FONS ELDERS

THE SEVEN KEYS OF ANY WORLD VIEW

INTRODUCTION

A study of world views forces the researcher into the role of a traveler who discovers that his city map does not give him any clue as to the real life of the city—its subterranean layers, its power structure and so on. The subject *World View* calls up images of the Generic City by the Dutch architect Rem Koolhaas: The Generic City is seriously multiracial, on average 8 % black, 12 % white, 27 % Hispanic, 37 % Chinese/Asian, 6 % indeterminate, 10 % other. Not only multiracial, also multicultural. That is why it comes as no surprise to see temples between the slabs, dragons on the main boulevards, Buddhas in the CBD (central business district). The Generic City is always founded by people on the move, posed to move on. This explains the insubstantiality of their foundations.

Koolhaas' plastic description destroys the notion of the city as a center of traditional culture, with stable, democratic institutions. The Generic city is an urban nomadic event, a melting pot of innumerable movements, dreams and conflicts. "Each period in history that witnesses a struggle of paradigms in which the borders between fiction and non-fiction lose their transparency, needs the dialogue. Only the dialogue form has the communicative freedom to confront and sometimes embrace the divergent philosophical and psychological positions, or the novelness as dialogue in Bakhtin's studies." [1]

DIALOGUE VERSUS APRIORI KNOWLEDGE

The powerful dynamics of permanent change cuts deep in the life-force of the vital-psychic order, the *longue durée* of Henri Bergson or Fernand Braudel. This urban crossing point of the old and the new confronts the various world views with the question of mutual negation, leading to fundamentalist tendencies, or with the choice of mutual investigation and open dialogue. The readiness for dialogue implies the recognition that we consider the various world views first of all as a human manifestation, regardless of their philosophical or spiritual background. Moses, Buddha, Sappho, Socrates, Jesus, Mohammed, Russell and Gandhi, as symbols of various world views, are confronted with each other in the Generic City. If their followers and admirers are willing to discuss their views with each other, then we will see a splendid manifestation of human insights and life strategies. The public will be inspired by their wisdom, their love for truth and the common good. If, however, the followers and admirers already know the answers before the questions are raised, we are in trouble. If they present their belief system as divine, absolute knowledge, then the communities of believers will start to distrust each other, followed by possible mutual exclusion. The Generic City will develop ghettos and nightmare-like dreams.

If we try to imagine how various world views enter into a dialogue about the question of good and evil, or about the expectations of some kind of life after death, we will witness a manifestation of "alterities." Such a manifestation will raise the question how much alterity a person, a community, and finally, a city can absorb before it enters the phase of becoming a generic city, in which the explosion of alterities destroys the roots of the past or drives them literally underground.

THE HUMAN DIMENSION IN DIFFERENT LIFESTYLES

New York City is a good example of a fragile balance between the past and the future. The Big Apple has an ongoing love affair with the apple in the garden of Eden. It cultivates the moment as a lifestyle, and with this lifestyle, its continuity among all the changes.

I believe that the capacity of human beings to absorb cultural differences is rather limitless, although there is ample evidence pointing to the opposite thesis. I remember a visit to New York that took place shortly after a stay of more than a month with the Dogon people in Mali. The difference in lifestyle between the Dogon and the New Yorkers was profound—so profound that I wondered in those days whether I could still apply the category of humankind to both New Yorkers and the Dogon. By humankind I do not mean an abstraction but rather a concrete, living notion. Can the gap between lifestyles become so deep that we are unable to recognize the human dimension in those different lifestyles?

My answer is both no and yes. No, if we are unable to recognize the human dimension, by seeing only the surface level of a different lifestyle. In such a case, our world view and psychology exclude *a priori* any serious attempt to understand an ontology and anthropology that differ from our own. Yes, if we are able to recognize the human dimension in a different lifestyle, by experiencing positively its aesthetics, because the aesthetic manifestation of a different lifestyle is the way *par excellence* to reveal its qualities. The great variety of lifestyles in a cosmopolitan city explains, at least partly, its attraction upon millions of people. It is especially in such an urbanized environment that each generation faces new challenges and will invent some answers, different from those of the former generation. I am convinced that the world views which contain the notion of transformation in the center of their ontology and anthropology, will be more capable of providing some orientation in an age of rapid changes than the world views with a static foundation. If this is true, Buddhism must be better equipped to deal with the actual transformations than Christianity. Some Buddhists experience the exile from Tibet since the Chinese occupation of 1959 as positive, because the exile challenges them to reflect upon the meaning and truth of the 'dharma', the Buddhist teaching, within the parameters of Western values.

NOMADIC AND SEDENTARY STATE OF MIND

In order not to overestimate the hermeneutical possibilities of *The Seven Keys of Any World View* I mention a dream I had in Turkey after traveling with my family for half a year in Europe, Africa and Asia. In this dream I realized that within a few weeks I would cross the borders of the Dutch state. It felt as if I were entering a prison: a clean, well-organized system but nevertheless a system. After that strong feeling of borders, I suddenly saw the origin, the hidden deep structure of my book *Analyze Decondition, an introduction to systematic philosophy*, written and presented for Dutch Educational TV (Teleac), two years earlier [2]. I had been writing this book on systematic philosophy, while sitting on a chair. It belonged to the sedentary culture, as was the case with all the philosophy books I had been studying. The dream didn't disappear, after I woke up. It was crystal clear to me that traveling through the Sahara, crossing many borders and living at the feet of the Himalaya in Kashmir, had changed my perception. Suddenly I realized the limits of the sedentary culture and the fundamental difference between a sedentary mind and a nomadic mind. I understood the fear of the dwellers for the nomads, the worldwide struggle of the states against the people who travel without documents, the ugly treatment of the gypsies by peasants, citizens and authorities. The nomadic state of mind differs from the sedentary state of mind as the hunter from the peasant, a traveler from a citizen, a river from a canal. This text *The Seven Keys of Any World View* springs from a sedentary state of mind, as all written reflections on methodology do.

DEEP STRUCTURE AND SURFACE STRUCTURE OF A WORLD VIEW

To grasp the deeper meanings of a world view, we have to cross some borders in our own consciousness, as if we are visiting the invisible cities of Italo Calvino. In order to understand the assumptions and aims of a world view, I need to introduce a theoretical distinction between the surface structure and the deep structure of a world view. In doing so, I use the terminology of Noam Chomsky's transformational grammar. The surface structure refers to the grammar, i.e., to the sensible manifestation of a world view, while the deep structure refers to the meaning of this manifestation. "Grammatical" expression and its implicate meanings are never identical. There is always the necessity of interpretation and, therefore, of the possibility of misinterpretation from the point of view of the supposed or assumed "original" intention and meaning. But any search for an "original" intention, can never escape the status of an interpretation, because of this creative tension between surface structure and deep structure. By studying the surface structures, the "forms", of world views, we will be able to study both their differences and similarities. I use the word "similarity" and not the words "unity," "one-ness" or similar terms, because of the sheer logical, semantic and artistic impossibility of 'expressing' or 'designing' the notion of oneness, a key-term in all kinds of philosophical and religious monism. Such an intuitive notion of oneness can only be expressed in paradoxical, metaphorical polarities, never as such in itself.

A METHODOLOGY OF WORLD VIEWS

A methodology of world views has to start with a number of questions. The choice of these questions will determine whether the questions do relate to fundamental, unavoidable "fields of experience" of a world view, or do not relate. The questions must be both congenial in order to reveal the basic pattern of a world view, and relevant to as many as possible world views.

At the same time, the questions we are posing may not guide the content of the answers, although I realize that this methodological approach is not value-free. The foundation of this methodology belongs to the analytic-rational tradition in Western philosophy. I hope, however, that the proposed method is open-minded enough to be attractive to students of comparative studies of world views. I distinguish seven questions of which the first six can be interpreted as three different coins, each having two sides. Questions one and two deal with the epistemological coin, viz. language and truth; questions three and four deal with the ontological coin, viz. man and cosmos; questions five and six deal with the normative coin, viz. ethical and aesthetical values. The seventh and last question deals with the *methodos* of a world view, i.e., the introduction and education of newcomers into a world view. Stated succinctly: the epistemological, ontological and normative "fields of experience" merge into the methodos as in a nutshell.

I believe that the answers to these questions can shed some light on the assumptions, the internal consistency and the reach of application of a world view. The (hidden) assumptions, the coherence and the reach of application are three different perspectives which we can use as meaningful and even normative criteria to estimate the quality of a world view. For example, if an assumption of a world view coincides with the conviction of the a-priori privileged position of the 'own' people, tribe or race, the coherence and the reach of application of such a world view in terms of humanity as the category or class of all humans, will lack consistency and universality.

We have some experience with this method at the University for Humanist Studies in Utrecht in a program called "Buddhism, Christianity and Humanism." The method seems helpful for understanding the mutual differences as well as the differences within the "same" tradition by clarifying the evolution and transformation of certain assumptions.

THE STATUS OF LANGUAGE

The first question deals with the function and the position of language. What is the status of language within Christianity? This status is clearly different from the one in Buddhism or Humanism. The language of the holy Scriptures carries, for an orthodox Christian, an 'absolute' or in any case an existential truth. The language contains revelations of God and the God-man Jesus. The deep differences among Christians can be understood, at least partly, by their different interpretation of the status of the language of the Bible, varying from strictly metaphorical, human language to a literal, time-and-spaceless language. The "anthropological" interpretation of language considers religious language primarily as a world of meanings; the "theological" interpretation of language sees its *own* religious language as a world of true statements due to its divine origin. "This book will not be a history of the ineffable reality of God itself, which is beyond time and change, but a history of the way men and women have perceived him from Abraham to the present day." [3]

Buddhists appreciate language as a liberation in so far it provides the possibility to communicate and to share a domain of meanings, including knowledge about the hidden structures of the laws which govern our existence. But at the same time, Buddhists will say that language is a prison. The reason for this critical approach is the insight that language has its own self-defining, limiting logic, based upon categories and generalizations. Language categories stand in the way of an all-embracing experience. Therefore, language has to give way to silence in the practice of meditation. "Language is on the one hand, an excellent tool and without it we would not be able to do anything and yet, on the other hand, language is also the great barrier to an understanding." [4]

Humanists have a special relation to language because books as the symbol and content of critical examination belong to the heart of the humanist tradition. There is, however, not one book that can be singled out as the principal source of knowledge. In the humanist tradition every text can be criticized. Humanists will stress the "anthropological" dimension of language, not the "theological" one. This is why *The Seven Keys of Any World View* belongs to the humanist tradition, and why the proposed methodology is not value-free.

THE VARIOUS SOURCES OF KNOWLEDGE

Analogous to the role of language within the Buddhist, Christian and Humanist traditions, one can distinguish similar differences in relation to the sources of privileged knowledge in each of the three traditions. The Buddhists stress the importance of study of their ancient texts, including the commentaries of the texts until today. In that sense there is no striking difference between the Christian and the Buddhist approach. The difference is in the status of the textbooks. Even if, tomorrow, someone could prove that the Buddha is a legendary figure who didn't exist at all but whose historical existence has been the result of a collective pious fraud, it would not diminish the truth of the Buddhist teachings. The insights of the Buddhist way of life do not depend on an "external"-Messianic cause but are the result of an internal process of insight into the roots of our existence and into the grids through which we perceive human existence. This explains that silent meditation and the practice of the teachings are more important for 'real' knowledge, i.e., insight than the study of the texts. Insight and concrete experience are inseparable.

Many Christians, especially in the monastic tradition, will sympathize with these Buddhist insights with one important exception. The non-existence of Jesus or the negation of his godlike nature with its special relation to the divine Trinity of Father, Son and Holy Spirit, would not destroy the value of his teachings in their general wisdom, but it would destroy the existential source of the message. Jesus is a Messiah; Buddha is not. This also explains the different status of these messages for Christians and Buddhists.

If we ask ourselves where and how humanists hope to find some truth, we face a whole array of questions because of the philosophical complexity of the humanist tradition. Perhaps the main distinction we have to introduce with regard to the question of truth within Humanism, is whether truth is a discovery or an invention. The Greek word for truth, *alètheia*, refers to discovery, the unveiling of what is already there, e.g., the radical correspondence between the (human) microcosm and the macrocosm. The (post)modern interpretation, however, regards truth mainly as a human invention. It implies a strong personal approach ending in the construction of truth statements and ethical values. This process runs parallel with the refusal in (post)modern humanism to allow for the existence of a spiritual reality, in contrast to Renaissance humanists, Rosecrucians or Freemasons, which stress the importance of the intimate relationship, within the human reality, of an immanent and transcendent reality. Because of the important distinction between a philosophical-materialistic, and philosophical-spiritual orientation, the notion of truth in the humanist tradition resembles a Janus face. "It makes sense to make a distinction between pagan humanism, and a Christian humanism. Within the pagan humanism one can distinguish between a philosophical-materialistic tradition and a spiritual tradition. To the philosophical-materialistic tradition belong the philosophies of Protagoras, Stoa, Averroes, Pomponazzi, Voltaire and many philosophes, B. Russell, but also Marxist-inspired forms of humanism; existentialism and postmodernism. To the spiritual tradition belong hermetic Gnosis, kabbalah, neoplatonism, freemasonry, various trends in feminism and deep ecology philosophy." [5].

Although the above-mentioned distinction is an important one, it does not destroy the inner coherence of the humanist tradition, as long as humanists agree upon some basic values, such as: the human search for knowledge and insight; the art of conscious not-knowing, also called the Socratic attitude; human dignity as a life project; a never-ending responsibility for personal actions and "beliefs." I use the word "belief" in a broad sense because of the unavoidable necessity for humans to make certain assumptions about themselves and reality in general. As Wittgenstein said: One can have doubts about everything but not about everything at the same time.

MAN AND COSMOS

In order to show the usefulness of the proposed methodology of the seven questions, I want to sketch briefly some ontological and anthropological characteristics of Buddhism, Christianity and Humanism.

Christianity

With more than 2,000 Christian denominations existing today, it is not easy to describe some basic characteristics of the Christian religion. But if the task were completely impossible, any reference to Christian tradition would lose its meaning. Orthodox Christian belief is rooted in the idea of a transcendent Creator outside time-and-space, for Itself fully independent of His creation; and in the Messiah Jesus, the Savior who is both god and man, the second person in the Trinity of Father, Son and Holy Spirit. This special knowledge is due to the revelation of both the old and new Testament. Important within the context of this discourse, is the idea of a divine revelation, also characteristic of Judaism and Islam. The acceptance of a revealed "truth" places the believer in a special position toward the non-believer: there is, in principle, an unsurpassable gap. The non-believer can belong to the category of atheists or, for example, animists, but also to the category of the believers of the "wrong" belief. The rejection by Christians of the other monotheistic beliefs, and vice versa, is proof of that. But one could also argue in favor of the idea that a religious believer who accepts the responsibility for his or her belief, enters the community of all those human beings who realize that their world view is based upon one or more hypothetical assumptions. This could create a common feeling that human beings are all believers, one way or the other. However, some important differences remain. One is the formal feature of the Christian belief in a revelation of one kind or another with all its consequences for ethics, human identity, including notions as soul, an eternal life after death or at least a *Jenseits* orientation, and so on. Changing such a belief-system must always be a painful process because of the original time-and-spaceless orientation. Even starting to doubt certain aspects can easily become suspect, although not necessarily.

In the Christian tradition, man is never alone, just as the cosmos is never alone, because there is always the eternal God, a loving Father for some Christians, an unapproachable Greatness for others. Nature has been placed subjacent to humans, not on an equal footing with humans. The ecological crisis of today cannot be separated from this traditional unequality between nature and humans. Nature is of a lower status than the divine order. Man is located between God and nature. His primary goal is individual salvation, according to Jacques Maritain. His mortality is relative because of his immortal soul, created by the Creator. The created soul makes each man unique and an image of his Creator. It is this thought that has become an inspiration for a more immanent, mystic tradition in Christianity, but always at the edge of the institutionalized orthodoxy, both Catholic and Reformed.

Buddhism

Buddhism chooses its point of departure in a specific experience, viz. that all sentient beings suffer. Everything that exists falls under the law of dependent origination. In Western philosophy we would call it the all-embracing law of cause and effect. Buddhist tradition is imbued with the notion of impermanence. Dependent origination is the source of *karma*. Insight into our *karma* is the beginning of a process of liberation. There are different interpretations about the relationship between *karma* and *samsara* at one hand, and *nirvana* or enlightenment at the other hand. Nagarjuna, the founder of Mahayana Buddhism, considers *samsara*, the circle of innumerable determinations, and *nirvana* as the liberating insight into this process, to be identical. *Nirvana* is, for him, the perfect way of perceiving human reality: a deep grasp of our reality is the first condition for enlightenment. Our liberated actions will subsequently influence reality to a certain degree.

In Buddhist anthropology and ontology, there are at least two striking characteristics. One is the notion of impermanence with regard to all reality, including the human. This explains why Buddhism focuses on reducing the human inclination to develop an "ego" as a fixed frame for individual identity. Buddhism considers the need for an ego as the basic problem, the source of confusion, passion and aggression, including warfare. Hajime Nakamura speaks of the theory of the non-self. The second characteristic is the process of self-liberation through the realization that desire is the root of our pain. The paradoxical result of such a process of self-liberation is a deep joy that emanates total peace, whether a Buddha is alone or together with *prayna*, known as *shakti* in Hinduism, *yap-yum* in Tibetan language. There is fundamental equality between man and woman, not as x is x, but as x and y, a unity in polarity. Every human being, female or male, is potentially Buddha. Nature, e.g., a flower or an animal, is also Buddhahood in many Japanese haikus. The reason for this universal Buddhahood is that Buddhism does not know the notion of original sin as Christianity does, neither with regard to nature nor to humans. In the so-called beginning, there is no disobedience to God, only lack of insight into the human condition. As I mentioned before, there is no external authority, no revelation in Buddhism and, therefore, a different conception about the role of language and about the road to insight. The four questions show their internal link when we try to answer them, the same as in Christianity.

Humanism

Humanist tradition, as I previously stated, holds knowledge and, therefore, books in high esteem, but not one book. In Roman times, knowing just one book was considered more dangerous than not knowing a book at all. So we may have a problem in determining the ontological and anthropological characteristics of the humanist tradition, especially since the Enlightenment humanism has interpreted time and subsequently history as an irreversible process with a surplus value toward the future, while Renaissance humanism kept the ancient past in high esteem.

If we trace the humanist tradition to the Arabic and Italian Renaissance, and with the Renaissance humanists again to the pre-Christian Greek and Roman antiquity, it is fair to say that the history of philosophy and the history of the humanist tradition coincide to a great extent. However, if they coincide fully, humanism and philosophy would submerge into each other. That is one step too far, or a step not far enough. It would deny the right to philosophy of defending a theoretical and practical nihilism, and to humanism to accept certain basic values and guiding ideas. Skepticism may be in high esteem among many humanists, but they do not want to push it too far. Arne Naess, the Norwegian philosopher and humanist, may have written a beautiful book in the defense of Pyrrhonian skepticism, but his skepticism does not prevent him from arguing for a total view and from inventing "deep ecology," without being inconsistent. An analogous distinction is valid for the classic polarity between the adherents of empiricism and rationalism, both movements understood and interpreted in many different ways. The empiricist Alfred Ayer and the rationalist Noam Chomsky can defend different philosophical positions while both are humanists. Just because of the intimate historical connection between various philosophical movements and the humanist tradition, Humanism may rather be defined as a hierarchy of values or a set of normative ideas than as a specific epistemology or cosmology. One can defend the thesis that the so-called humanist anthropology and cosmology can trace their origin equally well to the triple goddess as to the Greek myth of Narcissus who fell in love with his own image, as an act of self-reflection in the double meaning of the word 'reflection.' There is neither a revelation nor a clear beginning of the human endeavour, symbolized in Diana and Dionysus, Apollo and Athena. Pico della Mirandola will define human nature in his *Oration on the Dignity of Man* (1486) as open-ended, not defined by bounds or fixed limits. According to Pico, human nature can mold itself in every possible direction, because its final form depends on the own decisions, for better or for worse. So, the chances are equal. Humanists do usually agree about this indeterminist point of view with regard to the human nature. With Cicero, humanists adhere to the notion of human nature as a statement of fact and as a statement of value. Cicero developed his notion of *humanitas* from a double perspective: all human beings belong to "humanity" as a matter of fact, and have therefore—and this is the second perspective—to promote the consequences of that fact, viz. humanhood. The French philosopher Foucault, denying the universality of human nature, is a philosopher, even a good one in the classic sense of the word, but does not want to be called a humanist. Foucault is right in his refusal. The reason is his 'reduction' of the 'irreducible' human nature to the bourgeois or proletarian class-nature. It is the denial of "the infinite value of every human being", a statement of Arne Naess in Amsterdam's Paradiso during the Rushdie symposium in 1990.

Many contemporary humanists consider the Enlightenment, not the Renaissance, to be the true beginning of a humanist world view, because of the supposed irreversible development of a scientific world view and the liberation of man from every religious or transcendent orientation. Secularism and humanism were becoming Siamese twins. Immanuel Kant, with *What is Enlightenment*, made a strong appeal to break the chains of an imprisoned judgment. He tried to define the borders of human knowledge with regard to the old metaphysics. Kant believed so much in Newton's physics and in its underlying physical determinism that his entire philosophy became one great effort to prove human freedom within a world which is inanimate and determined solely by the universal law of cause and effect. Although Kant was a humanist and Christian, it was a relatively small step to accept his notions of theoretical knowledge, freedom and ethics without any Christian connotation.

The forthcoming dichotomy between nature and man since the Enlightenment, nature being deterministic and man being free, and the replacement of book Genesis by Darwin's evolution theory, seemed to open the road for an identification of humanism with an all-embracing belief in scientific-technological and ethical progress. Sir Julian Huxley is a good example of this conviction and aspiration: "We have only recently emerged from the biological to the psychosocial area of evolution, from the earthly biosphere into the freedom of the noösphere. Do not let us forget how recently: we have been truly men for perhaps a tenth of a million of years—one tick of evolution's clock ... Our feet still drag in the biological mud, even when we lift our heads into the conscious air. But unlike those remote ancestors of ours, we can truly see something of the promised land beyond. We can do so with the aid of our new instrument of vision our rational, knowledge-based imagination" [20]. I can only hope that Julian Huxley is right, but the fact that he wrote his text in 1961 makes me wonder why he believed in progress amidst of the ordeals of the 20th century, both politically and ecologically.

In its mainstream, (post)modern humanism has embraced two ontological assumptions. The first one is the above-mentioned dichotomy between nature and man. This dichotomy is the continuation of the Christian subordination of nature to man. There is no bridge between nature and man: nature and mind seem to be condemned to live forever a separate life. Our freedom rests upon controlling nature, not upon understanding and experiencing nature as our "alter ego." The second assumption is the denial of a transcendent or spiritual reality, as if the rejection of the Christian belief in transcendence has to coincide with the rejection of any spiritual reality. From a philosophical perspective, one can defend transcendence just as well as immanence. There is not any logical or empirical necessity to identify a transcendent reality with a monotheistic belief. Humanism and secularism do coexist quite well, but so do humanism and spirituality à la Ficino in the 15th century or à la David Bohm today. The belief in progress is a historical misunderstanding of the linear notion of time, formulated for the first time by Isaac Barrow, the master of Isaac Newton, in his *Geometrical Lectures.* Linear time has become an ontological and historical idea, instead of the mathematical concept it used to be for Barrow. My conclusion is that the unilinear notion of time is a mystification of a mathematical idea into an ontological category determining our vision of past, present and future as an absolute irreversible process.

These two ontological assumptions: the dichotomy between nature and man, and the denial of any transcendent or spiritual reality, seem to condemn the humanist tradition to coincide to a too great extent with the (post)modern world of science, technology and secularism, so that we are in danger of forgetting that humanism is primarily an open attitude to life, with some elementary, universal values as guiding ideas. These elementary values are, in my judgment, not so different from the values that Socrates practiced 2500 years ago.

ETHICAL AND AESTHETICAL VALUES

My last remark about values is an illustration of how value judgments subterraneously enter the domain of anthropology and ontology.

Humanism

Humanism defends the idea that values stem from an autonomous source, i.e., human reality itself. The old riddle whether the origin of values, ethical or aesthetical, is only the result of education and a specific environment, or stems also from an innate value-consciousness, is not a topic for this article on the methodology of world views. As said above, the humanist tradition does not coincide with the empiricist or the rationalist answers in this domain. The conviction that the choice and practice of values are the sole responsibility of the human person, is of vital importance in the humanist world view. But equally important is the insight that so-called facts and so-called values do not exist in separate domains as might be clear from my discourse on humanist anthropology and ontology. The preference of the humanist ethos for care and ethical questions, more than for the aesthetic realm, is probably a consequence of the need to demonstrate that secularism is not devoid of an ethical conscience and ethical responsibility. The ethical tradition in humanism can refer to the Ciceronian notion of *humanitas* as the center of their ethical commitment. The notion of humanity has to balance between at one hand a purely relativistic approach of ethical codes, and at the other hand a static, universalistic one. A helpful and essential criterion in determining the right balance between the two poles is attention for the means with which people try to realize their ethical aims. The means in ethical behavior correspond to their aims as the forms and designs in the field of aesthetics correspond to their meanings. The means in ethical behavior may never be subordinated to the so-called aims, because the means are the surface structure and visible realization of the aims, not different of the aesthetic forms which determine the possible meanings in the domain of the arts and design.

Parallel to the above mentioned dichotomy between humans and nature, another dichotomy has evolved during the last three centuries, viz. the separation of sciences and arts. Within a closed-off domain, i.e., the realm of the aesthetic experience, the arts in modern societies have reached an outstanding position. Museums, concert halls and theater houses flourish as never before. The price the arts are paying for this high-standing position, is their exclusion from daily life. One can describe this dilemma of humanist aesthetics as a dilemma of autonomy and isolation. This dilemma can only be solved if aesthetic freedom breaks through the walls of the museum. Artists such as Joseph Beuys or Richard Hefti were fully committed to integrate their art into daily life as an expression of their personal and social world view. Humanists can be at the forefront of this movement on the sole condition that they are willing and able to integrate the world of arts into their private and public life.

Christianity

Christian ethics derives its inspiration from the ten commandments and the moral codes of the various churches. The source is not autonomous, as in humanism, but heteronomous: the revelation by God and by God's intermediaries. Again, many Christians wrestle with the so-called timeless origin of their ethical values. The tension between an ethical rule from "outside" and the personal conscience, is often stressful. Clarification about the inner sense of ethical behavior seems to me to be the only answer to this problem. Christian aesthetics arises from a twofold experience, viz. the redemptive suffering of Jesus with his double nature: divine and human, and the presence of a transcendent, divine majesty. Among many Christians in (post)modern societies, there is a lack of concrete experience of the sacred. The result is that one can hardly find contemporary versions of Christian art, while much of the ancient liturgical traditions has been lost. Christian art is in crisis. The collective experience, needed for the creation of spiritual art, is absent, with the exception of those religious communities where a musical tradition, such as the gospel songs, is still alive.

Buddhism

Buddhism seems to have less problems than Christianity with regard to ethics and aesthetics. Buddhism does not suffer as radically from the dichotomy between man and nature as does humanism, and does not have to wrestle as tragically with the question of origin of ethical codes as does Christianity. In Buddhist tradition, ethical behavior is closely connected with the spiritual path. The spiritual path of self-liberation in the double meaning of this word, liberation by itself and from itself, is the point of departure for ethical behavior. The ethical prescriptions do not differ greatly from the Christian ones in their content, but the justification of those prescriptions is derived primarily from the intention with which they are practiced. There is no external, objective authorization.

A similar analysis is valid for Buddhist aesthetics. Nature is good and beautiful within its own order. The capacity of seeing and experiencing this beauty is linked to the notion of transcendence, not as an independent substance of nature or man but as the inherent, hidden dimension in everything that exists. Buddhists call it the spiritual dimension. The Buddhist transcendence is known as *drala* in our reality, as revealed in this haiku of Buson:

In the spring
relighting a candle flame
with a candle flame

The word *drala* is a Tibetan term for the poetic luster that glows over reality. Analogous to the Christian communities, we can wonder to which degree the Buddhist communities will be able to continue a rich aesthetic tradition. The answer will depend on their capacity to continue a sacred, aesthetic tradition among often very different cultural conditions, trying to find a new balance between the traditional icons and the metamorphosis of those icons into an contemporary idiom. Because of the assumption of the impermanence of all sensible reality and of the introspective strategy, i.e., do not project internal images upon outward reality, the Buddhist tradition has a strong point of departure for a renewal of its aesthetic tradition.

METHODOS

My last question regards the *methodos* of a world view. With the word *methodos* I pose the question of how a specific world view introduces children and outsiders into its tradition. Such a question only seems to be interesting from a pedagogical point of view. But this is misleading. I would like to defend the thesis that the methodos of a world view reveals more about its real intentions than any of the doctrinal and moral teachings. Education is the guidance of a child on the basis of values and normative ideas. Normative ideas of this kind could be: respect for authority, the eye of God, respect for parents, the value of money, the struggle for life, the nation-state but also respect for truth, to know what you don't know, respect for all life-forms, an open eye for the inherent beauty of nature, and so on.

Humanism

Humanist education will stress the importance of a personal judgment to strengthen the sense of moral autonomy, to let the child feel and understand the beauty of life but also its harshness. There is no *a priori* truth to teach except the respect for truth as the ultimate criterion for all humans to be able to live together without fooling themselves and each other.

Buddhism

Buddhist educational practice differs from the humanist one, but not essentially. Buddhists will use specific exercises, some rituals and songs according to their tradition, as is the case with Christianity. A student of Buddhism will accept a master as soon as the student decides to become a Buddhist. But the master has the explicit task of guiding the student so that he can bid the master farewell when time has come. The humanist idea and ideal of personal insight and autonomy is also the aim of the Buddhist education. The roads are different but not the aim.

Christianity

The Christian education will follow the path of the teachings of the Bible in order to introduce the child or outsider into the mysteries of the belief. The notion of salvation by Jesus and finally by the Father and the Holy Spirit has to become an existential experience. Salvation may take place by purification of (mortal) sin and by the surrender to God and Jesus, or also to the Holy Mary as the superb mediator in the Catholic belief. The road of the Christian ends in the identification with the death and the resurrection of Christ. There is no self-liberation in Christianity as in Humanism, or liberation of the self as in Buddhism, but rather salvation by surrender to God as the Creator of all there is.

CONCLUSION

The seven questions are various ways of looking and interrogation. The questions themselves are perhaps not even the most profound questions one could ask. At the heart of every world view, there are two fundamental ones: space and time. Space- and time-categories belong to the deep structure of the human mind and perception, and probably of all reality. The concrete answers to space- and time-questions reveal the essence of a culture and of a person. For that reason, they are difficult to deal with. I have chosen a more elaborate method via the seven questions. But everybody who studies the ontological and anthropological dimension of a world view, will discover that the space-time conceptions determine the final "identity" of a world view, even if that world view tries to escape every conceptualisation.

The attempt to uncover some aspects of the deep structure of a world view—assuming that it is possible to do so—leaves the researcher, in this case myself, with the feeling that any description of the deep structure through the concrete surface structure, is empty-handed. Life cannot be grasped in words; it can only be grasped by life itself. Every Buddhist, every Christian and every humanist can reject what I have been trying to elucidate. They are right. Personal experiences can never be reduced to the categories of my seven questions. Nevertheless, if one is not afraid to look into the mirror, everybody will discover, sooner or later, that the identity of humans cannot be divorced from nature and culture. Nature and culture are the two sides of one and the same coin, viz. the human being. Each world view is a specific expression of that consciousness, one way or another.

AFFILIATION

Fons Elders
University for Humanism
Utrecht
The Netherlands

REFERENCES

[1] **Elders, F.**, "Dialogue and Universalism. Towards Synenergy of Civilizations." *Dialogue and Meaning*, Polish Academy of sciences, Warsaw, Vol. **VI**, 5-6, 1996, p. 39.

[2] **Elders, F.**, *Analyseer Deconditioneer. Een inleiding tot de systematische filosofie*, Van Gennep Teleac, Amsterdam, 1971.

[3] **Armstrong, K.**, *A History of God. Introduction*, Mandarin, London, 1994, p. 4.

[4] **Guenther, H.V.**, "The Challenge of Proces Thinking", *Vajradathu Sun*, Vol. **12**, 3, 1990, p. 8.

[5] **Elders, F.** (ed.), *Humanism Toward the Third Millennium*, VUB-PRESS, Brussels, 1993, p. 10.

[6] **Huxley, J.** (ed.), *The Humanist Frame*, Allan & Unwin, London, 1961, p. 20.

FRANS GOETGHEBEUR

BUDDHISM AS PART OF A GLOBAL UNDERSTANDING OF REALITY

WHO IS SPEAKING?

One of the many relays in the network which is carrying out the "transfer" of Buddhism from the East to the West. Besides my work as a teacher of French in secondary school, for the past nineteen years I have been at the service of the Tibetan Institutes in Belgium. In my capacity of chief administrator I have, together with my life partner and under the spiritual direction of Tibetan Lamas, deliberately chosen to set up an infrastructure (on the administrative, financial and relational levels) which should guarantee in all its effectiveness an oral transmission assured by authentic Buddhist masters.

To set aside a place for the living word of Buddhism which flows from the source is not easy: the dangers stemming from unbridled, over-hasty and thus reducing translations or approaches are not imaginary and the primacy of the written over speech in the West is a fact.

Why Choose an Oral Direct Transmission in Preference to Anything Else?

The social sciences which are specifically faced with the presence of "the other" (anthropology, psychology, history, etc.) have a fundamental principle to teach us: if one wants to get to know the true face of a civilisation one must listen to the voice which rises from the depths of its heart and of its imagination and which follows "the strategies of meaning" [1] illustrated by its own way of facing reality.

This is why I have chosen to translate the public teachings of the Tibetan Lamas rather than giving any myself. Furthermore, Tibetan culture in general and Buddhism in particular are themselves largely based on oral transmission. And as Jullien puts it: "Here, speech does not tend to define (to erect a universality of essences) but rather to modulate to embrace the circumstance" [1].

A third reason is at the root of my presence here and now: the resonance of Buddhism allows Westerners, at least we think so, to breathe a gust of fresh air. In the desolate landscape and the heavy atmosphere of a dull and phlegmatic post-modernism [2] this refreshing seems indeed welcome to us. The tear-exciting effect and the hypochondriac nature of the air of the times we live in are quickly dissolved by the contact of Buddhism: indeed the latter evaluates precisely the lightness of existence and suggests to individuals as to societies an approach at once pragmatic and dynamic to reality, which allows them to conceive of it no longer in a stagnant form but as a living entity [3].

"Conditioned production" as conceived of in Buddhism helps us to recognise that the world (inner and outer) is a limitless network of variations and of relationships in a state of continuous evolution ... and that furthermore without them the world would simply not manifest at all.

TO SAY WHAT?

If I speak here on my own behalf instead of translating the words of the Lamas, as I usually do, it is for purely practical reasons: twenty minutes is short if one must include a consecutive interpretation.

Now, What is the Subject at Hand?

That of the West which is still too full of itself not to prefer to continue to ignore the obvious, namely Buddhism, which frightens it a little?

Perhaps, but my real aim is above all to attract the attention of the luminaries gathered on this occasion of the twenty-fifth anniversary of the VUB to the fact that Buddhism, in spite of the restraining influences it encounters here in the West [4], is developing everywhere, as it did before in the East, without generating shock waves and that the global understanding of reality to which it can lead has in the final analysis nothing to do with the superstitions, the simplifications or the negativisms that are sometimes made of it by Western readers and academic researchers.

I am not a scholar and I have nothing more to communicate to you than this.

You will reply concerning the first point (the development of Buddhism) that you will take note of it and concerning the second that "the global understanding of reality" is neither original nor specifically Buddhist since it has always been the goal of every scientific theory. You will add that furthermore to speak of a global understanding seems rather presumptuous since the times of Wittgenstein and Gödel: had we not agreed that the whole of the knowable lies outside our understanding, in fact? That reality necessarily escapes our grasp and that of our scientific instruments at least as long as we continue to separate the subject from the object of the analysis? And that in any case the pretension of wanting to take the omniscient point of view of God is outdated?

Later, we will state the meaning of the adjective "global" in order to show that in this discussion we are charting a "middle way", which is typically Buddhist in fact. Was it not a very sane modesty that inspired Einstein when he wrote as early as 1936 in an article called "Physics and Reality" that "all science is nothing but a refinement of our everyday thought"? So we do not have to adopt extreme views at all.

In the meantime we hope these words may be a homage from a Buddhist to the virtues of freedom of thought as practiced here at the VUB in Brussels.

BUDDHISM

It may be useful to know this tradition which has, like Christianity in the West, been the cultural support of the development of a great number of countries in the East.

The history of Buddhism is continuing since its meeting with Western culture is taking place before our very eyes. We already have a gracefully written history of the encounters of Buddhism with the West during the past 2,000 years: Stephen Batchelor's, "The Awakening of the West", (Parallax Press, 1994). Thus it can be noted that it took ten centuries for Buddhism to establish itself in Asia and twenty-four centuries before it reached the ears (for the most part, muddled [5] and incredulous) of Western scholars and the (greedy) hands of Western antique dealers before touching the hearts and minds of practitioners [6].

UNITY IN DIVERSITY

Buddhism first developed along the great commercial axes (the silk route) which crossed Asia, near urban agglomerations, and thanks to the support of kings, emperors and members of other classes of society.

Its extraordinary flexibility allowed it to integrate within itself the native traditions [7] in all the countries where it was transmitted. This resulted in the existence of a great diversity of points of view, of emphases, of artistic forms and of customs in countries which, although they are very different from each other, are all Buddhist.

The basic message of the Buddha, who said that human beings can liberate themselves from suffering, and awaken themselves from the confusion which binds them to the cycle of existences, has been translated into all the languages of Buddhist countries and is found in all the schools. The methods used to reach this goal differ greatly: one school will insist more on ethics, another on philosophical reflection, yet another on meditation. Different tendencies are also seen inside each school.

This explains the thousand faces of Buddhism and its unity in its very great diversity [8].

MODERNISM

It is interesting to note that at the organisational level the Buddha determinedly chose a structure without a hierarchy. As for the doctrine, it rests on a logic based on the primacy of reasoning and of experience in the analysis of reality as in the assessment of the results of spiritual practice. One could not be better adapted to our current times which value so much the criteria of logic and of empiricism. Buddhism certainly contributes to a liberated or liberating understanding of reality: it is more global in fact, in the sense that it lifts taboos and does not fear the blank spots which come from conscious or unconscious mechanisms in human experience, and which are called in the technical vocabulary of Buddhism "the veils of the knowable". As Jullien puts it: "By decompartmentalizing our apprehension of things, all these diversions of language make us join again the foundation of the real, *where nothing is excluded anymore*". Buddhism, everything can be envisaged from all points of view as long as the authenticity of the path of wisdom is not betrayed. There is not enough time for us here to describe precisely the latter, but it means not only to see reality as it appears but also as it is by nature ... and this wisdom fears nothing, needs no taboo to protect itself nor any certainty to impose on others; characteristics that are rather rare in spiritual traditions, at least as they evolved through the ages and are practised nowadays.

THE CONFRONTATION

The confrontation of the West with Buddhism is not solely interesting for the history of ideas; it could become that much more important as a vector of a profound transformation in several fields. We would like to emphasise one specific aspect which might interest the audience: as Jullien expresses it: "there must be an elsewhere to illuminate where we are. Thanks to that, the gaze looking at the question can be more global; especially, one can go back to what conditioned it, fathom its hidden biases—consider it more radically.". In this way we feel Buddhism can lead us to uncover certain structures (of the Western logos) up to now conscientiously denied (Verneinen in German).

This sleight of hand strikes what I will define provisionally in this way: the experience of the interdependent multiplicity of reality ("conditioned production") called "emptiness" in the technical vocabulary of Buddhism. Here emptiness is understood in its sense of "sparkling void" [9]. This experience is quite staggering for a person who has anchored his essence in the logos, which is unilateral and linear in nature.

The gradual dismantling of the bastion defended by the forces of monolithism (whether they stem from religions, from ideologies or from scientific circles) will not easily be accepted in the West.

ASTONISHMENT

Many attitudes will be strongly put to the test when certain key concepts of Buddhism (such as the relationship between interdependence and emptiness, the clinging to the existence of the ego) really come into play:

- The habit of conceiving of the origin of the universe as necessarily emanating from a single source [10] while according to Buddhism nothing exists by itself. There is always a series of causes and conditions at the origin of all phenomena which explain their presence,
- The habit of trusting the methodology which consists in limiting reality to what can be experienced, and truth to what is the result of quantified approach of a physical and mathematical type,
- The habit of projecting what surpasses us for the moment onto an external (divine) force which is above suspicion,
- The habit of considering that proselytism is only acceptable when it does not come from others etc..

This confrontation with Buddhism runs the risk of discountenancing Westerners a little. Which recalls in some sense the astonishment of the polytheist Romans faced with the first rumours of the monotheism of the early Christians.

In spite of all that, in the space of a few decades Buddhism has managed to enrich the quality of life of many individuals (indicative of this is the fact that our retreat centre in the south of Belgium receives the visits of nearly ten thousand people a year), that it impregnates all levels of society and that it influences scientific thinking [11]. It is enough to see it to believe it.

We believe that we are witnessing here a major phenomenon of the civilisation of the end of the second millennium which Einstein and Toynbee, to quote but these two eminent men of science, had actually foretold [12].

PLURALISM

We have the feeling that a true pluralism (similar to the tolerance which characterises the Buddhist when he considers tendencies within Buddhism different from his own or other religious traditions altogether) is a conditio sine qua non to understanding Buddhism. Otherwise the latter will appear as an obscure whole or one will take the risk of talking nonsense (about Buddhism).

Thus, the time is not too far away when people will agree to accept just as easily the conviction according to which theism can be interpreted as a superstition, just as this judgement is now accepted when it is aimed at anything that does not correspond to monotheism.

AUTONOMY

Why does atheism, as it is found in Buddhism and which has nothing to do with that anti-clerical reflex of a primary "down with the skull-cap" which it so often becomes in our countries, constitute a true form of autonomy?

We will limit ourselves to some sketches of an answer:

1 - Since Buddhism does not have the hierarchical structures of an institutionalised church founded by a supernatural being, geopolitical considerations (with all their accompaniment of persecutions and coups d'état) are less influential in its history.

I would prudently add that this remains true to the extent that Buddhist authorities do not set up such a hierarchisation!

Furthermore, the risk of oppressive or obscurantist claims emanating from dogmatisms (which are always positions taken when contact with reality itself has been lost) will become rarer.

2 - The proofs of the existence of God are one of the forms that the search for a minimal indivisible entity can take (whether on the material, the temporal or at the level of consciousness). Buddhist philosophers of the Madhyamaka school will consider these attempts as mistaken views of reality which arise from a tendency which is deeply anchored in us: it consists in wanting to find at any price a solid or permanent reality and disregards the emptiness which is the basis of everything that exists. Thus there exists an unconscious knot of "fundamentalism" in the literal sense of the word, at the very heart of any theistic approach.

3 - Symptoms of "mental decolonisation" [13] or true emancipation which is discussed here emerge in different ways: in the form of all that comes from science, philosophy or ethics and which follow the direction of what

Francisco Varela calls "transforming approaches to experience ... which are concerned not with escaping the world nor to discover a hidden, true self, but to deliver the everyday world from its avid grasp and its desire for an absolute foundation." [14] ... The ease with which the Dalai Lama uses the concept of God, apprehending it and denying it successively, and this without giving offence to others' convictions, illustrates well, we think, the true emancipation which we wish to allude to here.

INTERRELATIONS

* Buddhism is not preached from the heights of the Chair (of the papacy or the professorship), nor from the rooftops, besides. The tools of understanding which it puts forth (relative truth and absolute truth; the dialectics of interdependence and emptiness; the non-existence of the ego) make it possible not to take refuge in a transcendence (isolated from physical reality) nor to lace up reality with the help of a partial understanding. Buddhism, as Lama Karta explains: "goes against our tendency to believe that objects and beings exist such as we imagine them: as solid and substantial entities, single and separated one from the other, eternal and unchanging. The true nature of reality, the basis of all that exists is not what manifests itself but emptiness. When one comes to a profound understanding of reality one realizes that phenomena which manifest themselves are but pure relative apparitions, which have arisen from emptiness ... We start reasoning from the principle that the table, the mountain, the self, the other, good and evil, exist by themselves; that they correspond to a substantial and independent reality.

"Conditioned production" teaches us that this conception does not correspond to the truth: phenomena do not manifest themselves by themselves, just like that ... They are connected between themselves, composite and dependent and for this reason they are relative (impermanent) by their very nature. In fact they would not even exist without the presence of the other elements of "this reality". This interrelation was therefore recognised as the conditio sine qua non of existence itself".

* We propose also as a theme of reflection the shades of meaning which separate the two following statements: "the knowable finds its limit in the unlimited" (the profession of faith of all theistic mysticism) and "the knowable is unlimited" ... which would rather be ours as Buddhists.

* Likewise, it would be interesting to compare the similarities between system theory and Buddhist epistemology which one finds for instance

in these words of the great master Dogen "to forget oneself is to be recognised by the entire universe"!

Its way of establishing facts by looking for their causes, of being attentive to the interrelations which connect phenomena, to postulate that relationships are more important than the objects, the interdependence it discovers between manifest, scientific images of reality and the ultimate reality of those images, corresponds well, we think, to the principles of the theory of systems. One could for example evaluate in what measure Buddhism forms an integral part of these "fluid and adaptable communication networks" which Joël de Rosnay speaks of in his latest book.

CONCLUSION

During Einstein's and Bohr's debate between 1927 and 1955 (the year of Einstein's death), a debate which centred upon the nature of physical reality and its recognition, Einstein felt uncomfortable with the idea of the indetermination of atomic motion such as Bohr proposed to see it. Just as Einstein did his utmost to listen to Bohr, the West should not fail to listen to Buddhism with the same intelligent benevolence.

To imagine that all choices are absolute and that one is either for or against Buddhism; that in the former case, one should behave with servile obedience towards its principles: [15] this is a mistake.

In this context I could also suggest the following question, concerning what I would call univocal mimetic recovery: *why do Christians who practice Zen meditation in their monasteries never wonder why the Buddhists they have met never feel the need to meditate upon Christ without this being in any way a sign of a lack of respect on their part?*

CRUMBLING

In our times, everything seems to be becoming disconnected. In our opinion this crumbling arises from our distancing from the source of the ethico-spiritual base which connects all the aspects of living things, their evolution and their interpretation. To realize how "conditioned production" and "emptiness" relate to one another prevents the subject from clinging to working hypotheses interpreted as certainties and brings him to what one might call "the nature of things" [16].

Faced with this phenomenon of disconnection we put forward the idea of "universal responsibility" which the Dalai Lama has been advocating for the past thirty years. This humanistic idea is founded on a scientific approach and a global understanding of reality.

Another idea which the Dalai Lama has put forward recently and which goes in the direction of the depolitisation (politicisation often implies making use of egocentric patterns) of the decision-making entities is that of a United Nations of the Wise ... a little like what is happening right here.

To translate the key point into Buddhist terms: "The realisation of the emptiness upon which phenomenal reality is based is called the wisdom of the non-existence of the self of phenomena. The realisation of the emptiness upon which inner reality is based is called the wisdom of the non-existence of the self of individuals".

May all sentient beings aspire to realise this wisdom.

This wisdom is everywhere, as it corresponds to the nature of our mind and of reality.

AFFILIATION

Frans Goetghebeur
Karma Sonam Gyamtso Ling
Tibetan Institute
Belgium

REFERENCES

[1] **François Jullien**, *Le détour et l'accès. Stratégies de sens en Chine et en Grèce*, Grasset, 1995.

[2] "The West is behaving as if it were repressing its own distress, ignoring the void which it knows it carries", **Jean-Claude Guillebaud**, "*La trahison des lumières*", Seuil, 1995.

[3] "Our reasoning in the face of complexity remains analytical, our vision of the world disciplinary, our knowledge, of an encyclopaedic nature. We continue to extrapolate linearly the data of the past, whereas the developments which we are experiencing are non-linear, exponential, and in a state of constant acceleration.", **Joël de Rosnay**, "*L'homme symbiotique*", Seuil, 1995.

[4] Sometimes incorrect definitions of key-concepts of Buddhism are put forth: a confusion between the deterministic meaning of the word "karma" as it is used in Hinduism and the dynamic position which this word assumes in the Buddhist vocabulary is a common example of this. Simplifications are also encountered, for instance: to claim that the non-attachment to the ego which is discussed in Buddhism means that one must be without emotions, sometimes, it is further claimed, that one must be without character or personality ...

There is also the common deformation which consists in limiting Buddhism to an apophatic or negative mysticism. Not only the pope, but others, the Jesuits for example, have no scruples in perpetuating this sort of confusion knowing that there is no danger of a Buddhist fatwa. These reticences are brought about by the insecurity which the dominating ideologies feel following their meeting with Buddhism.

[5] Academic writings on Buddhism often tend to put a veneer of typically Western concepts on the original terms of Buddhism, thus deforming their meaning. In spite of all the good qualities of the book, **J.P. Carrière**, in his bestseller entitled "*La force du bouddhisme*", Laffont, 1994 (a title which furthermore could not be less Buddhist) cannot but admit that "the Dalai Lama was listening to me" whereas one would suppose that it was he who was listening to the Dalai Lama. Analysing the cover photograph of the book tells us already enough. Roland Barthes would have no difficulty in drawing certain conclusions from the frontal look of the Dalai Lama directed to the reader and the sidelong glance of the author thrown from above his interlocutor.

[6] This lesson of history may teach patience to impatient European Buddhists who imagine that everyone can reach the state of Enlightenment the day after their first teaching on Buddhism.

[7] Mouism in Korea, Shintoism in Japan, Confucianism and Taoism in China, Bön in Tibet, and in other countries spirit and ancestor cults.

[8] Bacot speaks of a: "gigantic plant with a hundred boughs, whose many different leaves have all sprung from the same source. Although they are dissimilar to the casual eye, all the branches are nevertheless nourished by the same invisible sap". The non-Buddhist neophyte sees this diversified whole which Buddhism represents to his eyes as a multiplicity of philosophical and pragmatic forms and approaches. It is normal that his consecutive visits to a Zen, Thai and Tibetan temple and the completely different customs and usages should disorientate him to some extent. He is well aware that all of these manifestations, which go from the most stripped-down, extreme atheism to an explosion of colours, of forms and of innumerable divinities, are tolerated by a large majority in Buddhism; but he himself finds it difficult to adapt to this and to discover the means of functioning of such an intrinsic plurality, insofar as he is always looking for a magical formula or a single basis for everything. As Jullien says: "the proper function of the logos is to grasp the object as closely as possible". o.c. p. 53.
This discomfort explains the fact that the multidirectional movement of Buddhism which is penetrating into our hemisphere in all its forms (Zen, Tibetan, Burmese or others), remains for the moment imperceptible to

the eyes of the monolithic. At any event it pretends to see nothing for the time being.

[9] **Michel Cassé**, *"Du vide et de la création"*, Odile Jacob, 1993. "the universe in its champagne version sparkles with innumerable autonomous foci of laws", p. 205.

[10] Even **Roger Garaudy** who inveighs so much against "market monotheism" remains under the influence of a certain idea of oneness which should take upon itself the function of a foundation. On page 31 of his last book "*Vers une guerre de religion*" he writes: "to seek a true universalisation of culture which brings about a unity no longer hegemonic, colonial or imperial, but symphonic through the contribution of each culture to the universal culture." We contrast with this the slightly different idea of Varela: the world as a symphony without a conductor and his notion of "autopoiesis", the autonomy of living structures.

[11] Indeed, scientists set a certain value on Buddhism because critical analysis based on experience is used as a valid criterion of knowledge. Its views on the functioning of the mind, experiences and the relationships between mind and body contain information of interest to sciences such as physics, medicine, neurobiology, etc.

In this context there are three books which are the result of meetings at a high level between the Dalai Lama on the one hand and Western physicists, neurobiologists, mathematicians and chemists on the other, "*Gentle Bridges*" by **Jeremy W. Hayward** and **Francisco J. Varela** (Shambhala, 1992) and "*Mindscience*", Harvard (Wisdom Publications, 1993), *Healing Emotions* by **Daniel Goleman**, (Shambhala, 1997).

[12] "The religion of the future will be a cosmic religion. It will transcend the personal God and will abandon dogma and theology. Including the natural as well as the spiritual, it will be founded upon the spiritual meaning of the experience of the unity which connects all phenomena. Buddhism answers these criteria. If there is a religion which corresponds to modern and scientific needs it must surely be Buddhism." **Einstein**. "One of the most prominent events of the twentieth century will be the arrival of Buddhism in the West", **Arnold Toynbee**.

[13] This way of putting it comes from Professor **Pinxten** in his latest book "*Culturen sterven langzaam*". Here we adopt also the idea expressed by Garaudy "L'intégrisme premier, c'est le colonialisme occidental" o.c. p. 31

[14] **Francisco Varela**, "*L'inscription corporelle de l'esprit*", Seuil 1993, p. 336.

[15] The philosopher **Glucksmann** analyses in his book "*L'innocence*

dangereuse" . . . this contemporary habit of hiding behind the role of the victim or of the innocent.

[16] In the Chinese tradition the nature of things is called "the law of heaven". It goes without saying that in its translation of this expression the West will not scruple to introduce a verticality instead of respecting the immanence expressed in the Chinese word.

TETSUNORI KOIZUMI

SYNTHESIZING FOUR IDEALS OF HUMANITY

1. INTRODUCTION

Herbert Read talks about four ideals of humanity—Primitive, Greco-Roman, Byzantine, and Oriental—in his *The Meaning of Art* [1]. Read derives these four ideals of humanity from his study of the history of art on the presumption that art, in one sense or another, involves creation of beauty, of pleasing forms, and that the sense of beauty which prevails in a specific artistic tradition is inseparably associated with the notion of an ideal type of humanity widely accepted in that artistic tradition.

While creating beauty, or pleasing forms, may not capture the gamut of motivations behind artistic creation, there is no denying the fact that the artist has always used a painting or a sculpture as a medium of expressing a specific humanistic ideal. In many cases, the image of an ideal humanity expressed in a work of art has performed a social function of eliciting a sense of identity among the viewers and/or of inducing them to emulate that ideal. It is in recognition of this social function of art that we accept Read's typology of humanistic ideals as representing the four archetypal images of humanity. In other words, Read's four ideals will be treated here as representing the four archetypal images of humanity in the sense that they appear and reappear in different places and different times. Indeed, these four archetypal images of humanity can be regarded as representing four "world views", for each image reflects a specific way of looking at the world and a set of ideas derived from it as to what man's relationship ought to be with respect to self, society and nature.

The purpose of the present discussion is, first, to examine the psychological and physiological implications of Read's four ideals of humanity as the four archetypal images of humanity and then to explore the possibility of synthesizing the four ideals in light of recent insights obtained in cybernetics, systems science and other scientific disciplines. The role of art education as a vehicle for learning about, and experiencing, the unified world view which is suggested by our synthesis is also considered.

2. FOUR IDEALS OF HUMANITY

One's ideas about an ideal image of humanity can be construed as reflecting one's ideas about what the world is like and what one's conduct ought to be in the world, i.e., one's world view. While this point was not pursued in detail by Read, it is instructive to examine what his four ideals of humanity mean as world views in light of recent advances in psychology, physiology, and other related scientific disciplines. Before embarking on this task, however, let us first reproduce Read's four ideals of humanity as they are described by him.

The Primitive ideal, according to Read, is "a propitiation, an expression of fear in the presence of a mysterious and implacable world". The Greco-Roman ideal expresses a sense of beauty associated with something which is "perfectly formed, perfectly proportioned, noble and serene". The Byzantine ideal is described as "divine rather than human, intellectual and anti-vital, abstract". Finally, the Oriental ideal is characterized as "abstract, non-human, metaphysical, instinctive rather than intellectual".

Read arrives at these summary descriptions of the four ideals based on his careful analysis of the representative works of art belonging to these four artistic traditions. The cave paintings in Dordogne, France, and Altamira, Spain, represent works of art belonging to the Primitive tradition; *Aphrodite of Melos* and *David* by Michelangelo represent the Greco-Roman tradition; the apse mosaics in San Vitale in Ravenna and San Marco in Venice, the Byzantine tradition; and the landscape paintings by Chinese and Japanese masters such as Wu Chen and Sesshu, the Oriental tradition.

Each of these four ideals will be treated here as representing an archetypal image of humanity, for it appears and reappears not only in each of these specific artistic traditions but also in different places and different times. In fact, the use of the term "Primitive", for example, is somewhat misleading in that the so-called "Primitive ideal" also appears in "modern" Western Europe in works of art by such artists as Paul Gauguin, Henri Rousseau, Paul Klee, and others. The Primitive ideal can indeed be regarded as a world view with universal relevance as it is, as Piaget pointed out, the kind of world view the child holds before it starts to develop "higher" cognitive faculties [2]. The same can be said of the other three ideals as they, too, have universal relevance beyond the confines of the original artistic traditions.

While Read's four ideals of humanity can no doubt be interpreted as having four corresponding world views, establishing that correspondence involves leaps of interpretation beyond what Read does in his book; Read, as a matter of fact, makes no explicit attempt to apply insights of modern science to his analysis of art. The first leap of interpretation, as was just suggested, involves treating Read's four ideals of humanity as representing the four archetypal images of humanity. To be more specific, this leap of interpretation involves inferring what Read's description of each ideal might mean in terms of the manner in which the individual who embodies that ideal utilizes his psychological and physiological faculties.

Read's description of the Primitive ideal suggests the image of man as a spontaneous creature of nature *à la* Rousseau's "noble savage". The Primitive man is hardly a sedate and reflective being. On the contrary, judging from vivid images of him chasing animals as portrayed in cave paintings, the Primitive man is full of action and vitality. The Primitive man can thus be interpreted as utilizing his psychological faculty of "sensation", in the sense the term is used by Jung, as a guide for his conduct [3]. After all, he would have to find out that something is out there for him to chase. However, he would not calculate the trajectory of his arrow in order to chase down and catch his prey; he would instead collaborate with others guided by his instincts. Physiologically speaking, therefore, the Primitive man can be interpreted as being ruled by the kind of brain which Jaynes termed "the bicameral mind" [4].

The Greco-Roman ideal, in contrast, suggests the image of man as a free and independent individual whose conduct is ruled by reason. The Greco-Roman man is capable of abstracting from concrete things he observes in the world around him the notion of idealized form, which is intellectually conceived as having certain proportions like the Golden Section. The main psychological faculty he relies on would, therefore, be "thinking". And the logical manner in which he goes about conducting his affairs, not unlike the manner in which one goes about proving theorems in geometry, suggests that the "left hemisphere" is the dominant part of his brain.

The Byzantine ideal suggests the image of man as a humble creature made in the image of God. As such, the Byzantine man is not a measure of all things. However, the Byzantine man also needs to conduct his affairs of life just as any other man. This he does with the use of his psychological faculty of "feeling" as opposed to "thinking", for determining what is important for him is not an intellectual task. The symbolic, as opposed to realistic or idealistic, representation of human figures in Byzantine art, which is done with splendid sense of balance and proportion like a mosaic depicting St. Mark and Boatsmen in San Marco in Venice, suggests the "right hemisphere" as being the dominant part of his brain.

Finally, the Oriental ideal suggests the image of man as an insignificant speck in nature. Nature represented in the Oriental landscape paintings called the mountain-water paintings, or in the design of Oriental gardens, is distinguished by the absence of intellectual conceptions such as geometric forms and symmetric arrangements. Instead, nature is represented as it is—as an all-embracing entity which contains man as a participant. There is indeed fusion of man and nature in the Oriental ideal. Thus, in stead of relying on the power of his intellect to analyze natural phenomena, the Oriental man would be guided by his "intuition" to grasp the ways of nature. As far as the physiology of the brain is concerned, the Oriental man would need "the unicameral brain", i.e., the totality of the physiological faculties of the brain—in contrast to the "bicameral", or unconnected, brain of the Primitive man—if he is to grasp the ways of nature by becoming one with nature.

3. FOUR IDEALS AS FOUR WORLD VIEWS

The four ideals of humanity derived by Read, interpreted as we did as the four archetypal images of humanity, have definite psychological and physiological implications as suggested above. However, a meaningful discussion of the image of humanity must include a discussion of what man's relationship ought to be with respect to his fellow human beings and the things in the natural environment, i.e., of man as a social being and as a biological being. Expanding Read's discussion of four ideals into these areas is the second leap of interpretation that we take in order to establish correspondence between humanistic ideals and world views. Before doing this, we first need a definition of a world view.

We define "a world view" as "a conception of what the world is like and a set of ideas derived from it regarding what man's relationship ought to be with respect to self, society and nature". While some may consider the second part of the definition to be redundant, it is actually an essential aspect of a world view. What we are suggesting with the above definition is that one's world view is more than just one's conception of what the world is like, for how one looks at the world is reflected in the way one goes about conducting the affairs of one's life as an individual, as a member of a society, and as a member of the natural world. In other words, a world view must be treated as comprising both the mode of perceiving the world and the art of conducting affairs in the world.

The Primitive image of man as a spontaneous creature of nature suggests a society organized around kinship among members of a family and extended families. In such a society of "hunters and gatherers", an individual engages in social transactions as a matter of course, as a spontaneous act of cooperation, and not as a matter of civic duty or out of his self-interest. No formal language would be needed to facilitate social transactions among Primitive men as everybody instinctively knows what is expected of him. With their superior senses of hearing and sighting, communication by "sounds and gestures" would be sufficient to induce individuals into spontaneous cooperation, as is vividly captured in the scene of bull hunting on a wall painting in Catal Hüyük, Turkey.

For the Greco-Roman individual, a society is basically a contrivance, an emergent entity, which is formed by the aggregate will of autonomous individuals in order to protect and promote their individual interests. A society operates as an efficient system of social transactions by exploiting specialization and the division of labor among individuals. One necessary consequence of such a functional society among autonomous individuals is that society tends to get divided and fragmented into different social groups, necessitating the use of formal language to facilitate social transactions, i.e., a shared set of "words and signs" incorporated into laws, rules and regulations.

The Byzantine man lives in a community of believers, among those who share the same faith. As long as men lead their lives according to the commandments and dictates of their faith, social transactions need not be enforced by a secular authority, be it in the form of a monarch or a representative government. Social transactions spontaneously spring up in and around the town square which serves as the center of religious as well as secular activities. Moreover, to the extent that the people believe in the value of a specific social undertaking such as the building of a cathedral, they would voluntarily participate in that undertaking. The mode of communication here is in terms of "images and symbols", as exemplified by mosaics and frescoes of medieval churches and cathedrals.

The Oriental ideal emphasizes connections and interdependencies among individuals. Some of these connections and interdependencies are in terms of blood relations and family ties but others are in terms of different positions and stations individuals occupy in society—the ruler versus the ruled, the teacher versus the student, the employer versus the employed, and so on. Men are made what they are by society, and it is a specific social context that defines what man's proper conduct ought to be to his fellow human beings. Social contexts are usually defined in terms of "customs and conventions", which indicate not only what man's place is in a society but also how man ought to relate himself to other individuals. As such, verbal communication plays only a supportive role in fostering social transactions.

What we have said above about the image of man as a social being can also be said about the image of man as a biological being, i.e., regarding what man's relationship ought to be with nature and the things in it. Thus, the Primitive ideal depicts man as a spontaneous creature of nature because nature, to him, is an implacable monster who overpowers and dominates everything, including human affairs. The natural world is seen as a living creature, full of vitality and energy. And if the world is ruled by animistic spirit which is to be found in every stone, every tree, every river, and every mountain, it is only by the power of magic that man hopes to appease the indomitable world. Hence, we can detect the idea of submission of the world by magic behind the Primitive ideal as a world view.

The Greco-Roman ideal, in contrast, suggests the view of the world as a clockwork. And as long as the world works like a clock, its mechanism can be understood by the power of reason, by the rational inquiry of the human mind as represented by science. The Greco-Roman world view is thus a mechanistic world view in that the world is seen as ruled by predictable laws of causality. Moreover, to the extent that man is able to discover these laws of causality, he is capable of "mastering" the world, not only in the sense of understanding how it works but also in the sense of manipulating it to suit his wishes and desires. While the Primitive man tries to appease the spirits in nature by magic, the Greco-Roman man tries to submit the world to his will by a different sort of magic, i.e., technology [5].

The Byzantine man sees the world as a product of divine providence, full of miracles and mysteries. How the world works is, therefore, beyond human comprehension, for the power of reason can only take us so far. This does not mean, however, that man is powerless before the world. He can hope to find salvation of his soul and transcend the world by entrusting himself to divine grace or by leading a life worthy of a true believer.

The Oriental ideal suggests the image of man as an insignificant speck in nature because the world is seen as an organic whole characterized by connections and interdependencies among all things, including humans.

The Oriental ideal as a world view is thus an eco-systemic view in that the world is seen as a space of ecological interactions among living as well as non-living systems. Since man is an inseparable part of this organic whole which is always evolving, man cannot hope to master the world in the same way the Greco-Roman man can; he can only hope to establish communion with the world by enlightenment, i.e., by an intuitive grasp of the nature of eco-systemic interactions, which is obtained through certain types of bodily exercises or meditation.

The four ideals of humanity by Read interpreted as four world views, each expressing a set of ideas about what man's relationship ought to be with respect to self, society and nature, are schematically summarized in the Appendix.

4. SYNTHESIZING FOUR WORLD VIEWS

We have developed the world views as embodying the four archetypal images of humanity by making the differences among them as sharp and distinct as they can be made. This we have done in order to make each world view a stylized view of the world whose relevance is not limited to a specific society during a specific period of time. In fact, the specific names used for these world views—Primitive, Greco-Roman, Byzantine, and Oriental—are, in a way, historical accidents and should not be treated as permanent features of specific societies. After all, both the Greco-Roman and Byzantine ideals have coexisted in most societies of the West. And the influences of both the Primitive and Oriental ideals are distinctly present in the art of modern Western societies. All this suggests that the four world views do not necessarily represent four sets of separate and incompatible ideas, but rather four sets of interrelated and complementary ideas which can potentially be integrated into a unified world view. Exploring the possibility of synthesizing the four world views into a unified perspective seems to be crucial if we are to gain insight into the nature of global change taking place in the world around us, where diverse groups, organizations, nations, and nation-states operate in the space of complicated webs of actions, reactions and interactions.

There seem to be at least two ways of going about the task of synthesizing the four world views. First, the insights offered by cybernetics and systems science may be employed as bridges between these world views. Second, art education may be reconstituted as a vehicle for learning about, and experiencing, the psychological and physiological implications of the four ideals of humanity and, therefore, four world views.

One of the most important insights that has come out of modern science in general and cybernetics and systems science in particular concerns the role of the scientist. Today we no longer subscribe to the classical notion that the scientist is a neutral observer of a phenomenon or an event which he studies. Rather, the scientist is regarded as a participant/observer who gains knowledge of the world around him by participating in a phenomenon or an event. This change in our perception of the role of the scientist has come about because of an increasing awareness on his part that the very act of observing is inescapably enveloped in the space of interaction around that phenomenon or event, as typified by Heisenberg's uncertainty principle in quantum mechanics [6]. This recognition of the role of the scientist—or the artist, for that matter—as a participant/observer, which is marvelously captured by Velázquez in his masterpiece, *Las Meninas* (1656), is of course the central insight of both cybernetics and systems science. In cybernetics, this insight underlies "second-order cybernetics", or the cybernetics of observing systems [7]. In systems science, the same insight is often expressed by the statement: a system is in the eye of a beholder [8].

As for art education, we may appeal to recent research findings about the functioning of the brain in neuro-physiology and other cognitive sciences. For example, it is now possible to test for one's brain dominance [9]. Moreover, it is also possible to develop the weaker part of the brain by engaging in certain types of activities. Many brain researchers have confirmed that the right hemisphere of the brain is mostly responsible for what we regard as "creative insight" in art as well as science. To the extent that this is the case and to the extent that fostering creativity is a desirable thing, the development of the right hemisphere can be stimulated by engaging in certain activities such as "drawing on the right side of the brain", as Betty Edwards puts it in her highly successful book on art education [10].

If it is possible to experience and learn about the different modes of operations associated with different parts of the brain, it is also possible to do the same about the different types of faculties associated with different psychological orientations of the mind. In fact, the four psychological faculties of "sensation", "thinking", "feeling" and "intuition" are, as Jung pointed out, the faculties that we need to orient ourselves in the world: we need "sensation" to establish that something is out there, "thinking" to interpret what that something is, "feeling" to evaluate what that something is worth, and "intuition" to determine where that something is in space-time relationships. The development of these four psychological faculties can be facilitated by developing the habit of always seeing things in perspective.

The first point about the scientist as a participant/observer suggests that it is possible to infer why and how we tend to embrace a specific world view, for the message here is that the way we look at the world, even in a supposedly objective activity such as science, cannot be divorced from our world view. As James Burke puts it, "[w]hen we observe nature, we see what we want to see, according to what we believe we know about it at the time" [11]. In fact, is this not the message Magritte tries to remind us in his *Les objets familiers* (1927-28)? In any event, the fact that world views are "relative" is not a novel discovery. What is important is that, once we understand why and how people come to embrace a specific world view, it opens up the possibility of establishing communication between peoples who hold different world views.

The second point about art education suggests that it is possible to put ourselves in the position of someone who embraces a different view of the world from our own. The message here is that we should be able to develop other ways of looking at the world by consciously trying to suppress and block out our customary mode of looking at the world. And to the extent that we are prepared to appreciate, if not fully embrace, other world views, communication between the peoples who embrace different world views will be greatly facilitated.

5. CONCLUSION

As Toynbee has shown, the history of humankind is very much a story of challenges and responses among civilizations, with each civilization built upon a specific world view [12]. While civilizations rise and fall, the past couple of centuries has witnessed the rise of a civilization built on what is characterized here as the Greco-Roman world view. This is the world view which sees individuals as free and independent beings whose conduct is ruled by reason, a society as a nominal aggregate of autonomous individuals with specialized skills and functions, and nature as an instrument which can be exploited to promote selfish interests of individuals.

There are many signs that the civilization built on the Greco-Roman world view faces a serious crisis today—as exemplified by identity confusion, ethnic conflicts, global disparities in standards of living, environmental degradation, and others [13]. While individuals are supposed to be ruled by reason, the world is full of ghastly stories of raw human emotions. While a society is supposed to reflect the aggregate will of free and independent individuals, the reality is that it is ruled by special interests of powerful groups of individuals rather than by the sum total of enlightened opinions of individuals. Most of all, the Greco-Roman ideal has failed miserably when it comes to maintaining harmony with nature, as nature has been viewed almost as an adversary to be conquered and exploited with the magic wand of technology.

It should be clear from our discussion in the preceding pages that, while somewhat stylized in its characterization of each world view, the linear, expansionary and masculine implications of the Greco-Roman world view need to be complemented, if not replaced, by the nonlinear, interactive and feminine implications as embodied in the other world views [14]. Needless to say, whether we succeed in redirecting the world from straying further into the pitfalls of the Greco-Roman world view depends on how much we as individuals can experience and embrace the unified world view as developed in this discussion. This means that we must get used to the idea of stepping out of our customary mode of looking at the world. When it comes to looking at the world from a different perspective, we can learn a great deal from artists who tend to anticipate a new view of the world which scientists open up to our perception later on. While we pay tribute to Einstein for his revolutionary insight as a scientist, artists such as Monet, Manet and Cézanne were already suggesting, before the turn of the century, a new view of the world as implied by his special theory of relativity [15]. In this sense, Magritte's work such as *La clef des songes* (1930, 1936) may well hold the key to unlocking the worlds of perception which have hitherto escaped us.

AFFILIATION

Tetsunori Koizumi
Department of Economics
Ohio State University
United Sates

REFERENCES

[1] **Read, H.**, *The Meaning of Art*, Faber & Faber, London, 1968.

[2] **Piaget, J.**, *The Construction of Reality in the Child*, Basic Books, New York, 1954.

[3] **Jung, C.G.**, *The Structure and Dynamics of the Psyche*, Princeton University Press, Princeton, 1969.

[4] **Jaynes, J.**, *The Origin of Consciousness in the Breakdown of the Bicameral Mind*, Penguin Books, Harmondsworth, 1980.

[5] **Bronowski, J.**, *Magic, Science, and Civilization*, Columbia University Press, New York, 1978.

[6] **Heisenberg, W.**, *The Physical Principles of The Quantum Theory*, Dover, New York, 1949.

[7] **von Foerster, H.**, *Observing Systems*, Intersystems, Salinas, 1981.

[8] **Klir, G.J.**, *Facets of Systems Science*, Plenum, New York, 1991.

[9] **Hermann, N.**, *The Creative Brain*, Brain Books, Lake Lure, 1990.

[10] **Edward, B.**, *Drawing on the Right Side of the Brain*, Tarcher, New York, 1989.

[11] **Burke, J.**, *The Day The Universe Changed*, Little Brown and Company, Boston, 1985.

[12] **Toynbee, A.**, *A Study of History*, Oxford University Press, London, 1954.

[13] **Koizumi, T.**, *Interdependence and Change in the Global System*, University Press of America, Lanham, 1993.

[14] **Capra, F.**, *The Turning Point: Science, Society, and the Rising Culture*, Simon and Schuster, New York, 1982.

[15] **Shlain, L.**, *Art and Physics: Parallel Visions in Space, Time and Light*, William Morrow, New York, 1991.

APPENDIX: SUMMARY OF FOUR WORLD VIEWS

	Primitive	Greco-Roman	Byzantine	Oriental
Individual				
Image	spontaneous creature of nature	autonomous being ruled by reason	humble creature with soul	insignificant speck of nature
Psychology	sensation	thinking	feeling	intuition
Physiology	bicameral brain	left hemisphere	right hemisphere	unicameral brain
Society				
Image	group joined by blood and kinship	aggregate of autonomous individuals	community of believers	network of relational individuals
Mobilization	spontaneous cooperation	functional specialization	voluntary participation	contextual collaboration
Communication	sounds and gestures	words and signs	images and symbols	customs & conventions
Nature				
Image	implacable monster	inanimate clockword	divine creation	organic whole
Operation	animistic	mechanistic	mysterious	eco-systemic
Ideal	magical submission	rational mastery	spiritual sublimation	mystic communion

HUGH CLAPIN

KANTIAN METAPHYSICS AND COGNITIVE SCIENCE

1. INTRODUCTION

Perhaps the most important attempt to reconcile a kind of idealism with a kind of realism in the history of philosophy was Kant's combination of transcendental idealism and empirical realism.

The cognitive sciences, by providing scientific explanations of mental phenomena are threatening to place within the unity of scientific realism the set of disparate experiences which are often thought to support relativism (e.g., cultural relativism).

The aim of this paper is to illustrate the connections between Immanuel Kant's thought and contemporary cognitive science, and to show how, in combination, these two approaches to the mind support a view about what there is which is intermediate between 'hard-nosed' scientific realism and radical relativism. Following Pat Kitcher's recent work [1] defending a psychological reading of Kant's *Critique of Pure Reason* [2], I argue that Kant's technical terms, the 'pure concepts of the understanding' and the 'pure forms of intuition' refer to aspects of what would now be called mind's functional architecture.

It is my view that Kant was drawing attention to important features of any possible system of mental representation, features that are typically considered peripheral in modern discussions. Once we accept that thought is representational, we are committed to a functional architecture which places important, and perhaps inviolable, constraints on what can be thought. This paper consequently shows that aspects of cognitive science provide support for Putnam's Kantian 'internal realism.' And unlike Kant we need not accept that the functional architecture is unchangeable.

TACIT REPRESENTATION AND FUNCTIONAL ARCHITECTURE

There are a couple of conceptual resources developed by cognitive science that are intimately connected with this interpretation of Kant. The first is Zenon Pylyshyn's notion of 'functional architecture,' the second is Daniel Dennett's technical use of the term 'tacit representation.' I take these two concepts to be closely related.

Zenon Pylyshyn [3] explicitly argues for the importance of the notion of *functional architecture* to distinguish between cognitive and non-cognitive levels of explanation. Systems of representation have basic operations and resources that serve to define the system.

Functional architecture is closely related to syntax. In a formal system, the various formal constraints which serve to define the system constitute its syntactic description: the nature of a symbol in the system, the sorts of operations that can be carried out, and so on. In an *automated* formal system, the syntactic constraints need to be embodied in some manner. In a computer, the electronic design of the central processing unit (CPU) fixes the nature of microcode symbols, and also fixes what can happen to those symbols (addition, shunting back and forth to memory, etc.). To a first approximation, then, the functional architecture of a system of physical symbols is described by the syntax of the associated formal system of symbols, where we understand the syntax to consist of the vocabulary, the formation rules and the rules of inference.

The functional architecture of a cognitive system—its *cognitive* architecture—is understood to be those resources that underpin cognitive representations. If the language of thought is first-order predicate calculus, then the syntax of the predicate calculus describes the mind's cognitive architecture.

Talk in terms of functional architecture provides a functional framework in which to understand systems of physical representations. The basic metaphor to keep in mind is that of a lock and key. If a representation is seen as a key, then the lock the key fits is analogous to the functional architecture. Both the key and the lock are required for the successful operation of either element. If the key is faulty, then it won't turn the lock, but conversely if the lock is faulty, then the key also won't work.

In his paper 'Styles of Mental Representation' [4], Dennett uses the term 'tacit' in a special way. He distinguishes between *explicit* representation—exemplified by sentences, program instructions, wffs of a formal system and so on—and tacit representation, which refers to content which is simply *embodied* by a system or organism. One sort of tacit representation, according to Dennett, is that found in the CPU of a digital computer. The CPU tacitly represents information about how to follow certain rules—microcode instructions, which are the lowest level of explicit representation in a computer. Thus another way of describing the relationship between the syntax of a given formal system and the functional architecture of a physical realisation of the formal system is that the functional architecture tacitly represents the syntax of the associated formal system.

But there is more that is tacitly represented by the computer's CPU. A CPU recognises memory addresses as being of a fixed length—for example 16 binary digits long. Since the maximum number that can be represented in 16 binary digits is 65 536, this is the highest possible memory address, and so the memory of a computer with a 16-bit CPU is restricted to 65 Kbytes. Using Dennett's terminology, the functional architecture of the CPU tacitly represents that the associated memory is less than 65K.

To take another example, in comparing the propositional calculus to the predicate calculus, we can say that the predicate calculus constrains potential models in a way that the propositional calculus doesn't. Any natural model of the predicate calculus must be able to map predicate letters and names, while a model of the propositional calculus doesn't. Conversely the predicate calculus provides insufficient resources to deal, for example, with modal information. That the predicate calculus lends itself to the coding of certain sorts of content, but not to others, suggests that the system's *syntax* has content-related significance. We might say that the syntax of the predicate calculus makes certain assumptions or presuppositions about the nature of any model. Thus the syntax of the predicate calculus can be said to *tacitly represent* certain broad assumptions about its model—that the model has the right sort of complexity to be carved up in a certain way. If implemented as a *physical* symbol system the functional architecture of that system would tacitly represent those same assumptions.

In general, it seems that any implementation of a symbol system will add further constraints. For example, the logical calculi typically permit indefinitely long strings as well-formed-formulae, whereas any computational implementation of such calculi will impose some finite upper bound on the length of symbols.

KANT'S PROJECT

Kant's analysis of human cognition, as interpreted by Pat Kitcher in her recent book *Kant's Transcendental Psychology* [1], offers important insights into foundational aspects of cognitive science. In establishing a link between Kant's own project and modern concerns Kitcher has argued that the *Critique of Pure Reason* ought be understood, at least in part, as an exercise in theoretical psychology. In the past there has been much resistance to this view. This can partly be explained by Kant's own rejection of what he termed 'empirical' psychology, and by the perceived non-philosophical nature of psychological concerns. Kitcher argues, however, that a clear understanding of Kant's distinction between the terms 'psychology' and 'logic' shows that the term 'logic' mostly refers to what we would now understand to be philosophical psychology (see pp. 5-14 of [1]).

Taking Kitcher's work as my starting point, I will review Kant's discussion of pure *a priori* concepts, and of the form of perceptual states, in relation to the notion of functional architecture. I will argue that Kant holds that cognitive states tacitly encode information in addition to the content they express explicitly. Thus the functional architecture of the faculty of perception (which Kant calls 'sensory intuition') tacitly represents the phenomenal world to be spatial and temporal, while the functional architecture of conceptual thought ('the understanding') tacitly represents the categories. Kant seems to have put his finger on a notion of non-explicit mental representation which has been afforded little attention until recently, but which is of crucial importance in understanding cognitive systems.

Kant claims that there are necessary preconditions on there being experience, and some of these preconditions can be understood as concepts. These preconditions for knowledge are themselves epistemological. The two key sections of the *Critique* which introduce both the forms and the categories (the 'Transcendental Aesthetic' and the 'Transcendental Logic') uncover these preconditions in terms of necessary types of processes that the mind must employ for there to be experience. Assuming some sort of naturalised epistemology, we can see the relevance of Kant's work to the philosophy of mind and cognitive science in general—he is taking seriously the consequences of understanding the mind as a representational system. The structural features of systems of representation—their syntax or grammar, roughly—cause the system to treat some principles as *a priori*. Kant is investigating what it is that *our* system of mental representation, by virtue of it structure, takes to be *a priori*.

Thus Kant is using the term *a priori* in a complex and (to us) somewhat unconventional way. Certain propositions are knowable *a priori* because they are an expression of the architecture of the mind—in particular, an expression of those aspects of the architecture that are necessary for there to be experience at all.

There are two areas in which Kant makes important claims of this sort. The first is in his discussion of the pure forms of intuition, the second is his discussion of the pure concepts of understanding. I shall argue that the way to understand both space and time (the pure forms of intuition) and the categories (the pure concepts of understanding) is as descriptions of the functional architectures that underlie perceptual and cognitive systems of representation respectively.

THE TRANSCENDENTAL AESTHETIC

To begin with the Transcendental Aesthetic; how might we understand the forms of intuition in contemporary terms? Kant uses two terms in discussing the contents of sensible intuition which will be of importance here—'appearance' and 'intuitions.' The latter Kant uses to refer to mental objects in the faculty of sensation (i.e., in the faculty of sensible intuition) which may relate to things outside the faculty of intuition, and which are only consciously accessible after they have been 'brought under' concepts. An example which might help give the flavour of an empirical 'intuition' is the effect created by the turning on of a bright light when one has just woken up. There is no doubt that the resultant mental state is some sort of perception, and it is even a perception *of* something (e.g., the walls of one's bedroom), however, it is not the perception of a thing understood as that thing. All the subject gets is a mass of incoherent visual information. I take it that this is the sort of thing Kant has in mind when he refers to intuitions.

For Kant, however, intuitions *per se* are not *experienced*. Experience involves the bringing of intuitions under concepts, and so before an intuition has been associated with appropriate concepts, it is *ipso facto* not able to be experienced. This reading of Kant is not committed to the existence of a kind of mental state which is the product of intuition but not yet an intelligible thought. We need only recognise that by the time an intelligible thought comes into being, it has been constrained in certain ways by the processes which have created it.

'Intuition' is a broad term, since it refers to pure contents of the sense faculty as well as empirical ones (e.g., appearances). In the first instance 'pure' means 'devoid of empirical content,' and refers to cognitive states or properties which do not have their origin in any external sense influence. In general, it is meant to refer to thoughts or other mental states which owe no part of their existence to the impact of objects on the senses. It will be important for our purposes, however, to sharply distinguish between those contributions of intuition to thought which are the result of the ordinary operations of the sense faculty, and those which are pure, since the latter are (I will argue) not *products* of the faculty of intuition in anything like the same way that appearances are.

An appearance, Kant says, is an 'undetermined object of an empirical intuition' (A20 = B34). (Following convention, I will differentiate between the first and second editions of the *Critique of Pure Reason* ([2]) by the letters A and B respectively). Thus appearances are things in the world—with the important qualification that they are 'undetermined'; that is, that they are not conceptualised, and thus not yet possible objects of experience.

A key passage for our purposes is the following:

> That in the appearance which corresponds to sensation I term its matter; but that which so determines the manifold of appearance that it allows of being ordered in certain relations, I term the *form* of appearance. That in which alone the sensations can be posited and ordered in a certain form, cannot itself be sensation; and therefore, while the matter of all appearance is given to us *a posteriori* only, its form must lie ready for the sensations *a priori* in the mind, and so must allow of being considered apart from all sensation. ([2], A20 = B34, emphasis Kant's)

The form of an appearance is that which allows it to be ordered, compared and combined in relations; however, it is distinct from that aspect of the appearance which can be accounted for by interaction with the object. Kant assumes that the information made available to the senses by contact with objects could not in itself be enough for us to determine what the sensed object was, nor to account for the content of appearances in general. Note that Kant's claim is that the information available from the object is insufficient to identify it, and thus must be supplemented by the subject.

We should pause to recognise *why* Kant thought that the information available to sensation is not enough to account for the content of our sensible intuitions. That is, why did Kant think that there had to be a semantic contribution from the mind?

Kitcher argues that we can best understand Kant's argument here if we acknowledge the preceding discussions of spatial perception among philosophers. How we perceive space had been a live issue since at least Descartes. The problem Kant is dealing with was recognised by Molyneux: information about distance cannot be directly available to the sense because '*distance* of itself, is not to be perceived, for 'tis a line (or a length) presented to our eye with its end toward us, which must therefore be only a *point*, and that is *invisible*' (William Molyneux in the *Dioptrika Nova*, 1692, quoted in [1], p. 32). That is, the three-dimensional world around us is collapsed into a two-dimensional image on our retinas, and in that process information about the distance between objects and the eye is lost. Molyneux's challenge had been discussed by Descartes, Berkeley and Leibniz.

Thus within Kant's intellectual context it was generally recognised that there was a problem in determining how we can see three-dimensionally when we only have two-dimensional information available. While Kant doesn't argue against the various answers presented to this problem, he does start with the assumption that there needs to be something other than the information available at the senses in order to explain distance perception.

Kant was aware of further problems relating to perception. For example, as Hume noted, sensory experience consists of a constant flux. This suggests the requirement that sensory intuition identify continuous objects in the rapid succession of changing sensations. In short, Kant recognised that for perception to succeed, information additional to that present to the senses was required. This is Kant's 'Isolation Argument': The pure form of sensory intuition is that information which cannot be traced to an empirical source. With regard to perception, Kant assumes the information required in that space is three-dimensional and Euclidean.

In the passage above, we find Kant distinguishing between the 'form' and the 'matter' of a sensory intuition. Because objects of perception cannot be identified using the information available to the sense, only a portion of the content of an appearance can be traced directly back to the contact between the senses and the object being sensed. The rest of the content of an appearance is contributed by the senses themselves, and is the form.

Why call this contribution by the mind 'form,' and the contribution of contact with the world 'matter'? Although the scholastic adherence to Aristotelian notions of form, particularly in the realm of the mental, was comprehensively rejected by thinkers in Kant's time, the general notion of distinguishing the form and matter of objects was clearly still alive. In general, the distinction between form and matter as it was passed through the scholastic tradition was importantly connected to the distinction between essential and accidental properties. The form of something is what makes it what it is, while the matter is whatever happens to make up that thing at a given time.

I take it that it is this distinction that Kant is using in distinguishing between the form and matter of sensory intuitions. Different intuitions have vastly different features depending on their contents, however, they must all have something in common by virtue of being intuitions—it is this common feature that Kant is attempting to identify, and which he labels 'form.'

Thus when Kant says that 'that which so determines the manifold of appearance that it allows of being ordered in certain relations, I term the *form* of appearance' he is suggesting, in part, that the orderability of sensory intuitions proceeds from their essence, and is not an accidental feature.

It is worth noting how close this claim is to a computational theory which posits explicit symbols as the media of cognition. Such symbols have as their (essential) form their shape and other physical features which allow them to be ordered and related to other symbols in ways determined by the rules of the symbol system. For example, well-formed formulae (wffs) of the predicate calculus have essentially features such as that they can be composed from atomic wffs according to certain rules. Regardless of the content of these formulae, they must have a certain *form*. Although we don't want to identify Kant's view with computational theories of mind, it is consistent with such theories in this respect. Thus exactly the sort of features we identify with functional architecture in the parade case of computational symbol systems appear to be what Kant refers to as the forms of intuition.

Kant's 'form' is a structure imposed on the incoming sensory information by the senses which is necessary for the information contained in the 'matter' (e.g., the retinal image) to be interpreted and made intelligible. Compare this to a computational system which uses n-tuples as the vehicles for representing appearances (e.g., (C, x, y, z, t)). If the architecture demands that all fields of this n-tuple be well-defined for the n-tuple to be well-defined, then each and every appearance (to use Kant's term) must carry spatial and temporal information. This would be a simple case in which the *form* of individual intuitions demanded that they carry spatio-temporal information. That is, the forms of this computational intuition are three-dimensional space and time.

Notice how this view of form would justify Kant's claim that 'We can never represent to ourselves the absence of space, though we can quite well think it as empty of objects' (A24 = B38). If your system of representation demands that there be spatial coordinates for every appearance, then all appearances must exist in space.

Another strong connection between Kant's own description of the form of intuition and the modern notion of functional architecture comes from his polemic against Eberhard's criticisms of the first edition of the *Critique*. In defending himself against charges of nativism, Kant makes the following précis of his view:

> The *Critique* admits absolutely no divinely implanted or innate *representations* ... There must, however, be a ground in the subject which makes it possible for these representations to originate in this and no other manner ... (Quoted in [1], p. 38.)

That the forms are 'grounds' and not explicit 'representations' provide clear support for my suggestion that they reflect aspects of the mind's functional architecture, especially considering that this ground ensures that the subsequent representations originate in 'this and no other' way. That is, the functional architecture imposes very strong, inviolable constraints on just how perceptual representations are able to represent objects.

The actual nature of the forms (i.e., space and time) is not what an intuition represents, in the ordinary sense of 'represent.' Something which has time as its ordinary representational content—for example the readout of a digital clock—is a representation *of* time. Rather, the form of the intuition is the *way in which* something in the world is represented, and must be represented, in our faculty of sensation. For Kant, the elements contained within the faculty of intuition must always carry temporal information.

According to Kant then, sensible intuition supports a system of appearances, and that part of the content of appearances which is supplied by the functional architecture is the spatiality and the temporality or the appearance—the very nature of appearances is that they represent spatial relationships and temporal relationships. In contemporary terms we understand this as the information tacitly encoded by the functional architecture—and that is that the world of sense is three dimensional and exists in progressive time. The functional architecture (tacitly) represents this information because something can't be an appearance unless it represents its object *as being* in some comparative time frame, and *as being* in some spatial relationship.

THE TRANSCENDENTAL LOGIC

The task Kant sets himself in the Transcendental Logic is to show that there are 'pure' concepts of understanding, and that these can be known *a priori*. He is looking for concepts whose content does not presuppose experience and which continue to shape our thought after we do have experience. As in the above discussion, we will understand the *a priori* as concerned with the functional architecture of thought. We understand Kant to be investigating the features of our mind which allow us to think, but which, at the same time, constrain our thought. Thus concepts are *a priori* if their existence in thought is due simply to the architecture of the mind, and not to any experience. The project here differs from that in the Aesthetic in that he is now concerned with conceptual thought, and not appearances. I will be attempting to better understand the nature of the categories by treating them as an expression of the information tacitly represented in the processes responsible for the syntheses required for us to have thought. In doing this I will look at the deduction of the categories in the first edition (the 'A' edition) of the *Critique*.

Kant begins in the A deduction by noting that the way to find pure concepts is to investigate the conceptual requirements of mind—the '*a priori* conditions for the possibility of experience.' If he can show that a certain set of concepts are necessary for us to think about *anything*, then this would be a sufficient justification for treating them as a priori. It is important to Kant's project, however, that these concepts not be innate, in the sense that we are born with them. As we have seen, the middle road can be trod.

The primary responsibility that needs to be accounted for in understanding is that of synthesis, and it is here Kant looks for the categories. Since such *a priori* concepts are to be understood as expressions of the functional architecture of the understanding, and they consist of information that is added to or imposed upon information coming from the senses, then the natural place to look for them is in the processes carried out on the products of sensory intuition which create the cognitive states in the understanding—these processes are termed processes of 'synthesis' by Kant.

In the A deduction, Kant enumerates three sorts of synthesis in the mind (A98 - A110). The first is the apprehension of representations in sensible intuition. This is the basic process by which the confused jumble of information being dealt with by sensation is interpreted as a single, unified representation which can be considered by higher faculties of the mind; that is, as something which can be thought about.

The second sort of synthesis is the reproduction of ideas by imagination—this is the process which explains why those ideas which are not the direct result of current sensation occur to us as unified, coherent ideas, since they are composed from a range of different perceptions and previous thoughts. In Kant's terminology, the faculty of imagination is also responsible for the ordering of thoughts. The rules by which ideas are associated are the responsibility of the imagination. The synthesis provided by imagination has a role both in the immediate products of intuition—what is currently being perceived—and in those thoughts which are not the result of current perception.

The third sort of synthesis is that which allows me to recognise a thought as my thought. Every thought I have is unquestioningly recognised as mine. This unification is required for me to recognise successive thoughts as related. This unity is furnished, says Kant, by a concept, and so this synthesis is termed by him the 'synthesis of recognition in a concept.' This third sort of synthesis is perhaps the most obscure. What can be seen clearly, however, is that in the A deduction Kant places a heavy emphasis on the creation or derivation of unity from diverse sources. According to Kant, the need for *a priori* pure concepts of understanding—the categories—is shown by the need for processes of unification (the various syntheses).

Understanding synthesis, says Kant, is the road to inferring the categories. According to Kant:

> Pure synthesis, *represented in its most general aspect*, gives us the pure concept of the understanding. (A78 = B104, emphasis Kant's)

The link between synthesis and the categories is explicated thus:

> What transcendental logic ... teaches, is how we bring to concepts ... the *pure synthesis* of representations. What must first be given—with a view to the *a priori* knowledge of all objects—is the *manifold* of pure intuition; the second factor involved is the *synthesis* of this manifold by means of imagination. But even this does not yet yield knowledge. The concepts which give *unity* to this pure synthesis ... furnish the third requisite for the knowledge of an object ...
> The same function which gives unity to the various representations *in a judgement* also gives unity to the mere synthesis of various representations *in an intuition*; and this unity ... we entitle the pure concept of the understanding. (A78 - A79 = B104 - B105, emphasis Kant's)

Thus the categories are what allow the synthesis to provide unity. All three elements are required before we have thoughts—the manifold which is to be synthesised, the act of synthesis, and the categories which ensure the synthesis produces the required result.

The connection is that the category expresses in a conceptual way what it is that the act of synthesis adds to the manifold. The categories, says Kant, are the 'original pure concepts of synthesis' (A80 = B106). The only way to understand this relationship, as I see it, is that the functional architecture, of which synthesis is a component inasmuch as it is one of the basic mental resources, tacitly represents information about the sort of conceptual realm that is appropriate to the objects of conceptual thought. Because this is tacitly represented in the functional architecture, this information is imposed upon the cognitive states that the functional architecture makes possible.

My reading of Kant's talk of synthesis can be exemplified as follows. The full task of perception requires that objects and their relationships be identified from the sensory field. In the case of vision, for there to be a conceptual recognition—say the identification of a cricket match—a series of light patterns on the retina needs to be first analysed in terms of objects and their relationships, and then these objects need to be understood in terms of concepts that the perceiver has.

The first process requires solving problems like shape from shading, background/foreground distinctions, distance measurements, motion detection and so on. As already noted, the perceptual system needs to make certain assumptions in order to analyse the retinal information in these terms.

The second process involves taking the objects—red spherical objects, white human-shaped objects—and applying concepts like 'person,' 'cricket team,' and so on. The process by which perceptual objects are understood in terms of concepts is what Kant terms 'synthesis.' Once again, this synthesis requires that the mind make certain kinds of very basic assumptions, for example that objects can have causal relations to each other (that the ball flies off the bat the way it does *because* of the way the bat strikes the ball), or that objects have properties (that the thing holding the bat has the property of being a batsman, and the property of wearing white clothes).

I take it, then, that the category of 'inherence and subsistence' which allows the mind to make categorical judgements is an attempt by Kant to express the mind's assumptions that the world is carved up into objects with properties. Similarly the category of 'causality and dependence' which allows the mind to make hypothetical judgements is Kant's attempt to express the assumption that objects can be causally related. If the mind didn't assume these very basic things, it wouldn't get very far in applying sophisticated empirical concepts like 'cricket game.'

Expressing exactly what it is that is tacitly represented in a functional architecture is not an easy task, and Kant offers the tacit content of human functional architecture in two ways. First, in the form of the types of judgements that must be possible of all thoughts (the table of judgements at A70/B95). The second is in the form of concepts which can be applied to all thoughts (the table of categories at A80/B106). For example take the affirmative and negative judgements, and see how they relate to their corresponding categories (reality and negation). Judgements in the understanding can be affirmed or denied. In affirming a judgement we are claiming that what it says is true, that is, that it corresponds with reality. On the other hand, having the concept of reality allows us to ask of a statement if it corresponds to reality, and thus to affirm or deny it. Thus the ability to affirm a judgement is materially equivalent to having the concept of reality. In denying a judgement we are saying that what it says about the world is false, and so we must have the concept of negation in order to do this.

Thus Kant sees the table of judgements and the table of categories as expressing the same thing—the information *tacitly represented* in the functional architecture of the faculty of understanding by virtue of the required act of synthesis. Of course, Kant is operating at the most abstract level he can, however, a more psychological example might help explicate this idea a little further. If Kant had subscribed to Fodor's language of thought hypothesis [5, 6], then his categories are an attempt to capture some of the syntactic constraints on the language of thought. The tables of judgements and categories set up constraints on any possible syntax—they determine abilities that the representational system must have in order to *be* capable of thought. For example, the system must be capable of expressing universal and singular statements, of affirming and denying claims, of modal statements, and of disjunctive, hypothetical and categorical (i.e., subject-predicate) statements.

Kant's whole discussion of the categories isn't nearly this clear, of course, and it is worth reflecting on why. Kant's goal was to ground metaphysics, not cognitive science. His major contribution was to recognise that the mind is structured such that we can only think in certain ways, and that this structure is specifically designed for thinking about the world via a perceptual apparatus.

Kant was claiming that the phenomenal world can only be as the mind presents it, and that the mind must present the world as being a particular way, to the exclusion of other ways. To reach his goal, however, Kant had to go further than this and actually enunciate what *knowledge* is assumed by virtue of our having a cognitive architecture. To do this, he needed to extract what he took to be tacitly represented in the mind's functional architecture in conceptual form—he had to make it sound like pieces of knowledge. This is an immensely difficult task.

The pure concepts of understanding, on this analysis, are known *a priori* and unconsciously. Kant was right to single them out and treat them quite differently to other concepts. He may have been wrong, however, to treat them as concepts at all. If this analysis is right, then the categories are merely an attempt at expressing what is presupposed, and thus added to thought by, the functional architecture underpinning thought.

COGNITIVE SCIENCE AND METAPHYSICS

We find, then, that the modern language of functional architecture and tacit representation provides us with powerful tools in understanding Kant. A representational system's functional architecture constrains what can be represented within that system. On the assumption that thought is mediated by a system of representation, thought is also so constrained. Attempts within this system to represent contents which are at odds with the tacit information of the system's functional architecture can only lead to grief. A simple example might involve trying to include concepts of quantification into propositional calculus without changing the functional architecture. We might try and introduce a proposition X that means 'all persons are mortal,' and another Y which means 'Immanuel Kant is a person,' and another Z which means 'Immanuel Kant is mortal.' But with only the functional architecture of the propositional calculus (i.e., without any additional premises), we can never show that the conjunction of X and Y entails Z. More importantly, the functional architecture of the predicate calculus would not only allow us to prove this, it would prevent us from proving its contradiction. The propositional calculus allows us to represent these three propositions, but will not prevent us from proving all sorts of nonsense about their relationship.

The lesson Kant himself drew was that there were questions we could ask, but to which we could never find an answer (e.g., 'Is space finite?'). The reason for this is that these questions challenge the assumptions tacitly encoded in the forms of intuition or the categories. The faculty of understanding—understood as a system of representation—is designed to operate with ideas derived from, and which relate to, experience. Attempts to use these thoughts in ways in which they were not designed will cause the system to break down.

Kant's arguments are not conclusive, however they are plausible and sensible given that he was trying to draw attention to aspects of the mind's functional architecture. The semantic contribution of functional architecture is often difficult to clearly express. It is the result of the form of representations, both inasmuch as it results from the physical shape of representation in its syntactic context, and inasmuch as it is content held *essentially* by a representation by virtue of its status as a representation.

Kant saw a deep connection between the nature of mind and metaphysics. Kant's metaphysical conclusions include that the world as we see it (the 'phenomenal' world) must be distinguished from the world as it actually is in itself (the 'noumenal' world), because the phenomenal world is necessarily presented to us as being in space and time (i.e., presented to us in conformity with the forms of intuition), and is conceptualised according to the categories. Because the forms and categories are absolutely necessary for any kind of thinking, we cannot understand the world in any way other than as existing in space and time and as determined by the categories.

In a nutshell, Kant is arguing that the noumenal world is in important respects unknowable. That is, we can never know whether the world as it is in itself, independent of our perception of it, really is spatial and temporal. Simply in order to think about the world, we necessarily apply the forms of space and time, and so we cannot think of a world outside of space and time. It is here that Kant's metaphysics takes hold. Epistemological claims about our knowledge of the world as it is experienced by us are termed 'empirical' by Kant.

Epistemological claims about our knowledge of the world as it is in itself are termed 'transcendental.' Transcendental claims go beyond the limits of experience, whereas experience is limited by the forms and the categories.

Empirical realism is thus the claim that the world as it appears to us considered empirically (i.e., within the bounds of experience set by the forms and the categories) is real and objective. *Transcendental idealism* is the claim that the world as it appears to us considered transcendentally (i.e., outside the bounds of experience, and so as it is in itself) is shaped by subjective features of human cognition and is thus 'ideal.' In particular, it is shaped by the categories and the forms.

Transcendental idealism is not a simple-minded idealism because the forms and categories are *necessary* for thought. The forms and categories are universal among thinkers, and thus (e.g.,) scientific discoveries retain their objectivity and universality. It also has bite over other idealisms because it acknowledges noumena behind the phenomena (but claims we cannot know anything about the noumena *per se*).

Kitcher's psychological reading of Kant, in combination with a close understanding of representation, delivers an understanding of Kant's metaphysics which comports, on the one hand, with contemporary cognitive science, and on the other with Putnam's metaphysics (where we 'shift ...our way of thinking to the extent of regarding "the world" as partly constituted by the *representing* mind' ([7], p. 162, emphasis mine)).

Putnam regularly draws the connection between 'internal realism' and Kant's empirical realism [7-9]. I hope to have shown that there exists a convergence between the Putnam/Kant metaphysics and one aspect of contemporary cognitive science. Recognising the role of functional architecture in enabling a system of representation to represent has important metaphysical consequences when the system of representation in question is human thought. In particular, it means that there are tacit assumptions about the nature of the world which pervade thought. While it is possible that cognitive architectures (and thus the tacit content of those architectures) are changeable, there must nonetheless always be *some* assumptions.

While there is significant convergence between Kant's theory of mind and the view drawn from cognitive science that systems of explicit representation must also involve tacit representation, there is an important area of divergence. Kant argued that the forms and categories are necessary for and universal with respect to thinking. From the point of view of cognitive science, however, the tacit content of a functional architecture could be changeable. Different systems of representation tacitly represent different content.

Take the example of the computational perceptual system, which, we noted, tacitly represents that space is 3-dimensional by virtue of the form of the representations involved (*viz.*, (C, x, y, z, t)). If this system were to change into one in which the form of the representations demanded four spatial dimensions (e.g., (C, w, x, y, z, t)), then the *form* of space would have changed for this system. Such changes to systems of representation are, of course, possible. In computer programs such changes are often made by programmers while developing a program when it becomes apparent that one sort of data structure is inadequate. Such changes could occur in humans, either via evolution, or through individual learning or via some kind of environmentally driven change.

Thus, from the perspective of cognitive science, there is nothing *necessary* about forms and categories. It could be that in order to communicate we must share common thought structures, however it is conceivable that individuals from different cultures, for example, have different forms and categories.

It is from this possibility that we gain a metaphysical middle ground between a hard-nosed scientific realism and an 'anything goes' relativism. As Kant argues, from within a certain cognitive architecture that which is tacitly represented (e.g., 3D space, causality, etc.) cannot be questioned—it is empirically real. But if persons from different cultures do, in fact, have different cognitive architectures then their world view could be just as empirically real. Thus the western world view which sees a landscape of trees, animals, rivers, rocks, etc. could be correct, and at the same time an Aboriginal world view which sees the same landscape in terms of ancestor spirits, song lines and cultural taboos could also be correct. On the other hand a pluralist empirical realism does not ascribe equal value to all possible conceptual schemes. Kant is probably right that in order to give rise to something recognisable as thought, a conceptual scheme must be tightly constrained.

So the metaphysical promise of a Kantian metaphysics seen through the eyes of cognitive science is that we could discover that a true metaphysical pluralism obtains. There could be a number of equally viable, but incommensurable, cognitive architectures which understand the world very differently. This discovery would not require, however, that we stop valuing scientific knowledge.

One last comment. Cognitive science is a *scientific* paradigm, and its language and preconceptions are on the same metaphysical footing as the rest of science. So the scientific knowledge (of functional architectures, neural systems, and the causes of behaviour) which justifies the metaphysics at the same time prevents the metaphysical claims from being truly universal. This raises difficulties for the metaphysics in that what appears to be a genuinely universal (indeed *noumenal*, if you like) claim also *cannot* be, by its own lights, because it is justified from within the perspective of the scientific world view. The flip side to this is that, as I have shown here, the scientific world view which is often taken to presuppose a single correct theory of the world in fact provides reasons for believing that there is not a *single* correct theory.

AFFILIATION

Hugh Clapin
Department of Philosophy
University of Sidney
Australia

REFERENCES

[1] **Kitcher, P.W.**, *Kant's Transcendental Psychology*, Oxford University Press, New York, 1990.

[2] **Kant, I.**, *Critique of Pure Reason*, (trans. N. Kemp Smith), St. Martin's Press, New York, 1929.

[3] **Pylyshyn, Z.P.**, *Computation and Cognition*, The MIT Press, Cambridge, Massachusetts, 1984.

[4] **Dennett, D.C.**, *Styles of Mental Representation*, Proceedings of the Aristotelian Society, **3**, 1982.

[5] **Fodor, J.A.**, *Psychosemantics*, The MIT Press, Cambridge, Massachusetts, 1987.

[6] **Fodor, J.A.**, *The Language of Thought*, The Harvester Press, Hassocks, Sussex, 1975.

[7] **Putnam, H.**, "The Place of Facts in a World of Values", in: *Realism With a Human Face*, 1979. First published in: **Huff, D. and Prewett, O.** (eds.), *The Nature of the Physical Universe*, New York, John Wiley and Sons, 1979, pp. 113-140.

[8] **Putnam, H.**, *Reason, Truth and History*, Cambridge University Press, Cambridge, 1981.

[9] **Putnam, H.**, *Representation and Reality*, The MIT Press, Cambridge, Massachusetts, 1988.

JOHN MINGERS

SYNTHESISING CONSTRUCTIVISM AND CRITICAL REALISM: TOWARDS CRITICAL PLURALISM

INTRODUCTION

The discipline within which I work is generally known as *management science*. It aims to assist the management of organisations using a broadly-defined scientific approach. Within management science, I have worked particularly in *operational research* (OR) and *information systems* (IS). My work is generally informed by a *systems* approach. Historically, management science developed with a very positivistic view of science [1]. OR, for example, was concerned largely to develop mathematical and computer-based models of the quantifiable aspects of problem situations in order that these might be "optimised", while systems methodologies were based on "hard", engineering principles.

However, it was found increasingly in practice that such methods had only limited success as they could not adequately capture (or even acknowledge) the richness of human social and political interaction [2] [3]. This led to a major fissure within the discipline (similar to that which has occurred in other disciplines such as sociology and philosophy) based on a critique of the limitations of positivism, and the development of alternative, interpretive or phenomenologically-based methodologies such as *soft systems methodology* (SSM) and *cognitive mapping* [4] [5] [6]. These focused on the inherently *meaningful* and subjective nature of social reality, and the necessity of understanding and expressing individuals' different perspectives and world views. Such "soft" approaches, however, brought with them problems of relativism and subjectivism [7] [8]. If it is people's perceptions and constructions of problems that matter, and these are seen as all equally valid in that there can be no "objective" standpoint from which to evaluate them, then any connection to an external, constraining, "real-world" is lost.

A third paradigm arose known as *critical systems* or *critical management science* [1] [9] [10] that attempted a synthesis. This was based largely on the social theory of the Frankfurt School, and particularly the critical theory of Habermas [11] who argued that human beings had a *technical* interest in predicting and controlling nature (hence empirical science), and a *practical* interest in mutual understanding and communication (hence an interpretive science). A third, *emancipatory,* interest in self-development and freedom from constraint and dogma, leads to a critique of existing social arrangements and power structures (critical science).

This historical sketch serves to locate my own long-term research programme that is concerned with reconceptualising the philosophical foundations of management science, particularly in terms of the validity claims of knowledge (epistemology) and action (praxiology). In the following sections of the paper I shall first summarise the main theoretical sources on which I am drawing, and then outline the main stages of the resulting framework in terms of a hierarchy of four levels of inquiry—philosophical, concerning questions of ontology and epistemology; methodological, concerning the nature of science and particularly the relations between the natural and social worlds; cognitive, concerning the nature of human cognition and linguistic interaction; and praxiological, concerning our active intervention and research in real-world situations.

CENTRAL PROBLEMATIQUE

The central problematique that my research addresses is the resolution or reconciliation of the two main paradigms outlined above. Recognising, on the one hand, that our experience and description of the world (and even more so the social world) is inevitably constructed by the subject or observer, and yet, on the other, that there is an external world that exists prior to and independently of, human beings. And further considering the implications of these ontological and epistemological questions for an essentially action-oriented discipline such as management science.

In trying to overcome or dissolve this duality, my work has aimed at synthesising a number of theoretical developments.

i) Maturana and Varela's [12] [13] [14] [15] theories of *autopoiesis* and cognition.

Autopoiesis is a theory concerning the nature of living systems. It claims that all living systems are *self-producing* systems and that they are therefore intrinsically autonomous and organisationally closed. Autopoiesis provides a comprehensive and consistent theory explaining the phenomena of living systems from the organisation of a single cell to the consensual domain of human language. The latter is of particular importance. It is argued that it is a mistake to see language as *denotative*, that is objectively indicating and pointing to an external world. Rather linguistic behaviour is *connotative*, based on the ground of a pre-existing, implicit consensus and agreement. The observed communication of meaning and the practical efficacy of language do not stem from the words and terms themselves but reflect similarities in the organisms' structures developed through their history of interactions.

Living systems, because of their autopoietic organisation, are inherently structure-determined. Their behaviour is determined by their own structure—other systems/organisms can only act as triggers. However, organisms which interact recurrently with each other become structurally coupled. They develop behaviours which reciprocally trigger complementary behaviours, and their actions become coordinated so as to contribute to their continued autopoiesis. Language developed from linguistic acts which coordinated other activities in the practical daily lives of early humans. Moreover, the particular behaviours or conducts are divorced from that which they connote—they are symbolic and thus essentially arbitrary and context-dependent. They only 'work' to that extent that they reflect agreement in structure and this is what Maturana means by a domain of consensual action. Communication is thus only possible to the extent that organisms share the same structure, and is thus inherently intersubjective.

In terms of the positivist/interpretivist debate, while phenomenology may be useful as a *description* of our experience, it cannot claim to be primary since it takes for granted the existing, languaging subject which, it transpires, is generated inter-subjectively not individually.

ii) Bhaskar's [16] [17] formulation of a consistent and coherent non-objectivist version of critical realism.

It accepts the interpretivist criticisms that observations are theory-dependent and that we cannot have pure access to an independent world. It asserts strongly, however, that such a world does exist, and that it is populated by objects and entities, some of which may be in principle unobservable, that have causal powers or tendencies. It is these which generate the phenomena we experience. Bhaskar distinguishes these *intransitive* objects of knowledge from the theories and descriptions we employ—the *transitive* objects of knowledge. Science can no longer be seen as creating true theories, but proposing and identifying potential causal objects and structures, the descriptions of which are at least approximately true. Moreover, we can accept the fact that theories and descriptions are historically and culturally conditioned, *epistemic relativism*, without having to accept that they are all equally valid, *judgemental relativism*.

Critical realism has a well defined methodology which Bhaskar terms *retroductive*:

- Identify an effect or phenomenon to be explained.

- Postulate a hypothetical mechanism or structure which, if it existed, would generate the phenomenon.
- Attempt to demonstrate the existence of the mechanism by experimental activity leading to direct or indirect observation, and by eliminating alternative explanations.

Finally, in contrast to an epistemic dichotomy (or, for Habermas, trichotomy), critical realism maintains that it is essentially valid in both the natural and the social world although it recognises that social structures have different properties from physical ones.

iii) Habermas's [18] [19] theory of *communicative action*, that is social action oriented towards understanding and agreement. In analysing the use of language in a pragmatic (i.e., as actually used) sense, it is important not to take the line of individual intentionality, consciousness and meaning, but rather to start from the structure of linguistic expressions and the preconditions of communication in a formal sense. Habermas limits the domain of actions that he analyses to those intended to reach understanding or bring about an agreement (*communicative action*), as opposed to *strategic action* where an individual wishes to succeed personally in spite of or against another. He also excludes non-linguistic actions such as gestures and bodily expression. The justification for this is that communication for understanding is the original and primary mode of language use. Any form or coordination requires both understanding and agreement or acceptance. Other types of social action are derivatives or distortions of communicative action.

In communicative action between people the utterances, or speech acts, imply certain presuppositions or assumptions. The first of these is simply *comprehensibility*—the hearer understands what is said. However, from a pragmatic point of view, there are three wider *validity claims* which are raised by a speech act and which a hearer implicitly accepts. These are *truth, truthfulness* (sincerity), *and rightness* (legitimacy). By a validity claim is meant simply that a particular aspect of a speech act is in fact correct. It may not be—it is up to the hearer to accept the claims, or to dispute one or more of them. If claims are disputed then *discursive* action occur. The validity claim itself becomes the topic about which agreement is needed. Such questioning of the grounds of utterances is potentially endless and should ultimately be closed by the "force of the better argument" in an *ideal speech situation.*

The claim to truth means that the states of affairs implied by the utterances are, in fact, correct. That its propositional content, or existential presuppositions are true—what it assumes is actually possible or feasible. The claim to truthfulness or sincerity refers to the motivation of the speaker and the assumption that he/she genuinely and sincerely intends what they say. Finally, the claim to rightness or legitimacy concerns accepted social norms within a particular group and assumes that the actions or practices implied by the speech act are consensually valid. These three validity claims actually refer to three different "worlds" to which any utterance is related—the *objective world* of external nature which obtains or can be brought about; the *subjective world* of internal nature, constituted by peoples' private experiences; and the *social world* of society consisting of consensual, intersubjective practices and norms. *The* world, *my* world, and *our* world.

ONGOING RESEARCH PROGRAMME

This section will outline the main framework of the research programme as constructed so far.

Ontology/Epistemology—Critical Realist

The ontological and epistemological assumptions can best be summarised in the following quotation from an earlier work of mine:

> "There is a single, real, materially existing world. This has, through processes of evolution, generated organisms capable of creating distinctions, descriptions and constructs, subject only to their own internal structure. This leads, in such organisms, to a proliferation of domains of experience and interaction essentially free from dependence on, and determination by, the material world. One of these domains is that of science where observers cast their net of descriptions back onto the world itself. These are free human constructs, based always on subject-dependent experiences; yet the world that they relate to is independent of the observer's descriptions, and existed prior to them. To the extent that the mechanisms hypothesised within scientific descriptions do exist in the material world, then the praxis of observers based on these descriptions will be more successful. The contingent fact that science has been successful in enlarging our domain of interactions is, if not proof, at least strong evidence for such existence. What can, in any case, be shown is that the fact that our descriptions are always subject-dependent does not preclude the existence of a world independent of such descriptions." [15, p. 116]

This position represents a synthesis of Maturana's and Bhaskar's theories as outlined above. The nature of the relations between the observer and the external world are illustrated in Fig. 1, taken from Mingers [20] [21].

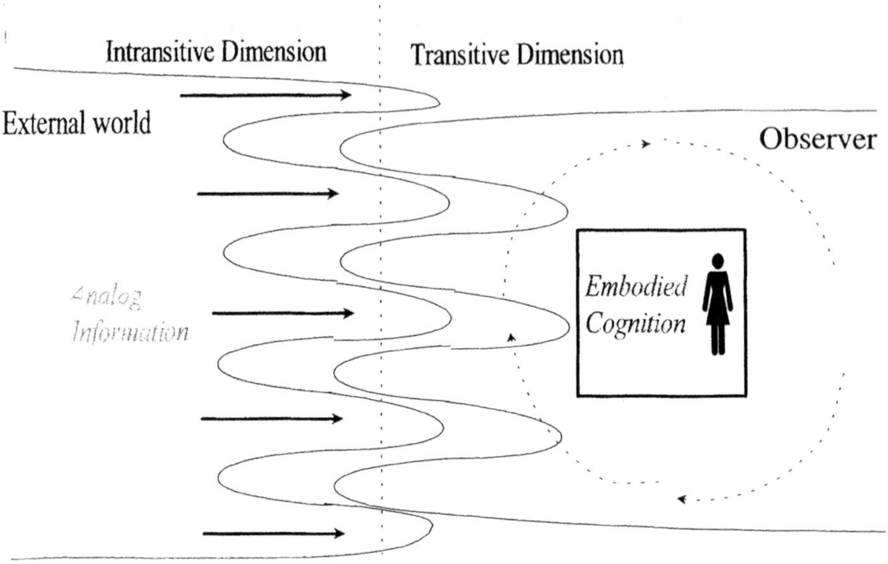

Relations of Structural Coupling

Figure 1. Relations Between the Observer and the External World

At the philosophical level, critical realism provides a post-empiricist ontology and epistemology. It is accepted that we can have no objective, observer-independent, access to reality but against constructivism it is maintained that there is an independent external world constituted by structures or entities with causal powers (the domain of the *real*). These are seen as generative mechanisms responsible for the events which actually occur, or which may not because of countervailing tendencies (the domain of the *actual*). A subset of these events may impinge on humans—the domain of the *empirical*. Epistemologically, science proceeds by hypothesising potential generative mechanisms that, if they did exist, would account for our observations and experiences. A distinction is drawn between the *transitive* and *intransitive* dimensions of science. The intransitive (ontological) dimension is the domain of the real objects of scientific knowledge, the transitive (epistemological) dimension is the domain of humanly-constructed cognitive objective of science such as theories, experiments and concepts.

At the biological level, autopoiesis provides an explanation of our interactions with the world. As living systems, we have a closed (autopoietic) organisation, but are interactively open to our environment or medium. The nature and limitations of such interactions are determined primarily by our own physical structure (structure-determined) rather than by the environment. External stimuli provoke or trigger a response, but the nature of the response is determined by the structure of the organism at that instant, not by the stimuli. Moreover, it is the structure that determines what can or cannot be a stimulus for the organism—organisms without eyes or the equivalent cannot be triggered by light. Given this under-determination of the organism by its environment, how is it in fact that responses in particular situations are generally appropriate? This is answered through the concept of structural coupling. The maintenance of autopoiesis through recurrent interactions with the environment or other organisms will lead to the generation of mutually adapted structures that can be said to be structurally coupled.

Moving to the level of information and meaning, events or signs in the empirical world, especially symbolic and linguistic ones, can be said to carry information about their origins, the causal mechanisms giving rise to them. Such signs will be triggers or perturbations for the nervous systems of organisms or human observers but, as shown above, cannot determine or control the reaction or internal compensation that it provokes. We can see this process as one in which the *information*, carried in analogue form by the sign, it transformed into *meaning*, expressed in digital form for the observer. The information is objective in the sense of being independent of the observer, but the meaning that it generates is (inter-)subjective and observer-dependent. This process of conversion from information to meaning, the digitalisation of the analogue as Dretske [22] calls it, carried out, largely unconsciously, by the body and its nervous system is *embodied cognition* (see the section on cognition below).

The Natural/Social World—Modified Naturalism

Having considered the fundamental philosophical questions of ontology and epistemology, we must move up a level to discuss a major debate that concerns the question of whether the methods of the natural sciences (whatever they are taken to be) are appropriate for the social sciences. Naturalists argue that the same scientific method (e.g., hypothetico-deductivism or falsificationism) should apply to all domains, while anti-naturalists (e.g., phenomenologists or interpretive social scientists) hold that the social world is constituted by humanly-constructed systems of meaning and so is inherently different and requires a quite different methodology.

The position adopted here is a modified naturalism. That is, that at a very general level the same retroductive Bhaskarian methodology

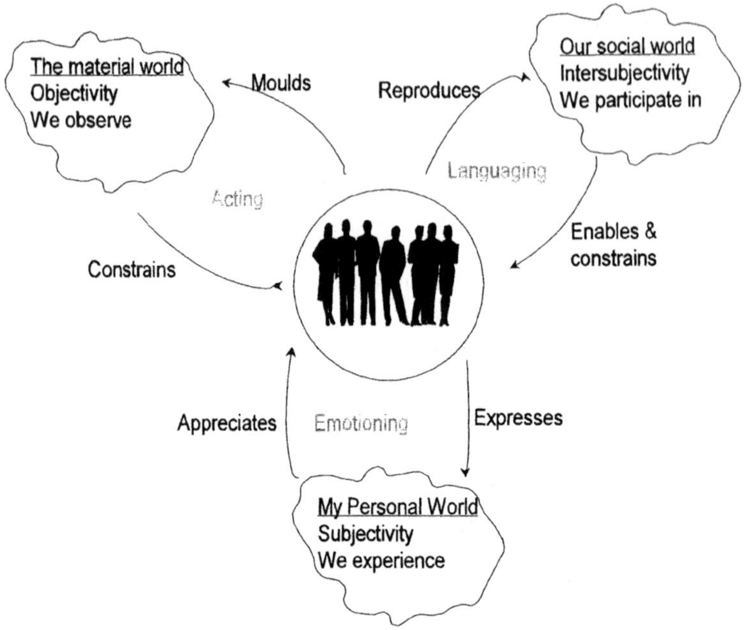

Figure 2. Habermas's Three Worlds

applies in all domains, but that the specific nature of the social world—its human construction and dependence, its self-referentiality, and the impossibility of laboratory-type experiments—alter the generation and validity of social science knowledge.

In fact, following Habermas, we would distinguish three worlds or domains of interactions to which we can relate in different ways (Fig. 2). This framework suggests that it is useful, ontologically and epistemologically, to distinguish three worlds—the material world, the social world, and the personal world.

Each domain has different modes of existence, and different means of accessibility. The material world is outside of and independent of human beings. It existed before us and would exist whether or not we did. We can shape it through our actions but are subject to its constraints. Our relationship to this world is one of observation (rather than participation or experience) but such observations are always theory and subject dependent. We can characterise this world as objective in the sense that it is independent of the observer although clearly our observations and descriptions of it are not, in this sense, objective.

From this material world, through processes of evolution, linguistically-endowed humans have developed capable of communication and self-reflection. This has led to the social and personal worlds. The personal world is the world of our own individual thoughts, emotions, experiences, and beliefs. We do not observe it but experience it. This world is subjective in that it is generated by, and only accessible to, the individual subject. We can aim to express our subjectivity to others and, in turn, appreciate theirs.

Finally there is the social world that we (members of particular social systems) share. Our relation to it is one of intersubjectivity since it is, on the one hand, a human construction, and yet, on the other, it goes beyond and pre-exists any particular individual. It consists of a complex multi-layering of language, meaning, social practices, rules and resources that both enables and constrains our actions and is reproduced through them. One of its primary dimensions is that of power [1].

The main conclusions of this analysis are:

- The social world is emergent from the linguistic and practical activities of human beings.
- It is inherently meaningful—we are participants rather than observers.
- Yet, social structure is external to us as individuals. It is both a precondition and a consequence of our consensual, cooperative activity [23].
- Power is of central importance both positively, enabling our activity, and negatively, constraining us.

Cognition—Embodiment

Having covered the general philosophical questions, we now move up another level to review the individual's cognitive and linguistic abilities. What is the nature of human cognition and what possibilities and limitations does this constitute for our observations and descriptions of the world?

The cognitivist program of representationalism, and indeed much of Western philosophy of mind, has been guided by the Cartesian dualism between mind and body—mind being a disembodied realm of pure ideas. However, the thrust of Varela's (and Maturana's) work is in the opposite direction—showing how our cognition and thought is inextricably bound to our *embodied* selves. This is a reversal that has also occurred in phenomenology, beginning with the difference between Husserl's [24] [25] analysis of pure consciousness and Heidegger's [26] focus on concernful, day-to-day, activity. This trend was continued by Merleau-Ponty

[27] [28], who took phenomenology down to the level of action and perception, emphasising the embodied nature of cognition. In considering perception, and cognition in general, Merleau-Ponty critiqued both what he called empiricist (realist) and intellectualist (idealist) modes of explanation. The empiricist simply takes the world as objectively given and sees cognition as a reflection of the world. Intellectualism recognises that the subject is involved in constituting the experienced world, but is too disembodied and mentalistic. For Merleau-Ponty, cognition is embedded in our body and our nervous system. It is the body which "knows how to act" and "knows how to perceive" on the basis of pre-formed readinesses and habits developed through its structural coupling with the environment. This is what Varela [29] means by "embodiment"—the idea that cognition necessarily occurs through and within our bodily structures which are themselves coupled to biological and social contexts.

Merleau-Ponty also analyses the relationship between the perceiver and the world, which he sees as a reciprocal relationship—the world does not determine our perception, nor does our perception constitute the world.

> "...The properties of the object and the intentions of the subject ...are not only intermingled; they constitute a new whole." When the eye and the ear follow an animal in flight, it is impossible to say "which started first" in the exchange of stimuli and responses. Since all the movements of the organism are always conditioned by external influences, one can, if one wishes, readily treat behaviour as an effect of the milieu. But, in the same way, since all the stimulations that the organism receives have in turn been possible only by its preceding movements which have culminated in exposing the receptor organ to the external influences, one could also say that the behaviour is the first cause of all the stimulations.
>
> ...it is the organism itself ...which chooses the stimuli in the physical world to which it will be sensitive. "The environment (Umwelt) emerges from the world through the actualisation or being of the organism—[granted that] can exist only if it succeeds in finding in the world an adequate environment." (Merleau-Ponty, quoted in [29, p. 441])

This is the basis of Varela's theory of enactive cognition which has two aspects:

- that perception consists in perceptually guided actions;
- that cognitive structures emerge from the recurrent sensory-motor patterns that enable action to be perceptually guided [29, p. 441].

The first point is that perception is neither objectivist nor purely constructivist—*pace* Maturana [30, p. 254]. Rather, it is co-determined by the linking of the structure of the perceiver and the local situations in which it has to act to maintain its self. There can be no fixed point independent of the organism, nor can the organism construct its own closed world. The organism's activity conditions *what can be perceived* in an environment, and these perceptions, in turn, condition future actions. Varela [30] [31 Ch. 8] assembles various neurophysiological evidence for this. For instance, in the area of perception, it is clear that colour and smell are by no means simple mappings of external characteristics. Rather, they are co-creations, dependent on the colour and smell "spaces" constituted by a particular organism's nervous system, and only triggered by external stimulation. Equally, our perception depends for its effectiveness on movement, as shown by Held and Hein's [32] kittens. Two groups of kittens shared the same, artificial, light conditions, but one group were allowed to be active while the other group were passively moved around. When released, the active ones were normal while the passive one acted as if they were blind even though their visual system was unimpaired.

The organism must interact with its environment for its self-continuation and so the question becomes, how does it happen that the world it carves out is one which permits its continuance? The answer lies not in the world, but in the relations between the sensory and motor surfaces of the nervous system. How is it that these are such as to enable effective, perceptually-guided action in a perception-dependent world? This brings us to the second of Varela's points—how action is selected and how the process generates higher cognitive structures.

Our behaviour is seen as a constant switching from one task or activity to another according to our readinesses for action. How is it that one is chosen rather than another? This seems to occur as the result of what might be thought of as a competition between different sub-nets or "agents" in the brain. Brain studies have shown that there are bursts of fast activity followed by more stable patterns as activities stop and new ones start. At each choice-point or breakdown there are many possibilities available, but eventually the historically conditioned structure leads to a selection and a new stability. It is next argued that this dynamic interplay linking sensor and motor activity gives rise to the higher cognitive structures. It does not determine them, but does both enable and constrain the more conceptual and abstract modes of thought. The key here is the emergence of the symbol or sign, and thus language itself, as a new domain of neuronal activity. Other work on language that is important in this context is Piaget [33] and, particularly, Johnson and Lakoff [34] [35] [36].

Intervention/Research—Critical Pluralism

Finally, we move up to the level of conscious and purposeful action, whether it is conducting research or intervening within problematic organisational situations. In particular, we shall consider the discipline of management science which is precisely concerned with such interventions. In recent years, the number of methods, techniques and methodologies within the broad field of management science (to include both O.R. and systems) has burgeoned. There is now an enormous variety of approaches all having very diverse characteristics and stemming from various paradigms based on differing philosophical assumptions. Whilst this plenitude can enhance practice [37] it also poses problems for practitioners who often tend to restrict themselves to one paradigm or even one methodology.

It is the contention of this paper that it is necessary to go beyond using a single (or on occasions more than one) methodology to always combining several methodologies, in whole or in part, and possibly from different paradigms, to make the most effective contribution in dealing with the richness of the real world. I argue for this use of *multimethodology* [38] both on theoretical/philosophical grounds, and on the practical grounds that practitioners are increasingly doing this already. However, mixing methodologies, particularly from different paradigms, does present serious problems—philosophically in terms of paradigm incommensurability, theoretically in terms of effectively fitting methodologies together, and practically in terms of the wide range of knowledge, skills and flexibility required of practitioners. In fact, this paper does no more than attempt to establish multimethodology as a broad interdisciplinary research programme.

It is encouraging to note that similar attempts to move away from paradigmatic isolation and conflict are occurring in other disciplines. In organisational studies there are strong attacks on the paradigm incommensurability advocated by Burrell and Morgan [39] [40] [41] [42]; in sociology a mixture of research methods is increasingly being advocated (triangulation) [43] [44] [45]; and in philosophy there is debate about theoretical and methodological pluralism [46] [47] [48].

The first main reason for adopting a pluralist approach is the multidimensionality of the real-world as already discussed above. Adopting a particular paradigm is like viewing the world through a particular instrument such as a telescope, an X-ray machine, or an electron microscope. Each reveals certain aspects but is completely blind to others. Although they may be pointing at the same place, each instrument produces a totally different, and seemingly incompatible, representation. Thus, in adopting only one paradigm one is inevitably gaining only a limited view of the problem situation, for example, attending only to that which may be measured or quantified, or only to individual subjective meanings and understandings. This argument is a strong one in support of multimethodology suggesting that it is **always** wise to utilise a variety of paradigms.

The second point is that interventions (and research) go through a number of phases and have different task requirements at each stage. A categorisation of different stages has been developed that is neither too general or too detailed, and that can be applied equally to the material, personal, and social aspects of a problem situation.

- i) **Appreciation**—the initial observation and structuring of the situation as a whole as it is entered or encountered by the agent—"here it is".
- ii) **Analysis**—trying to understand and explain the underlying factors that make the situation like it is—"this is why".
- iii) **Assessment**—going beyond what is there or how the situation appears to be towards framing alternatives—"this is how it could be".
- iv) **Action**—helping to bring about or facilitate appropriate change—"lets do it".

With these very general stages of the intervention process it is possible to examine various alternative methodologies and map out their strengths and weaknesses. A third reason is that practice is actually ahead of theory—management science practitioners are mixing methodologies together already to deal with their practical problems.

Ormerod [49] provides an illustration of how one methodology can be used as a framework to incorporate others. He uses Ackoff's Interactive Planning as a framework to incorporate other soft OR and soft systems methods in information systems strategy development. SSM has been used in a similar way in information systems development [50]. Traditional IS methodologies are 'embedded' within the SSM process as a means of better understanding the information needs of users. Alternatively SSM can 'front-end' information systems design methodologies.

The idea of 'front ending' exploits the relative strength SSM in relation to the different stages in the intervention process. Bennett [51], and Bennett and Cropper [52] adopt a similar line in combining conflict analysis, SODA and strategic choice. These authors show how the different methods may be most fruitfully employed at various stages in interventions. In a similar vein, Lane and Oliva [53] illustrate how SODA can help in the problem formulation stage of systems dynamics modelling.

The objective of combining methods to sustain particular purpose is pursued by Gains and Rosenhead [54]. They combine cognitive mapping with SSM in medical quality assurance. Likewise Ackerman and Belton [55] combine SODA and multiple criteria analysis to acquire, organise and make use of corporate knowledge.

To conclude this section, we have argued for the desirability and feasibility of multimethodology, i.e., mixing together a plurality of different methods and methodologies, from different paradigms, within a single intervention. What remains to be developed is *how* this should be done in an effective way. Here we must caution that such a pluralistic approach runs the risk of licensing an "anything goes", purely *ad hoc*, pragmatism. This is not, at all, the intention. Instead we wish to develop a theoretically-informed and *critical* pluralism in which self-reflective agents generate well-informed combinations of methodologies having a critical regard for the effects of their actions on all those involved.

CONCLUSIONS

This paper has sketched a framework for a research programme into the philosophy and practice of management science. The motivation for such a framework was the paradigm split within management science into "hard" and "soft", objectivist and subjectivist, that has only been partially overcome through critical theory. The framework that has been developed is based on an original synthesis of autopoiesis, critical realism, and the theory of communicative action. It addresses four complementary levels—ontology and epistemology, the nature of science, human cognition, and the methodology of intervention and research.

Much further work needs to be done both in linking these areas together more thoroughly, and in developing them, particularly the practical domain of critical pluralism. Ultimately, we engage in theoretical and philosophical reflection not for its own sake but to better inform our thought and practice, being, as we are, both *observers* and concerned, situated and embodied *agents*.

AFFILIATION

John Mingers
Warwick Business School
University of Warwick
United Kingdom

REFERENCES

[1] **Mingers, J.**, "Recent developments in critical management science", *J. of the Operational Research Society*, **43**, 1, 1992, pp.1-10.

[2] **Ackoff, R.**, "The future of operational research is past", *J. Opl. Res. Soc.*, **30**, 1979, pp. 93-104.

[3] **Checkland, P.**, "OR and the systems movement: mappings and conflicts", *J. Opl. Res. Soc.*, **34**, 1983, pp. 661-676.

[4] **Checkland, P. and Scholes, J.**, *Soft Systems Methodology in Action*, Wiley, Chichester, 1990.

[5] **Eden, C.**, Using cognitive mapping for strategic options development and analysis (SODA). In: Rosenhead, J. (ed.) *Rational Analysis for a Problematic World*, Wiley, London, 1989, pp. 21-42.

[6] **Rosenhead, J.**, (ed.) *Rational Analysis for a Problematic World*, Wiley, London, 1989.

[7] **Jackson, M.**, "The nature of 'soft' systems thinking: the work of Churchman, Ackoff and Checkland", *J. Appl. Sys. Anal.*, **9**, 1982, pp. 17-29.

[8] **Mingers, J.**, "Subjectivism and soft systems methodology: a critique", *J. Appl. Sys. Anal.*, **11**, 1984, pp. 85-103.

[9] **Flood, R. and Jackson, M.**, *Creative Problem Solving: Total Systems Intervention*, Wiley, Chichester, 1991.

[10] **Ulrich, W.**, *Critical Heuristics of Social Planning: A New Approach to Practical Philosophy*, Wiley, Chichester, 1984.

[11] **Habermas, J.**, *Knowledge and Human Interests*, Heinemann, London, 1972.

[12] **Maturana, H. and Varela, F.**, *Autopoiesis and Cognition*, Riedel, Dordrecht, 1980.

[13] **Maturana, H. and Varela, F**, *The Tree of Knowledge*, Shambhala, Boston, 1987.

[14] **Maturana, H.**, "Reality: the search for objectivity or the quest for a compelling argument", *Irish J. Psychology*, **9**, 1987, pp. 25-82.

[15] **Mingers, J.**, *Self-Producing Systems: Implications and Applications of Autopoiesis*, Plenum, NY, 1995.

[16] **Bhaskar, R.**, *Reclaiming Reality*, Verso, London, 1989.

[17] **Bhaskar, R.**, *Plato Etc.*, Verso, London, 1994.

[18] **Habermas, J.**, *The Theory of Communicative Action Vol.* **1***: Reason and the Rationalization of Society*, Heinemann, London, 1984.

[19] **Habermas, J.**, *The Theory of Communicative Action Vol. 2: Lifeworld and Systems—a Critique of Functionalist Reason*, Polity Press, London, 1987.

[20] **Mingers, J.**, "Information and meaning: foundations for an intersubjective account", *Information Systems Journal*, **5**, 1995, pp. 285-306.

[21] **Mingers, J.**, Embodying information systems. In: Jones, M., Orlikowski, W., Walsham, G. and DeGross, J. (eds.) *Information Technology and Changes in Organizational Work*, Chapman and Hall, London (in press), 1995.

[22] **Dretske, F.**, *Knowledge and the Flow of Information*, Blackwell, Oxford, 1981.

[23] **Giddens, A.**, *The Constitution of Society*, Macmillan, London, 1984.

[24] **Husserl, E.**, *The Idea of Phenomenology*, Martinus Nijhoff, The Hague, 1964.

[25] **Husserl, E.**, *Cartesian Meditations*, Martinus Nijhoff, The Hague, 1977.

[26] **Heidegger, M.**, *Being and Time*, Blackwell, Oxford, 1962.

[27] **Merleau-Ponty, M.**, *Phenomenology of Perception*, RKP, London, 1962.

[28] **Merleau-Ponty, M.**, *The Structure of Behaviour*, Beacon Press, Boston, 1963.

[29] **Varela, F.**, "Making it concrete: before, during and after breakdowns", *Revue Internationale de Psychopathologie*, **4**, 1991, pp. 435-450.

[30] **Varela, F.**, Whence perceptual meaning? A cartography of current ideas, in: Varela, F. and Dupuy, J. (eds.), *Understanding Origins: Contemporary Views on the Origin of Life, Mind and Society*, Kluwer Academic, Dordrecht, 1992, pp. 235-263.

[31] **Varela, F., Thompson, E. and Rosch, E.**, *The Embodied Mind: Cognitive Science and Human Experience*, MIT Press, MA, 1991.

[32] **Held, R. and Hein, A.**, "Adaption of disarranged hand-eye coordination contingent upon re-afferent stimulation", *Perceptual-Motor Skills*, **8**, 1958, pp. 87-90.

[33] **Piaget, J.**, *The Construction of Reality in the Child*, Basic Books, New York, 1954.

[34] **Johnson, M.**, *The Body in the Mind: The Bodily Basis of Imagination, Reason, and Meaning*, University of Chicago Press, Chicago, 1987.

[35] **Lakoff, G.**, *Women, Fire and Dangerous Things: What Categories Reveal about the Mind*, University of Chicago Press, Chicago, 1987.

[36] **Lakoff, G. and Johnson, M.**, *Metaphors We Live By*, Chicago University Press, Chicago, 1980.

[37] **Rosenhead, J.**, Diverse unity: the principles and prospects for problem structuring methods, in: Rosenhead, J. (ed.) *Rational Analysis for a Problematic World* Wiley, Chichester, 1989, pp. 341-358.

[38] **Mingers, J. and Brocklesby, J.**, Multimethodology: Towards a framework for mixing methodologies, *Warwick Business School Research Papers* **188**, Warwick Business School, Coventry, UK, 1995.

[39] **Burrell G. and Morgan, G.**, *Sociological Paradigms and Organisational Analysis*, Heinemann, London, 1979.

[40] **Willmott, H.**, "Breaking the paradigm mentality", *Organizational Studies*, **15/5**, 1993, pp. 681-719.

[41] **Jackson, N. and Carter, P.**, " 'Paradigm wars': a response to Hugh Willmott", *Organizational Studies*, **14/5**, 1993, pp. 721-725.

[42] **Weaver, D. and Gioia, G.**, "Paradigms lost: incommensurability vs structurationist inquiry", *Organizational Studies*, **15/4**, 1994, pp. 565-590.

[43] **Denzin, N.**, *The Research Act in Sociology: A Theoretical Introduction to Sociological Methods*, Butterworths, London, 1970.

[44] **Bulmer, M.**, *Sociological Research Methods*, Macmillan, London, 1984.

[45] **Blaike, N.**, "A critique of the use of triangulation in social research", *Quality and Quantity*, **25**, 1991, pp. 115-136.

[46] **Roth, P.**, *Meaning and Method in the Social Sciences: A Case for Methodological Pluralism*, Cornell University Press, Cornell, 1987.

[47] **Watson, W.**, "Types of pluralism", *Monist*, 1990, pp. 350-366.

[48] **Rouse, J.**, "Indeterminacy, empirical evidence, and methodological pluralism", *Synthese*, **86**, 1991, pp. 443-465.

[49] **Ormerod, R.**, "Putting soft OR methods to work: information systems development strategy at Sainsbury's", *J. Opl. Res. Soc.*, **46**, 3, pp. 277-293.

[50] **Stowell, F.**, (ed.), *Information Systems Provision: The Contribution of Soft Systems Methodology*, McGraw-Hill, London, 1995.

[51] **Bennett, P.**, Mixing Methods: Combining Conflict Analysis, SODA, and Strategic Choice, in: Eden, C. and Radford, J. (eds.), *Tackling Strategic Problems: The Role of Group Decision Support*) Sage, London, 1990.

[52] **Bennett, P. and Cropper, S.**, "Uncertainty and conflict: combining conflict analysis and strategic choice", *J. Behavioural Decision Making*, **3**, 1990, pp. 29-45.

[53] **Lane, D. and Oliva, R.**, The Greater Whole: Towards a Synthesis of SD and SSM, In *Proceedings of the International Systems Dynamics Conference*, Stirling, 11 - 15 July, 1994.

[54] **Gains, A. and Rosenhead, J.**, "Problem structuring for medical quality assurance", *London School of Economics Working Paper* No. **93.8**, LSE, London, 1993.

[55] **Ackerman, F. and Belton, V.**, "Managing Corporate Knowledge Experiences with SODA and VISA", *British Journal of Management*, **5**, 1994, pp. 163-176.

YOAV BEN-DOV

SCIENCE AND THE HUMAN: COMPLEMENTARITY AS RECONCILIATION

1. INTRODUCTION

In this paper, I try to sketch in general lines the idea of complementarity, which was first suggested by Niels Bohr as an answer to the conceptual problems of quantum mechanics, but later extended by him to other domains, both within science and outside it. In these applications, the idea of complementarity allows for the acceptance of different sets of beliefs and principles, which contradict each other but have all their value. In particular, it may help to ease the conflict between the scientific approach and human aspirations.

2. COMPLEMENTARITY

The idea of complementarity was first introduced by Bohr in 1927, in a lecture given at Como, Italy, and repeated at the fifth Solvay conference in Brussels [1]. Its aim was to come to terms with the new and conceptually problematic situation in atomic physics. On one hand, the new quantum theoretical formulations introduced by Heisenberg and Schrödinger in 1925-26 could account with impressive exactness for the atomic experimental results. On the other hand, these formulations could not be cast in a way that clearly identifies the quantum object as something which exists by itself, independently of the manner in which we choose to observe it. Thus, the new quantum mechanics fails to satisfy one of the basic tenets of physics, and science in general, as these terms were understood from the 17th century onwards: it does not give an objective description of an observer-independent reality. For Einstein, this was a good enough reason to reject the new quantum mechanics, and to look for a better theory (which, in spite of much effort, was never found). For Bohr, on the other hand, this situation called for the re-formulation of the scientific concept of "objectivity", or, in other words, for a re-evaluation of the relation between human language and reality. However, this re-formulation was perceived by him as pertaining not only to the understanding of atomic physics, but also to much broader issues.

In the original Como lecture, Bohr applied the idea of complementarity to the possible modes of description of an atomic electron. On one hand, we can in some cases know precisely the energy of an atomic electron, and apply concepts of conservation of energy to it. But in order for these concepts to be valid, the atom has to be kept isolated from any external influence. On the other hand, we may wish to have a space-time account of the detailed motion of the electron inside the atom. But in order to give such an account, we have to perform a position measurement, for example with the aid of a gamma-ray photon which collides with the electron. But now the atom is no longer isolated, and the atomic energy is defined only up to a relatively large degree of uncertainty. The space-time description thus excludes the energy-conservation description: both are possible, but not at once. In this sense, the two descriptions are "complementary": they exclude each other when applied simultaneously, but only their conjunction gives the complete story about the atomic electron.

Up to this point, one can understand Bohr's idea in terms of a classical image, in which the electron "really" has both a well-defined space-time location and a well-defined energy, but we cannot know them together because of the perturbation caused by the act of measurement. But Bohr later clarified his position, and stated explicitly that it is not a question of a mechanical disturbance of the particle by the measurement, but something deeper, which concerns the very definition of the physical phenomenon. One cannot speak of "the quantum object" as something which exists independently of its act of being measured, but rather one should give a full account of the experimental set-up which actually was implemented [2]. This description can be formulated only in classical terms. However, no single description in classical terms can exhaust all the knowable features of the quantum phenomenon, so that we have to use "complementary" descriptions. In other words, the quantum phenomena exhibit a richness and complexity which surpass our ability to give a full account of them in the language accessible to us. Instead of trying in vain to look for a better language, we should accept that any description of the quantum world is partial, and be ready for the possibility that different descriptions, which seem to be incompatible with each other, are all correct. A logical contradiction does not arise in practice, because at any given moment, only a single description is accessible. But this means that no description will be taken as "objectively true", independently of the practical context in which it applies.

3. APPLICATIONS OF COMPLEMENTARITY: QUANTUM PHYSICS

In the domain of quantum mechanics, Bohr applied the principle of complementarity not only to the different possible descriptions of the electron's state, but also to the different possible answers to the question "what is the electron". The electron, like any quantum object, behaves in experiments sometimes like a particle and sometimes like a wave. The classical concepts of particle and wave exclude each other. The wave is a physical entity which extends over space, and two waves can annihilate each other through interference, when the crest of one falls on the trough of the other. In contrast, a particle is located at each moment at a specific point in space, and two classical particles can never annihilate each other. The application of these concepts together to the same electron thus leads to a contradiction.

Again, one could look for a new language, in which terms the electron will be described using a single well-defined concept. But Bohr thought that such a language is not required. Physicists should not look for alternatives to the classical language. Instead, they should examine the circumstances in which descriptions formulated in classical terms are used. It is possible to construct an experimental arrangement in which the electron behaves like a particle, and another arrangement in which it behaves like a wave. But these two experimental arrangements cannot be realized simultaneously. The particle and the wave descriptions are thus applicable each in its own experimental arrangement, and never together in the same arrangement. Since in Bohr's view it is not possible to define the physical phenomenon outside its experimental context, the question "is the electron really a wave or a particle" is meaningless.

The statements "the electron is a particle" and "the electron is a wave" therefore stand in a relation of complementarity: excluding each other when applied at once to the same object, but giving together, along with the specification of their corresponding experimental arrangements, the complete account of the electron's entity.

One could, of course, ask about the relation between these two applications of complementarity. For an answer, it is important to note that complementarity is not a feature of the quantum object (although its description in physics textbooks sometimes gives the impression that it is). Rather, it is a general point of view which concerns the relation between language and the objects which it describes, or more exactly, between different descriptions in human language of what is taken to be the same object. Further, it is perhaps not even essential that complementarity always should involve only two descriptions and not more—

although Bohr did not discuss such cases. Thus, the two applications of complementarity described here are not directly related, and it is "the same" complementarity only in the sense that they both reflect the use of human (classical) language to describe quantum phenomena.

Bohr's approach can be extended to other formulations of the conceptual problems of quantum mechanics, for example to von Neumann's [3] formulation, which serves as a starting point for some of the new interpretations. In this formulation, the state of the electron is described by a quantum wave function. For this mathematical function, two modes of temporal evolution are assumed: one when no measurement is performed, and the other during a measurement. The coexistence of these two modes is usually considered as a serious problem, for it implies that a measurement is different from an ordinary process in which the electron is undergoing some physical interaction. It is therefore necessary to include the concept of measurement, which implies a reference to the observing agent, in the definition of the electron, which was supposed to be a physical object existing to itself. To solve this difficulty, von Neumann treats the measurement device as a quantum object, which has its own wave function. But now it has to be assumed that the wave function of the device evolves differently when nobody is looking at it and when somebody is looking, so that the (implicit) reference to the observing agent still remains.

Bohr did not share von Neumann's account, because he believed that the measuring device should be described in classical terms, so that no wave function is assigned to it. And as to the fact that the definition of the measured object includes a reference to the measurement procedure, as we saw, Bohr accepted it as unavoidable. Against Bohr's position one can claim that the measuring apparatus is made of atoms, which are quantum objects, so that the complete device should be describable in quantum terms of a wave function. But Bohr could point out that in order to demonstrate the quantum behavior of the atoms in the measuring apparatus, one should construct an experiment in which this apparatus serves as the object measured by another device. In such a case, we can treat it as a quantum object, but now it no longer serves as the measuring device. The role of the apparatus as a measured object and its role as a measuring device thus occur in different and incompatible experimental arrangements. Therefore, the description of the apparatus as a quantum object with a wave function, and its description in classical terms as a measuring device, are complementary, and it is meaningless to ask what the apparatus "really" is.

4. APPLICATIONS OF COMPLEMENTARITY: BIOLOGY

Bohr did not think that the principle of complementarity is limited to the new quantum theory. As he saw it, atomic physics was not even the basis of this principle, but only its clearest exemplification. Complementarity is a general epistemological principle, which should apply to all the domains of human thought and activity—from the hard sciences of physics and biology to the study of man, society and morality. Bohr tried to show the relevance of complementarity to problems in all these domains. As he insisted, these extensions from physics to other domains are not superficial analogies, but something fundamental.

Some of these extensions are perhaps related to his own background. His father, Christian Bohr, applied in his physiological researches the mechanistic view of the biological organism as a physical machine. But as he claimed, a description of the living creature in vitalistic terms, in which "life" is an irreducible concept, is also necessary, although not in the same observational context. The reason is that in order to study the living creature in mechanistic terms we have to dissect it, and for this we have to kill it. As Pais [4] points out, Bohr spoke explicitly about the need to apply the principle of complementarity to the mechanistic and the vitalistic descriptions of living creatures until the discovery of the structure of DNA in 1953. Then he changed his terminology, and spoke of a complementarity between a mechanistic description in terms of active factors and a finalistic description in terms of goal and purpose.

Regarding this position of Bohr's, one can claim that within the domain of biology, the application of the complementarity principle between a mechanistic and a finalistic description of the living organism seems unnecessary. The theory of evolution makes it possible to use finalistic expressions like "the eye is made for seeing", and still to regard them as short forms for a mechanistic description in terms of embryology and natural selection. The relation between the finalistic and the mechanistic accounts can therefore be a reduction, in which one of them is considered to be more fundamental then the other, and not complementarity, in which both have the same status. Thus, biologists can apply in their studies any combination of mechanistic and finalistic considerations, while their basic beliefs remain purely mechanistic. Clearly, Bohr did not think of such a "complementarity of practice", but of something more fundamental. After all, even Einstein was ready to admit that in actual practice, the physicist is an "opportunist" as regards conceptual systems, using whatever serves his needs at any moment.

If the principle of complementarity has any place at all in the description of living organisms, this place may lie outside the strict domain of biological research, and concern the perception of the living creature as a subject with which one can sympathize, identifying oneself with its feeling of purpose. For example, one can perhaps regard the description of a cat as a physical mechanism that can be dissected for research purposes, and its description as a living creature which suffers on the dissection table, as complementary. The key point here is the existence of consciousness in the living creature. As Bohr put it [5]:

"The fact that consciousness, as we know it, is inseparably connected with life ought to prepare us for finding that the very problem of the distinction between the living and the dead escapes comprehension in the ordinary sense of the word. That a physicist touches upon such questions may perhaps be excused on the ground that the new situation in physics has so forcibly reminded us of the old truth that we are spectators as well as actors in the great drama of existence".

5. APPLICATIONS OF COMPLEMENTARITY: MORALITY

The idea of a language that is imposed on us by our modes of communication but does not reflect reality as it is, as Bohr describes the classical language, bears a resemblance to the ideas of Kant. Bohr did not mention Kant as a direct influence on him. However, at the time when Bohr got his education, Neo-Kantian ideas were very much in the air, and it is difficult to assume that Bohr could have completely avoided their influence. Kant indeed argues in his *Critique of Judgment* that a complete mechanical description of living creatures are impossible, and that finalistic considerations should also be applied to their study. However, as we saw, Bohr's treatment of living creatures is problematic as long as consciousness is not mentioned, and consciousness can be ascribed with certainty only to humans. Therefore, even if there is a resemblance here between Bohr and Kant, in concerns a rather weak point.

Concerning humans, however, it is perhaps possible to say that Kant applies different kinds of considerations in a complementary way. We can regard the human person as a phenomenon, and study it with the scientific tools of physics, chemistry and biology. In this case, its description is limited by the principles laid down in the *Critique of Pure Reason*: we do not know another person as a thing to itself but only the phenomena related to him, and therefore the causal laws that we discover reflect the structure of our system of perception as much as they reflect the properties of the human object observed. On the other hand, as Kant states in the *Critique of Practical Reason*, each one of us perceives his own mind as a thing to itself, without the mediation of sense perception. In this self-perception, the contents of mind appear as finalistic. The same person can therefore be described in terms of two systems of concepts—the causal description as an object, and the finalistic description as a moral agent. In each of these systems, his laws of conduct obey different constraints: in the scientific description all his actions are determined by the causal laws of classical mechanics, while in the moral description, we have to assume the existence of free will. The antinomy between the assumption of determinism in classical mechanics (or the randomness in quantum mechanics) and the assumption of free will is thus solved, without a need to settle any contradiction: each one of these assumptions is valid in a different description of the same person, and the correct description at any moment depends on the kind of questions that we ask.

However, there is an important difference between Bohr and Kant. For Kant, the description of the person "from within," as he perceives himself, is completely subject to rational considerations, and Kant derives the principles of morality from such considerations. His analysis is abstract, and does not concern any feeling of human empathy. Bohr, on the other hand, describes the moral dilemma which arises in a concrete human situation, for example between the need to punish a beloved child and the tendency to forgive him [6]:

"Through the closest possible combination of justice and love presents a common goal in all cultures, it must be recognized that in any situation which calls for the strict application of justice there is no room for the display of love, and that, conversely, the ultimate exigencies of a feeling of love may conflict with all ideas of justice".

Since a similar dilemma arises in other circumstances, it is possible to say that Bohr uses here the word "love" in a somewhat Christian sense, as a kind of Grace. The feeling of love in this sense enables one to identify with the internal experience of the other person, and to feel a sort of communality with him. In contrast, the principles of justice are objective, and belong to the description of the person "from without". Thus, Bohr puts the duty to act according to some general rules in the external and objective description of the person, and human empathy in the internal experience, as complementary components of what can be called "morality". Kant, on the other hand, assigns the concept of duty to a moral imperative which is supposed to belong exclusively to the rational description of the person "from within".

6. COMPLEMENTARITY AND PHILOSOPHICAL MONOTHEISM

In the previous sections, we discussed several applications of the concept of complementarity to different domains, as suggested by Bohr. As can be appreciated from these applications, complementarity reflects a general epistemological approach, which can be applied on several levels. For example, within atomic physics, complementarity is applied to different descriptions of the quantum object. In a broader context of physics in general, Bohr mentioned the possibility of applying the atomistic and the energetistic description to thermodynamical systems [4]. And in a still broader framework, complementarity can be applied to the relation between the physical and the mental-teleological descriptions of living beings. This again stresses the point mentioned above, that complementarity is not a feature of "reality", but rather of the language used to describe reality.

The idea of complementarity thus runs counter to a very old tradition in western thought, which predates the scientific revolution. In fact, already in the 13th century one can find in this tradition the central idea that reality (or, if we prefer, "truth") can be given a single, complete and unambiguous description in human language. The fact that the content of this description has undergone very radical changes, for example from the Aristotelian-theological dogma of the late middle ages, to the mechanistic world-picture of the 17th-19th centuries, is secondary in this respect. The alternatives to mechanism reflected a similar tendency: for example, German idealism in the 19th century rejected physical materialism along with mechanism, but still tried to find the single set of principles which would account for all existence, for example Hegel's dialectical evolution of the spirit through history. One could continue this trend to 20th-century positivism, which rejected metaphysical accounts of "reality" as meaningless, but still looked for a single set of methodical principles which would generate the collection of all meaningful statements about the world. In this sense, as Henri Atlan calls it [7], western culture has kept a basic "monotheistic" obsession throughout its intellectual evolution.

The idea of complementarity suggests a different approach, much more modest, which is perhaps the only way to cope with the conflicting accounts of reality which today cannot be denied their claim to validity. One should not look for the single "True" account of reality, which should be cast in human language (either the existing language, or a new ideal language, as the positivists tries in vain to formulate). Instead, one should ask, for any given description of reality, what is its domain of useful application. In quantum mechanics, this domain is defined by the experimental set-up, in a more or less unambiguous way. In other domains, the question of useful applicability can be more difficult to answer. However, it is of the essence of the approach of complementarity that no clear-cut answers can be generated in advance to all possible questions (because there is no universally valid set of principles to generate these answers). One might therefore be called to make, in specific cases, a choice which is not dictated by a general set of principles. It is exactly in this choice that human existence as a moral agent is manifested. In this sense, complementarity reflects not only the ability to reach a reconciliation between conflicting views of reality, but also our reconciliation with the finite and limited nature of any form of human knowledge.

7. SCIENCE AND THE HUMAN

As an example, we can try to apply the concept of complementarity to a debate which cuts through the academic world, and consists of the relative priority of the scientific and the humanistic disciplines. The fact that such a debate exists, and that it is sometimes expressed in virulent terms, needs hardly to be pointed out: One could follow, for example, C.P. Snow's [8] concept of "the two cultures" to its recent manifestations in some scientists' attacks on the humanistic disciplines as "higher superstition". However, it is today clear that hostilities along the dividing lines between the disciplines can be dangerous and harmful, in a situation in which the central problems facing humanity involve both scientific and human aspects.

As long as we insist on the (supposed) existence of a single set of methods and principles which regulate all valid knowledge, a peaceful and productive coexistence between scientific and humanistic disciplines is impossible. The reason is that the two domains (Snow's "two cultures") rely on radically different basic assumptions. One of the central points of this cleavage involve the use of teleological considerations. In natural science, teleology is explicitly rejected. In the humanistic disciplines, in contrast, teleology is indispensable: one cannot arrive at any understanding of human affairs without recourse to concepts like volition and purpose. And in particular, one cannot formulate any acceptable precepts of morality without using such finalistic terms.

As we saw, Bohr's use of the concept of complementarity allows us to apply different sets of concepts, without asking which one is "absolutely true". In this issue, one should not ask whether there "really" are purposes in the world. Instead, one should ask in which domains the use of finalistic terms is productive. Therefore, scientific and humanistic approaches can coexist in the same cultural framework, without the need to choose one of them as the only "true" one.

Behind the mechanistic-finalistic distinction lies, of course, a deeper issue, that of the mind-body question. Ever since Descartes' separation of these two substances, western philosophy has been involved in a hopeless attempt to determine, once and for all, which of the two "really" exists, or, if both exist, what is the exact relation between them. In our terms, the body (brain included), which is made of matter, can be studied by the natural sciences, which involve no finalistic concepts. On the other hand, the most important features of the subjective experience of mind appear to itself as finalistic: I want, I choose, I hope. But the distinction between mind and body is not between two substances which exist separately in the world: in the living human, they always come together. Thus, we can regard the mind-language and the body-language as two different descriptions of the same human entity, to be applied in different circumstances. The basic assumptions of these two languages are contradictory, and therefore, they cannot be applied at once. but each one of them is applicable in a different context. Therefore, the relation between mind and body can be regarded as complementarity between two different descriptions, and not as some mysterious connection between two distinct substances.

The interesting questions arise, of course, on the borderline between the two languages. For example, at which point does a collection of cells become "alive"? More specifically, if we speak of a human embryo, it is sometimes necessary for practical and very definite purposes to decide whether or not it should be regarded as a human being. As mentioned above, an approach based on complementarity will not be able to give a clear-cut answer, which should be valid in any circumstances. What we have is two different languages—in one of them, we have a biological object, and in the other a human being—and none of them can determine where the borderline between their respective domains of applicability exactly lies. It is in such circumstances that we have to rely on our moral autonomy, and make a choice which is not dictated by a previously given set of principles. In other words, in such cases we cannot have recourse to anything external—God, science, categorical imperative or whatever—to relieve us from our moral responsibility. Thus, at least in this example, and maybe also in others, the approach of complementarity does not give a general solution to problems which should perhaps be left without.

AFFILIATION

Yoav Ben-Dov
Department of Philosophy
University of Tel Aviv
Israel

REFERENCES

[1] **Bohr, N.**, "The quantum postulate and the recent development of atomic theory", *Nature*, **121**, 1928, pp. 580-590.

[2] **Bohr, N.**, "The causality problem in atomic physics", in: *New Theories in Physics*, Nijhoff, the Hague, 1939, p. 11.

[3] **von Neumann, J.**, *Mathematische Grundlagen der Quantenmechanik*, Springer-Verlag, Berlin, 1932; English Translation: *Mathematical Foundation of Quantum Mechanics*, Princeton University Press, Princeton, N.J., 1955.

[4] **Pais, A.**, *Niels Bohr's Times In Physics, Philosophy and Polity*, Clarendon Press, Oxford, 1991.

[5] **Bohr, N.**, Naturw. **18**, 1930, p. 73, Cited in Ref. 4.

[6] **Bohr, N.**, in: *Studia Orientalia Ioanni Pedersen*, Munskgaard, Copenhagen, 1953, p. 385.

[7] **Atlan, H.**, *A tort et à raison*, Le Seuil, Paris, 1986.

[8] **Snow, C.P.**, *The Two Cultures*, Cambridge University Press, Cambridge, 1959.

VLADIMIR KUZNETSOV

CONDITIONS AND FEATURES OF UNITY CONCEPTS IN SCIENCE

1. UNITY CONCEPT IN TERMS OF SCIENTIFIC KNOWLEDGE SYSTEMS

There are many phenomena in different areas of human intellectual life that have been subsumed under the concept of unity and its relatives: integration, generalization, synthesis, unification and so on. Numerous authors have lobbied for one or another variety of the unity within traditional scientific disciplines (in mathematics, physics, psychology, etc.). Other have argued for the unity between the different natural sciences (for example, physics and biology) or between the natural sciences and mathematics. Still others have focused on forms of the unity between science and humanities. It seems that the scrutinizing these unity forms is a precondition of considering any meaningful form of the unity between science and art, science and other spiritual types of human activity.

The concept of unity may be investigated with reference to various phenomena and at different levels of generality and formality. Keeping in mind set-theoretical or/and categorical foundations of mathematics, one may speak of, correspondingly, the set-theoretical or/and categorical unity of mathematics. One may also speak of the unity of contemporary physics created by the ideas of relativity, complementarity, and self-organization. In these cases the unity is treated in terms of some particular concepts and ideas. It gives a room for scientific-like speculations that are sometimes very interesting and plausible, especially when they have been produced by outstanding experts in the field.

There is also a possible alternative to such a consideration of the unity concept in science. It is connected, first, with taking into account not the whole science, but only scientific knowledge. Second, it recognizes as some empirical fact that scientific knowledge has been split in various and interrelated systems. These have not been remained stable and constant. The development of science has been producing new knowledge systems as well has been leading to elaborating old ones and establishing fruitful and original relations between different systems. The supersystem of scientific knowledge systems is in a permanent state of flux. The unity here may be treated as some property defined on pairs of elements from this supersystem. From such a point of view the unity concept in science may be investigated in terms of scientific knowledge systems and relations between them.

To put it more definitely, one of the important forms of the unity in science is connected with the consideration of common structures that two distinctive knowledge systems possess. Examples of such unity may be found among different theories of classical physics; the nonrelativistic quantum mechanics and wave mechanics, quantum relativistic mechanics and special relativity theory; current programs of unification of theories of fundamental physical interactions; different models of concepts in contemporary cognitive psychology, etc.

It is necessary to note, that considering the unity concept in terms of scientific knowledge systems would be more efficient if one explicitly uses a certain methodological model or reconstruction of scientific knowledge systems. The modern methodology and philosophy of science has produced many such models. Among them are cognitive [1], erotetic [2], standard [3], semantic [4], structuralist [5], structure-nominative [6], etc. They put forward different pictures of the nature and composition of scientific knowledge systems. This has immediate consequences for methodologically possible modes of the unity of various knowledge systems.

Indeed, if one separate a specific list of structures in knowledge systems, then it means that she or he may speak of their unity in more concrete and detailed terms. Simply stated, if knowledge systems include a single type of structure, then one may describe their unity only in terms of this type. If knowledge systems include two different and interrelated types of structure the number of possible modes of unity grows essentially. As a first approximation, we may have two types of partial unity connected with either types of structures. Then, we have also the type of unity associated with both structure types. In each case there are some constraints on the unity arising from ties between both types of structure.

As compared with others, the structure-nominative reconstruction has separated inside of scientific knowledge systems much more various types of interrelated structures. From this it follows, that it has provided more opportunities for the analysis of the concept of unity of scientific knowledge systems and for the classification of possible types of unity.

2. THE STRUCTURE-NOMINATIVE VIEW OF COMPOSITION OF SCIENTIFIC KNOWLEDGE SYSTEMS

Let us suppose that we consider a variety of functions realized in scientific practice with the help of a scientific knowledge system (SKS). Among them are the naming (assigning special and manageable names), attribution (imposing properties and relations), description, explanation, classification, prediction, understanding, generalization (establishing patterns of behavior and changes) of entities, events, states, and processes in the domain of a SKS, etc. An SKS must have some structures by means of which these functions have been fulfilled. Including only logical structures in SKSs is not sufficient for their realization.

The very idea of an SKS presupposes that its structures are ordered in a way. Further still, there are many kinds of these structures and many methods of their ordering. On closer examination one may conclude that these structures create a huge system with many hierarchies and subsystems. The numerous case-studies of mathematical, physical, economical and social SKSs have shown the justifiability of such an approach [6].

It is possible to separate five subsystems at the highest hierarchy level of an SKS. These are logico-linguistic, model-representing, problem-heuristical, pragmatic-procedural subsystems and a subsystem of links between them. Their careful study shows that any of them may be represented in others only partially. These subsystems are not totally reducible to each other.

All kinds of knowledge associated with an SKS are expressed and logically ordered by means of its logico-linguistic system. Among resources of this system are concepts, alphabets of languages, word construction rules, vocabularies of assertoric languages, expression construction rules, assertoric languages of an SKS, selection rules for basic expressions. Evaluating some of them as true, we obtain a system of initial propositions (axioms, laws, postulates, principles) and enter to the logical part of a logico-linguistic subsystem. It includes also a system of rules of transformation (deduction, inference, argumentation) of initial propositions, and derivative propositions (statements, consequences, theorems).

With only a few exceptions, practically all experts in the current philosophy and methodology of science and many philosophically minded scientists identify an SKS with its logico-linguistic subsystem. Only adherents of the structuralist approach connect systematically an SKS with the kind of structure different from logico-linguistic ones. The elaboration and further development of their arguments led to the separation of a model-representing subsystem inside an SKS. Its constitutive structures are abstract properties and models. To a first approximation, the model-representing subsystem includes levels with the following subsystems. These are systems of: names of supposed entities from the domain of an SKS; names of properties and relations of entities from this domain; scales for properties and relations between domain entities; abstract properties as models of properties and relations of entities; model scale of an SKS; model forms of the representation of domain regularities; hierarchies of such forms.

In addition to these subsystems of an SKS, there are reasonable arguments for separating others. Indeed, numerous case-studies of SKSs (as they exist in scientific monographs, textbooks, papers, manuals and lectures) show that an SKS also includes other kinds of constitutive structures above concepts, languages, propositions, properties, models, and laws. These are problems and tasks specific for a given SKS, methods and operations of problem solving and model building. An important part of an SKS is associated with its various axiological structures. Among these are traditional logical estimations like the truth of propositions and the completeness of their system. Above this there are variety of other estimations as the adequacy of models, the complexity of problems, the universality and efficiency of methods used, the beauty of a whole SKS [7]. These constitutive elements are naturally distributed between the problem-heuristical and pragmatic-procedural subsystems. In turn, either of them consists of two subsystems.

The problem-heuristical subsystem of an SKS contains problem and heuristical systems. The basic constitutive structures of the first are problems, questions, puzzles and tasks as well solutions of problems, answers to questions, guess-works about puzzles and results of fulfillment of tasks. The basic structures of the second are hypotheses, ideas, effective but not well grounded assumptions.

It should be stressed that any set of structures associated with a given SKS alters during its development. For a problem-heuristical system it means that employing any nontrivial SKS, one is able not only to solve some known problems, but also to formulate and resolve new and original ones.

The constitutive structures of an SKS mentioned above may be called constructive or static ones. The introduction of so-called dynamic structures is caused by the following reasons. In scientific practice an SKS exists as a large and complex system of transformations, actions and operations with constructive elements. Possessing and applying an SKS, scientists are doing something with concepts, language expressions, models, problems, etc. In the framework of a given SKS these actions are not arbitrary and chaotic. Each kind of constructive structure corresponds to a specific system of operations over them. Operations over concepts include abstraction, idealization, combination, etc. Operations over propositions contain their truth estimation, deduction of some propositions from a certain set of them, confirmation, falsification of propositions, etc. When one deals with properties he or she connects operations of determination of property values with them. In an empirical SKS the realization of this determination frequently presupposes the use of the calculation and measurement operations and procedures.

There is a widespread tendency of reducing all kinds of operations and methods associated with an SKS to deductive and inductive ones. However, besides them, there are many other kinds of operations and methods in it. For example, the methods of solving mathematical equations in any SKS with mathematical apparatus are not reducible either to deductive or inductive ones.

The operations appropriate to the various constructive structures of an SKS are the main constitutive structures of the operational subsystem of a pragmatic-procedural subsystem of an SKS. It also contains the descriptions of these operations (procedures and algorithms), the rules of construction of new operations from the known ones, etc. It may be shown that methods are systems of some basic operations and the rules indicating the sequence of their fulfillment.

The second system of the pragmatic-procedural subsystem is an axiological one. Its main constitutive structures are estimations and evaluations of other structures of an SKS. Examples are the effectiveness and universality of methods used in an SKS, the practicality of scientific theories [7].

There are many important and poorly investigated relationships within any subsystem mentioned above. Along with this, all subsystems of the highest hierarchy level have many interdependencies and ties. By way of example one may take links between the logico-linguistic and model-representing subsystems. Languages used in the former essentially determine the types and nature of models that may be constructed by means of these languages. If a scientist uses only ordinary language, then she or he is not able to construct rather sophisticated and experimentally testable models. We mention in passing that (from the point of view of ties between the logico-linguistic and model-representing subsystems) the importance of mathematics in SKSs is connected not only with a need of conducting complex calculations and statistical processing of experimental data. Mathematical languages are also used for constructing models that could be experimentally tested. Only after such testing the models may be provisionally estimated as more or less adequate depiction of the domain of an SKS.

Similar considerations lead to the separation of the subsystem of links among other four subsystems of the highest hierarchy level of an SKS.

3. THE UNITY AS A FUZZY DICHOTOMOUS PROPERTY

Having rather detailed reconstruction of SKSs, we need some formalization of the concept of unity between them. Usually this concept has been considered as dichotomous one. It means that pairs of SKSs subsumed under it have been treated as possessing the unity or not. In reality there are many types of the unity between SKSs as well many degrees of the unity of a certain type. For instance, it seems that the types of unity between two mathematical theories are different from the types of unity between sociological theories. Moreover, the degrees of the definite type of unity between mathematical theories may depend on the pairs of theories chosen.

First of all, we consider a way of expressing an idea about the unity degrees. Then, we propose an idea of the unity sorts connected with different measures of the unity.

The unity under study may be studied as some abstract property [8; 9] defined on pairs of compared SKSs. It allows us to speak of the unity not only in terms of its presence or absence, but also in terms of its degree.

In more formal way, any sort of the unity of SKSs may be defined as follows.

Definition 1. The unity **U(S)**of the sort **S** of SKSs is a triple (**M**, **p(S)**, **L(S)**) where **M** is a set of all possible pairs of SKSs, **p(S)** is a unity function of the sort **S** and **L(S)** is a scale of the unity of the sort **S**.

More specifically, **L(S)** is a set of values of the unity of the sort **S**, **p(S)** is the function that assigns some value $\mathbf{v} \in \mathbf{L(S)}$ to a given pair of SKSs. For dichotomous concepts of unity, one takes the set $\{0, 1\}$ as the scale **L(S)**. The assigning of the value **0** to the pair of SKSs means that there is no unity between them. The assigning of the value **1** means that there is the unity between them.

It seems more realistically to take the interval **[0,1]** as the scale **L(S)**. In this case we may give some numerical characterization of the degree of unity between SKSs and may compare various degrees.

The unity function **p(S)** assigns some numerical value from **[0,1]** to any element from **M**. It allows us in principle to compare the degrees of unity between SKSs. For example, the SKS(1) may have the numerical value of unity equal to **0,8** with SKS(2), while the appropriate value may be equal to **0,5** for SKS(3). The way of determining value of **p(S)** on a given pair of SKSs is depended on the definition of the measure of unity at hand. Under this treatment it is naturally to interpret the value **(1 - p(S))** as the degree of deviation of SKSs.

Definition 2. The deviation **D(S)**of the sort **S** of SKSs is a triple (**M**, **d(S)**, **L(S)**) where **M** is a set of all possible pairs of SKSs, **d(S)** is a deviation function of the sort **S**, **L(S)** is a scale of the unity of the sort **S** and **d(S)** = **1 - p(S)**.

Let SKS(1), SKS(2) and SKS(3) be three SKSs. Their possible coupling includes pairs designated as (1,2), (1,3) and (2,3). We can assign or determine definite values of **p(S)** for these pairs: **p**(1,2), **p**(1,3) and **p**(2,3) and analyze various numerical relations between them. It is interesting to investigate how different measures of unity are comparable with our intuition about the unity of knowledge.

There are variety of possible measures of unity connected with the appropriate definitions of the unity function and the scale of unity. Our ideas about the compositions of SKSs play here the crucial role. Each of the possible measures introduces a specific sort of the unity between SKSs. Some of them may be found below.

Definition 3. The measure of nonspecific structural unity of two SKSs is the ratio of the number of their common structures to the sum of numbers of all their structures.

Definition 4. The measure of specific structural unity of two SKSs in relation to specific sort **S** of their structures is the ratio of the number of their common structures of this sort to the sum of numbers of all their structures of this sort.

In many cases it is more convenient to discuss the unity of SKSs in terms of sorts of their structures, but not in terms of their structures themselves. It gives the nonspecific and specific sortal measures of unity.

Definition 5. The measure of nonspecific sortal unity of two SKSs is the ratio of the number of their common sorts of structures to the sum of numbers of all their sorts of structures.

Definition 6. The measure of specific sortal unity of two SKSs in relation to specific sort **S** of their structures is the ratio of the number of their common subsorts to the sum of numbers of all subsorts of this sort.

These definitions are rather flexible and their application depends on many conditions. First of all, it is an objective splitting SKSs into subsystems and structures. Then, our ability to recognize and identify these components should be mentioned. These conditions are correlated with each other. However, there may be various contradictions between them. For example, an SKS may have very rich hierarchies, but we could not recognize their existence.

The deepness of our possession of SKSs is the next condition. Usually only a few scientists are well aware of both SKSs compared. It is possible that the deeper understanding of SKSs may reveal the unity lacking in the case of their superficial treatment. At least, but not last, one should mention the objectives during the study of unity of SKSs. They extend from general talking about SKSs to attempts to unite them in order to solve real scientific problems.

Because there are many reasonable measures of the unity, the problem of their compatibility and consistency appears. Its preliminary analysis revealed many inconsistencies in the common scientific intuition about the unity of SKSs. The results of this analysis will be given in the future author's papers.

It should be noted that to achieve the fuller presentation of the unity between two SKSs we may use the construction of composition of the measures of unity mentioned above. This construction is a partial case of composition of abstract properties [8].

Besides the idea of different measures and sorts of the unity between SKSs, we introduce an idea of types of this unity. This idea deals with internal splitting of SKSs.

4. THE STRUCTURE-NOMINATIVE CLASSIFICATION OF UNITY TYPES

One may use the structure-nominative reconstruction as some methodological frame of reference for discussing the unity concepts. From stated above about the structuring of an SKS it follows that one may speak of, at least, two major presuppositions of the unity of two SKSs.

First, they should reveal all their subsystems in distinctive form. In this sense all SKSs demonstrate the unity connected with their systematic and structural similarity.

From such a point of view, it is rather difficult to separate any nontrivial mode of possible synthesis of art and science or scientific and common knowledge. Under trivial modes we mean the usage, for example, the same alphabet of natural language and a huge list of the same words from its vocabulary by many SKSs and literary texts. These kinds of the unity may be called, correspondingly, the alphabetic and lexicographic. However, the proper specifics of SKSs began from the usage of artificial languages, especially mathematical languages with their symbolic alphabets and vocabularies of terms. As a rule, these languages have semiotic, syntactical and semantic rules that are different from those of natural language. Nevertheless, they may intersect. It gives the natural grammatical unity between SKSs and written and spoken arts.

Second, SKSs under comparison should possess about the same level of development of their corresponding subsystems. There is no sense to consider the unity between mathematical category theory with its highly developed logico-linguistic and procedural subsystems and some social theory that used ordinary language in its logico-linguistic subsystem and elementary rules of argumentation in its pragmatic-procedural subsystem.

Generally speaking, one may introduce the following principal types and subtypes of the unity of SKSs.

The logical-linguistic unity.

The compared SKSs possess the logico-linguistic unity when they share some structures from their logico-linguistic subsystems. The main subtypes of this unity are conceptual, language, logical and axiom unity. We may speak of them when SKSs share some structures from, correspondingly, conceptual, language, logical and axiom systems of SKSs. Let us consider briefly some illustrations. We may speak of the conceptual unity between geometry and most scientific theories that using the concept SPACE. Classical field theory and wave theory of light possess the language unity arising from the use of the language of the theory of differential equations in partial derivatives. Some scientists say that there is the logical unity between all classical theories because these theories use logical systems with two truth values. According to this view, the logical unity of quantum theories arises from their common many-valued logical systems. We may speak of the axiom unity of intuitionistic and non-intuitionistic logical systems having in mind that they share many axioms but not the law of the excluded middle.

The model-representing unity.

The compared SKSs possess the model-representing unity when they share some structures from their model-representing subsystems. The main subtypes of this unity are nominal, attributional, model and nomic unity.

The nominal unity takes place when compared SKSs use the same names of objects under study. There is a great degree of nominal unity between atomic physics and chemistry, especially in the field of naming atoms, chemical elements and their atomic structure.

The attributional unity emerges between SKSs by means of which the "same" properties are studied. In spite of all discrepancies of classical and relativistic mechanics these theories have a great value of attributional unity. This follows from their usage of such properties of investigated objects as mass, momentum, energy.

The model unity appears when SKSs use some common models. There is the model unity of most contemporary theories of elementary particles due to their usage of quark model.

The nomic unity appears when SKSs share some laws and principles. Practically all contemporary natural sciences share the law of energy conservation. There are varieties of modern theories of fundamental physical interactions that share the same supersymmetry principles. In both cases we may speak about the nomic unity of corresponding SKSs. In the last case the laws of SKSs may be different and simultaneously may obey the same supersymmetry principles.

The pragmatic-procedural unity.

The compared SKSs possess the pragmatic-procedural unity when they share some structures from their pragmatic-procedural subsystems. The main subtypes of this unity are operational, methodical and axiological unity.

The operational unity is connected with a usage of the same operations by SKSs. Having in mind such operations as abstraction, generalization, idealization, one may speak of operational unity of science and humanities. The subtypes of this unity are connected with using arithmetical, geometrical, topological operations, etc.

The methodic unity is connected with an application of the same methods. All contemporary natural sciences use experimental methods and methods of measuring. On this ground one may introduce the concept of experimental and measuring methodic unity of some SKSs. The methods of solving differential equations give the way for the corresponding subtypes of the methodic unity.

The axiological unity is associated with sharing the same collection of estimations by different SKSs. The best-known estimation is the truth evaluation. If we agree that one objective of any SKSs is the truth attainment, then we may assert the truth unity of science in the whole. The extraordinary important roles in the social sciences and humanities play some hermeneutic estimations. Among them is a requirement to provide the understanding of reality under study. In view of this, one may consider many types of the hermeneutic unity between various social and humanistic knowledge systems.

The problem-heuristical unity.

The compared SKSs possess the problem-heuristical unity when they share some structures from their problem-heuristical subsystems. The main subtypes of this unity are problem and heuristical unity.

The problem unity is associated with common problems formulated and resolved by SKSs compared. There are many kinds of problem generality. If we agree that one of the main problems of science is to obtain adequate and testable explanation and prediction of phenomena studied, then we may speak of the appropriate problem unity of all SKSs. On more concrete consideration, problems are separated by nature (mathematical, physical, social, spiritual, etc.) and forms (precise, formal, fuzzy formulated, etc.). Taking this into account, one obtains appropriate subtypes of the problem type of unity. In spite of widespread opinion even highly formalized mathematical and physical theories use many plausible assumptions, efficient but not well grounded methods, some general and empirically unverifiable ideas, heuristic argumentation, etc. In this sense there are many subtypes of heuristical unity between all SKSs.

Studying the real life cases of the unity of SKSs, one may meet practically all possible combinations of these basic sorts and types of unity. Frequently researchers concentrate only on some of them. This generates different opinions about the unity of any given pair of SKSs, because researchers consider only one or two unity aspects. Applying different concepts of unity, they may obtain different conclusions about concrete realization of the unity. For instance, keeping in mind the nature of mathematical concepts and basically deductive character of mathematical argumentation, some eminent mathematicians find only a low degree of unity between mathematics and natural science. In contrast to them, others point out the fact of using mathematics in physics and conclude that there is a great degree of unity. This may be explained as a result of the application of the different unity concepts.

5. UNITY AND UNIFICATION OF PHYSICS

Let us consider briefly the problem of unification of physics from the view point of the above analysis of the unity concept.

Almost always this problem is investigated as follows. It is supposed that in the future it will be possible to invent such an axiom and law system from which one may obtain axioms and laws of any other physical knowledge system as particular cases. This program rests obviously only on two types of unity: axiom and logical ones. Even though the building of this system will be possible, this would lead only to the partial, but not to the total unification of all physical knowledge systems. These contain not only common problems, models, methods, etc., but also specific ones. This means that the unity of the appropriate types with degrees close to 1 is practically unattainable. The lack of the complete reduction of any subsystem of SKSs to other means that all types of unity are irreducible to the axiom and logical unity.

The unification itself is not an ultimate goal of science development. It is only one of the tools of this development. It strengthens human power to solve new and more complicated scientific problems. It also gives us the interplay of different methods that leads to inventing more powerful and universal methods. It also reveals deeper and more fundamental structures, patterns and regularities of nature. It creates more effective and universal forms for nature understanding. From this point of view, one may conclude that only with the help of the axiom and logical unification, this tool cannot be put into action.

AFFILIATION

Vladimir Kuznetsov
Center for Philosophy, Logic and Science Theory
University of Munich
Germany

REFERENCES

[1] **Giere, R.N.**, *Explaining Science. A Cognitive Approach*, Chicago University Press, Chicago, 1989.

[2] **Meyer, M.**, (ed.), *Questions and Questioning*, De Grueter, New York, 1988.

[3] **Suppe, F.**, (ed.), *The Structure of Scientific Theories*, 2nd ed., The University of Illinois Press, Urbana, 1987.

[4] **Suppe, F.**, *The Semantic Conception of Scientific Theories and Scientific Realism*, University of Illinois Press, Cambridge, 1988.

[5] **Balzer, W., Moulines, C.U. and Sneed, J.D.**, *An Architectonic for Science*, Reidel, Dordrecht, 1987.

[6] **Burgin, M. and Kuznetsov, V.**, *The Axiological Aspects of Scientific Theories*, Naukova Dumka, Kiev, 1991 (In Russian); *The Nomological Structures of Scientific Theories*, Naukova Dumka, Kiev, 1993 (In Russian); *The Introduction to the Exact Methodology of Science*, Aspect, Moscow, 1994 (In Russian); "Scientific problems and questions from a logical point of view", *Synthese*, **100**, 1994, pp. 1-28.

[7] **Burgin, M. and Kuznetsov, V.**, "A Formal Aesthetic for Scientific Discourse. The Beauty Measures of a Scientific Theory", in: **Harré, R.**, (ed.), *Anglo-Ukrainian Studies in the Analysis of Scientific Discourse. Reason and Rhetoric*, Lewinston, Edwin Mellen Press, 1993, pp. 69-93.

[8] **Burgin, M.**, "Abstract Theory of Properties", in: **Smirnov, V. and Karpenko, A.** (eds.), *Non-Classical Logics*, Moscow, Institute of Philosophy, 1985, pp. 109-118 (In Russian).

[9] **Burgin, M. and Kuznetsov, V.**, "Properties in science and their modeling", *Quality & Quantity*, **27**, 1993, pp. 371-382.

PHILIP L. PETERSON

BEYOND 'EINSTEIN', 'MAGRITTE', AND 'QUINE'

I

Einstein was the discoverer of spatial-temporal relativity. Let "Einstein" (Einstein's name surrounded by quotes) be the name of him and his achievements together with other revolutionary scientists of the 20th century and their equally surprising discoveries (such as Bohr, Schrödinger, Heisenberg, Dirac, Fermi, etc. through S. Hawking). So, when I say that "Einstein" is a 20th century scientific achievement of a very peculiar nature, I am not talking just about Einstein, but also his results plus other noteworthy scientists and their results. Analogously, I will use Magritte's name with quotation marks added to refer not just to Magritte and his artistic achievements, but also to many other modernist 20th century artists and their achievements (e.g., Picasso, Duchamp, Pollock, Dali, but also Calder, Schönberg, Escher, Le Corbusier, Bausch, Maplethorpe, and many others).

By extending the labels "Einstein" and "Magritte" in this way, I can more efficiently express some general claims. My first claim is that "Einstein" has produced a permanently relativistic and indeterminate (even probabilistic) conception of what the physical world (which may be the only reality) is. Secondly, I claim that there is a mirroring of this event of scientific cognition in "Magritte"—especially, if we attend to the more exotic and/or obscure elements of this artistic era (e.g., Dali' s paintings, not to mention Magritte's, Cage's music, or performance art). For on one way of thinking about "Magritte" (not just Magritte), there is a kind of relativity and indeterminacy about what an artwork is—and what is "in" it (seemingly anything at all in the visual arts, music without discernible melody, apparently any movement being a dance).

My second contention is simply an extrapolation on Dali's famous "*The Persistence of Memory*". (Myself, I prefer to use Magritte's "*Time Transfixed*" for the same purpose.) It can be taken to represent a malleability to what is quite simply not malleable at all—time. But this couldn't be, could it? It's a nightmare (or science-fiction fantasy) of a nearly deranged artistic personality, isn't it? Maybe. But maybe not. Perhaps it's a more literal representation than one might think (even if still metaphoric), since time itself has turned out to be not so determinately fixed and unbending as we previously assumed. I assume that this interpretation of Dali's painting can be generalized to cover all the achievements of "Magritte" to some degree. They can be taken to mirror, at least loosely, the unraveling of absolutes and determinacies of our most basic sciences. How this mirroring came about historically is something I won't go into. Rather, my interest is, first, to make the link between "Einstein" and "Magritte" rather more explicit than this sketch. Then, after making it more explicit (via "Quine" below), I will show how to "overcome" it—or move beyond it.

20th century Physics, all by itself, inspires metaphysical relativity. (Thus, there is no need to make a dubious inference from some popular moral relativity to a matching metaphysical relativity.) That there is no privileged inertial system (frame of reference) seems to be the ultimate extension of Galileo's/Copernicus' dismantling of an earth-centered (meta)physics. Not only is the earth not the center (nor is there any reason for taking anything or any region to be spatially central to the natural universe), but the idea of a spatial-temporal framework itself being privileged has evaporated. This alone, even without the indeterminacies from quantum mechanics, is sufficient to produce a revolution in metaphysics. No philosopher was even needed. Surprisingly, we did have artists giving us visual (and other, e.g., atonal) images associated with this metaphysical revolution—most notably the Dali painting mentioned above. Some might take that to be a nightmarish image of a conceptual level horror, suggesting time itself could be bent or melted as the clock in it is. Of course, some others may not have found this image, or the metaphysical implications from physics which it is an image of, to be so horrible. A Zarathustra, for example, might be thrilled by the liberation from spatial-temporal fixedness.

Now extend the name "Quine" parallel to the two extensions above. Let the name surrounded with quotes label not just Quine and his results, but also many other 20th century philosophers and their results—results which fit together to some degree with Quine's (e.g., from Davidson, Rorty, Kuhn, Putnam, and some others—though not, by my lights, from Kripke who remains more in Aristotle's ambit than Quine's). This extension allows me to say things such as that Kuhn's concept of incommensurability of concepts across paradigms is as central a component of "Quine" as Quine's indeterminacy of radical translation, both contributing to a "Quinean" (as well as Quinean) metaphysical relativism (or ontological indeterminacy). "Quine" (perhaps even Quine alone gives us the explicit means for making the "Einstein"-"Magritte" link (i.e., the 20th-century science-art link) very explicit and transparently understandable. For "Quine" has raised to the level of language—and to the level of any concrete system of understanding or thought capable of expressing it—the new relativistic and indeterminate facts which are apparently the basic facts of nature (or the physical world). At bottom, physical nature has no specific foundation—of microparticles, or of spatial-temporal framework for any phenomena. Reality, if physics is to be believed (and why not believe it?), is not at bottom specifically this or that, or here or there. It's not just that we can't tell what the most basic physical foundations are. They aren't there to be told. Now, nature, or the natural physical world, exists with its characteristics (relative or absolute, determinate or not), but we can only understand it, or seek to rationally believe propositions about it (or simply have thoughts about it), by descriptions and explanations expressed in a language (or in some such system embodying, or amounting to, a conceptual system akin to a natural language). Quine (himself) transferred the kind of relativity and indeterminacy found in what we use languages to report, describe, and explain (such as the achievements of "Einstein") to the languages themselves. This transfer parallels applying relativistic and quantum mechanical effects to instruments and modes of measurement used (i.e., language as an instrument used) in demonstrating the relativistic and quantum mechanical effects in the first place—an application any "Einsteinian" would not object to, but rather expect.

Quine's final result—the heart of "Quine" (cf. Rorty's or Putnam's dogmatic mouthing of the demise of the analytic-synthetic distinction)—is ontological relativity, via semantic indeterminacy. The semantic indeterminacy (cf. [12] [13] [7]) results from the permanent opacity of meanings themselves. If meanings did exist and were non-circularly usable, then we could, at least in principle, determine analyticities, synonymies, and translations. That, in turn, could lead to semantic determinacy "at home" at least and the possibility of defending a determinate ontology for the language we speak and think in. But, according to Quine [10, Ch. 1 & 2], meanings don't exist. Further, if meanings did exist, we couldn't use them. For there is no non-circular or non-stipulative way to individuate them. They are like (or identical to, I would say, cf. [8]) universals that are attributes, unlike universals that are classes which we can individuate (as in [10]). So, referential indeterminacy follows from semantic indeterminacies of languages (in radical translation or at home, cf. [12] [13]). Referential indeterminacy is ontological indeterminacy (relativity). For since the kinds of things we say or think we are referring to are always revisable (e.g., with proxy-functions, [12]), we cannot be sure what in particular (the content, as well as its referents) we are saying or thinking IS.

How does this Quinean "Quine" illuminate the "Einstein"-"Magritte" connection? On one way of looking at paintings, for example, they can, and often do, give us visual representations of not just some immediate (or potential) realities—landscapes, people, human events, activities, and disasters, etc.—but also representations of our ideas, ideals, hopes, and fears about nature and ourselves. Our ideas or thoughts about love and desire can be portrayed in one lofty and refined way as feminine beauty by Botticelli in "*The Birth of Venus*". And our fears and horrors be depicted in another way by Munch in "*The Scream*". So, more than just the visual appearances of things and persons shows in the paintings. Our concepts and ideas, and our emotional and sensual feelings can show. Often, it seems that beyond the phenomenal appearance of things, it is mainly emotions and feelings (pain and suffering, for example; cf. Gericault's "*The Raft of the Medusa*") that in some way show in Western World paintings. I think the same is true though to lesser and more difficult extent, in arts like architecture and music—to a greater and easier to see extent in performance art.) What about ideas, concepts, and understandings we have—aside from emotional content or connection? How do they come into art? Do they come in at all? Yes. Recall, the touch of God in Michelangelo's "*Creation of Adam*" (Sistine Chapel). The point of that is deeply metaphysical (a theological version of the creation of humans). And consider B. Shahn's "*The Passion of Sacco and Vanzetti*" (legally and morally interesting as well as personally and socially saddening). So, a lot of artwork does encompass non-emotional, non-sensory ideas in one way or another. (The moral virtue of courage in heroism was encapsulated by Beethoven's "*Eroica*", wasn't it?)

Now to "Quine". Quine's concepts, ideas, or hypotheses are (in the main) ideas about what ideas, concepts, and hypotheses ARE, as well as what the realities are which instantiate, or are covered by, them. And the heart of "Quine" is Quine's conclusion that ideas, concepts, or other entities to the same ends do not exist at all. There are only languages, and their uses. And the languages are entirely inadequate for making clear what particular things (and kinds of things) there really are which we refer to with (and think about via) their expressions. That is, Quine's idea is that there aren't really any ideas and so, what ultimately exists (things, stuff, stages of things, numbers, anything you like) can't be learned—not because it's so difficult, but because "it" (ultimate reality) isn't anything in particular. Quine's solution—just a new form of traditional American pragmatism (which I consider to be merely "happy-face" skepticism, contrasting with the "foul mouth" skepticism of much postmodernism)—is practical. Choose whatever terminology, or conceptual frame-of-reference that "works for you", and you will be as right or knowledgeable as is ultimately possible. For there are no ultimate metaphysical "facts of the matter" to be right about. Dewey would be proud (which is no revelation—cf. [12], Ch. 2, first paragraph).

What visual art, if any, could mirror this kind of philosophical relativism? It certainly would be hard to say before 1900. But in this century, it has turned out to be easy. Here Magritte himself comes to the forefront of "Magritte". This surrealist clearly shows us how to produce an artistic creation that importantly encapsulates "Quine" (and maybe the specific ontological relativity of Quine himself). Look at "*The Natural Graces*" (1960). Are these birds that just look like leaves? Or vice versa? Does it matter? A leaf might BE a bird. Really? Yes, depending on how you look at it—or, rather, think about it via your analytical hypotheses for "bird" and "leaf". For how you think about it just amounts to the language you use (to think in, as well as to express your thoughts in). This object, or kind of object, I am viewing—or viewing a representation of in "*The Natural Graces*"—can be a leaf, if my analytic hypotheses (for the paraphrase lexicon of my language "at home") are chosen to "see" (think) it so. But there is nothing ultimate in that. Another lexicon (set of semantic hypotheses) is equally possible. That object, under the alternative lexicon of ideas and images, is now a bird. Which is it, really? There is no answer to that. For there are no such ultimate facts of the matter in terms of which it could be decided. Realities are relative to one's semantic (imagistic) hypotheses. A mountain (or immense rock) could be a cloud—as in "*Castle in the Pyrenees*" (Magritte, 1959). "It" (what we are gazing at and/or thinking to be portrayed) easily admits of either interpretation. And neither interpretation, or any alternatives (compatible with all empirical observations of our speech) to be discovered, are "the" ultimate truth. There are no ultimate truths or absolute metaphysical facts. Neither Einsteinian explanations nor Quinean arguments are needed to simply introduce the idea of "no fact of the matter" responses to some questions about what is ultimate or "really real". For consider this pencil on the table in front of me. Is it in motion? Or at rest? Well, relative to me, this table, and this room, the pencil is at rest—not in motion. However, relative to the sun, it is not at rest but (like me, the room, and the table) in both rotational and elliptical motion. If someone, after hearing these answers, persists and asks, "But which is it really or ultimately—in motion, or at rest?", then the right (and apparently only) response is that there is no such fact of the matter as the pencil being really at rest (or really in motion) except relative to the room, the earth, the sun, or some other arbitrarily adopted spatial temporal framework. So, we don't really [sic] need "Einstein" as basis for Quine's extension to language. Any relativistic phenomenon (such as the ordinary relativity of rest and motion) can serve as the source for analogistic extensions of "no fact of the matter" judgments (e.g., to what a predicate "really" means).

Perhaps my point can be most easily grasped by pretending that Quine's indeterminacy of translation thesis was produced by reflecting (perhaps unconsciously) on Magritte's famous "*Ceci n'est pas une pipe.*" On one way of interpreting Magritte's picture the sentence "in" it has a grammatical subject ("ceci") which is referentially ambiguous—on one reading it refers to the picture of a pipe (in which case the sentence is true, since a picture of a pipe is not a pipe) and on another reading it refers to the pipe which the picture depicts (in which case the sentence is false, since that object is a pipe—or, let's presume so here). Quine, then, thought as follows. How should I translate a similar French sentence—e.g., "Cette pipe est brune" (when said of an actual pipe I have, not a picture of one or of one depicted)? There seem to be alternatives for what "Cette pipe" might be used to refer to depending on alternative translations of the French word "pipe". For example, I could translate French "pipe" with English "pipe" and French "est brune" with English "is brown". But I could also translate the French "pipe" with the English "stage [or image, or aspect] of a pipe" and French "est brune" with "is a stage [or image, or aspect] of a brown thing". On either way of translating it, we get an English sentence (two different ones, which are seemingly about quite different things—pipes vs. stage [images, or aspects] of pipes) which is true under the same conditions as the alternative translation (or might easily be—especially if we allow for alternative translations in the same spirit of associated sentences with similar words in them). But given that this IS possible—mutually exclusive alternatives supported by appropriate adjustments in closely related sentence translations—which is the "right translation" ? Is it the one that gets the French MEANINGS of the words right so that those meanings are accurately expressed in English (so that, finally, we can figure out what we are talking about—pipes, pictures, stages, images, etc.)? To this question Quine gives a surprisingly Magrittish reply: You can take it either way you like (and there are probably more than just these two alternatives). Just as the leaf can be a bird—if you choose (or simply find yourself) "seeing" it that way—or just be a leaf (in Magritte's painting), so French "pipe" can MEAN what the English "pipe" does (i.e., be *translated* that way), *or not!* But there is no necessity here—or guiding "fact of the matter". French "pipe" might well be translated by "stage of a pipe" (and so "really MEAN" that), if appropriate adjustments (the root of Quine's later proxy function ploy) are made in associated translations (as just illustrated). THEN, to the question "But which is it REALLY?", Quine gives the relativistic answer (more typical of *pre*"Einstein" views on the relativity of rest and motion than of "Einstein" itself—viz., there is no thing which French "pipe" REALLY means, *no fact of the matter!* It only has a meaning and translation relative to a manual of paraphrases (in French) and translation (French-English). (Can anyone prove that Quine did *not* get his indeterminacy thesis in this way?) And this conclusion applies in retrospect to Magritte's "*Ceci n'est pas une pipe.*" What REALLY does "Ceci" in it refer to? There's no determining fact of the matter to answer that question. The right response is "Choose!"

The utility of Quine's philosophy for understanding Magritte's art in particular (as well as *vice versa*) is my model for how all of "Quine" illuminates all of "Magritte". (Of course, I don't expect every instance of "Magritte"—e.g., atonal music—to be quite so easy to "quintify" over.) The linking of "Magritte" to "Einstein" via "Quine" consists of two steps. First, "Quine" is an adoption of the first-level physical relativity and indeterminism ("first-level" since the physical phenomena are what the so-called "object-level" of language is used to describe or apply to) at the "second-level"; i.e., the object-level relativities and indeterminacies (the ideas of them occurring) are raised to the language itself in which they are described. Thus, not only the what's-described-and-discussed possesses crucial relativities and indeterminacies, but the means for discussing it—language itself—does too. Second, this linguistic indeterminacy—which results in a replacement of words for ideas and concepts leading to a referential indeterminacy amounting to ontological relativity—is encapsulated rather directly in the artworks of surrealism (surrealism in particular, and in other 20th century art forms analogously). "Quine" is a vital means—maybe the "ultimate and absolute" [sic] one—for seeing what surrealism (and the rest of) "Magritte" really is. Of course, you can hear the "Quineans" and "Magrittites" groan, "There isn't anything this art, or this philosophical theory, REALLY is. The lesson of both is to forget about satisfying such absolutist cognitive and metaphysical longings. Rather, choose your concepts as you like (from alternative systems compatible each with confirmed observations even if incompatible with each other). (As my criticisms below will suggest, I don't think the recommendation to choose your *concepts* validly follows for "Quineans".) For upshot of Quine's philosophy is that there simply do not exist any to choose. (Does this "simply choose" echo Sartre? And, is the Sartre echoed part of the artistic era? Or of "Quine"? Or both?)

II

A Zarathustra (as I suggested above) might find this relativity and indeterminacy—at the philosophical as well as scientific level, mirrored by the surprisingly creative artistic explosions of 20th century "modernisms"—to be absolutely [sic] liberating. There's no need to overcome it, or go beyond it, because it frees every thing, phenomena, or idea from being eternally fixed and sacrosanct. I, however, am not that kind of Zarathustra. "Einstein", "Magritte", and "Quine" ("EMQ") do seem to me to be a disappointing, even gloomy result. They do need overcoming. But how? My answer is that on closer inspection, "EMQ" is not what it appears to be. Some dismantling and re-evaluation—though not of the (self-)deceptive "de-constructive" sort—of the relativisitic/indeterminate picture of reality will provide the materials for going beyond it.

First, "Einstein" does NOT show every physical thing, event, or fact to be relative or indeterministic. That is a pervasive misunderstanding. (The misunderstanding is akin to discovering that despite the fact that there are notable moral differences between cultures, still (i) there is not a culture (nor could there be one) for each logically possible moral position or point of view and (ii) not every moral value or virtue (e.g., courage, wisdom) is found to be absent or reversed in some culture or other (i.e., some moral universals do exist).) The lesson of "Einstein" is that certain facts that we might expect to obtain, just don't—and among these are some we might have previously thought to be fundamental, absolute, and/or basic. For example, the fact that for any object at a particular time, it is absolutely in motion, or else at rest, has a certain mass or energy, and/or stands in the relation of simultaneity (or not) to others. Or, that each particle (even photons), at a point in time has both a specific position and a specific momentum. Absolute facts like these don't obtain. However, OTHER absolute facts do obtain. It's easy to mechanically produce one from any debunked alleged absolute of the pre-relativity type. Just add the qualifying conditions which undermined the absoluteness of the candidate fact to that fact to produce a new conditional fact (e.g., IF object a's velocity is measured by using the earth's inertial system as the frame of reference, then object a at time t has velocity v). Some of these conditional facts are absolute. And if any such proposed one, f, is found to depend on further conditions (where the proposed one is already conditionalized, of the form "if C, then O"), we can make it into an improved candidate by adding the further conditions, C', to it (to obtain a new candidate for absolute fact f': if C& C', then O). "Einstein" provides no motivation for thinking that some (at least) few absolute facts cannot be obtained this way. To logical atomists, this will be disheartening, since it would then appear to be possible that some basic, absolute facts of the physical world are not even potentially atomic facts. For these are inherently molecular (compound) facts—e.g., conditionals, or equivalent disjunctions, etc. That is admittedly surprising, but it does not motivate any retreat to "Quine". It may well be that some (possibly even all?) absolute facts are somehow disjunctive—of the form "p V q", but not true because just one or the other *alone* of "p" and "q" is (which logical atomism requires). Disheartening as this may be, still it is not incompatible with there being some absolute facts of nature.

Secondly, for absolute facts of nature we may have to admit that some of them are generic. That nature consists of physical objects in relative motion (possibly inherently very disjunctive objects themselves in that they are either matter or energy complexes or wave-particles complexes). Exact specificity of what such material objects are at a specific time and place (motion-wise, or particle-wave-wise) may be impossible. But that does not at all mean they don't (or might not) exist at all. Physical objects (in the face of macro-relativity and microindeterminacy) still EXIST—i.e., generically (without absolute specificity) they do (e.g., the earth, me or my body, the lunar lander still resting on the moon, elephants!, etc.). It is an absolute (non-relative, non-indeterminate) fact that physical objects exist. One might think of this genericness as a variation on conditionalizing alleged absolute facts to produce genuine absolutes that are conditional or disjunctive in form. But whether or not that is so, still some generic facts can turn out to be nonrelatively (determinately) the case.

Third, there are new and different facts which are absolutes, ones previously unsuspected and which are troublesome to use in any traditional way as a new foundation for physical reality and our knowledge of it. One absolute fact, in particular, stands out—that c (186,000 miles/sec, the speed of light) is constant across alternative inertial systems. There is an absolute for you! Granted it's new and it's difficult to see how to reconceive the physical world as based upon it (as foundational). But that, apparently, is what we must do (*metaphysically,* as well as physically)..

Fourth, about "Quine", I will point out just one mistaken feature of it (of Quine, in particular, though I am sure analogues can be found in Rorty, Kuhn, and others)—its explanatory vacuity. (For additional discussion of Quine's failure, SEE APPENDIX below.) "Einstein" rejects absolute simultaneity, any privileged inertial system, determinacy of absolute properties of particles (simultaneous position, momentum, etc.), pure-particle-hood or pure-waviness to photons, etc., because doing so has great explanatory value. Quine, however, does not reject a host of analogous propositions about language. Rather, he rejects just one—that determinate meanings exist. And the rejection does not lead to a fruitful new theory or picture of the phenomena in question (knowledge of one's language and/or use of it to communicate detailed information)—one which explains something. Quine the contrary. His rejection leads only to more and more relativities and indeterminacies, resulting finally in absolute [sic] ontological relativity. This nihilistic conclusion doesn't explain anything (except what follows *validly* from the assumption that meanings don't exist). In fact, Quine's ontological relativity is tantamount to never being able to have determinate explanations (relativistic or not) of anything—since conceptual systems in terms of which explanations are expressed and understood are radically indeterminate and can always be replaced with incompatible alternatives which are such that "no fact of the matter" exists that could make one or the other of the alternatives true. But we do have successful explanations in "Einstein" today! So, any *philosophy* of scientific explanation which ends up showing that they don't exist has to be mistaken. The relativism and indeterminacy of "Quine" prohibits facts—e.g., that c is absolutely [sic] the same across inertial systems—from obtaining which we know (or have scientific reasons of the most fundamental sort to believe to be) actual facts.

III

So, there are no reasons to think that "Quine" is either well supported by "Einstein" or is more plausible (or nearer the truth) than at least some other philosophical alternatives. There are errors and mistakes in the interpretations "Quineans" give to "Einstein", just as there are errors in the popular beliefs about "Einstein"—e.g., that the resulting space-time relativity has something to do, in the end, with moral or cultural relativity.

What now, then?

(1) Does this leave us with no "world view" encompassing "Einstein" and "Magritte" as "Quine" appeared to.

(2) If "Quine" is a failure and mistake (through vacuity and inept mimicking of "Einstein"), does this destroy the explication of the "Einstein"-"Magritte" link?

(3) Do we, thereby, leave "Magritte" hanging—seemingly unsupported and/or inexplicable.

My answer is "no" to all three of these questions. I'll explain them in reverse order.

"Magritte" is not left hanging, for it is no less important than it always was. For even without "Quine", there is a noticeable, if only metaphorical, connection of "Magritte" to "Einstein". And metaphors aren't nothing (nor useless). Quite the contrary, they (and other creative analogies) are usually at the heart of surprising proposals, discoveries, and artistic and scientific creations. Further, about (2), we do not lose the use of "Quine" for an explanation of the "Einstein"-"Magritte" link. We still have it. I urge that it is the correct explanation. What is incorrect about "Quine" is not its explicit utility for linking "Einstein" and "Magritte". That (like the validity of the forms of Quine's arguments) is correct. "Quine" does illuminate "Magritte" *vis-à-vis* "Einstein"—and even *vice versa.* But as for "Quine" *tout court* —that is mistaken (mainly via Quine's vacuity; see APPENDIX).

Finally, about (1), without "Quine" as the philosophical moral of 20th century science and art, are we left with no encompassing "world view". ("Hurray", we can hear Zarathustra shout. Or did he cackle, as Russell was alleged to?) Shouldn't we be just as gloomy and disappointed without "Quine" as with it?

My answer, again, is no. No, because scientific realism fills the vacuum. Perhaps a label other than "scientific" (can't you hear the postmodernists *contest* (as they say) it with "Positivism"?) for the required variety of realism is needed—such as "ultra-realism", "neorealism", or "super-realism". On the other hand, we may already have the needed label—"surrealism". The kind of realism required—which I shall prosaically deem "scientific"—is already present in (and at least implicitly adopted by) both "Einstein" and "Magritte" (though doubtless this needs to be shown). My diagnosis of the recent (though not unprecedented) panicky retreat to pragmatism, skepticism, and nihilism (such as by postmodernists and "Quineans" like Rorty) is that it was caused by the juvenile hope that knowledge could be God-like and reality perfect. But God (in all respects) IS dead.

There are *as many* errors and mistakes in the general understanding (by and among philosophers, as well as others, today) about what such a scientific realism is as there were in (i) the popular, commonsense interpretations of what "Einstein" is and implies and in (ii) the interpretation by "Quineans" of the lessons of "Einstein" and of Quine himself. On (ii), for example, one correct lesson to be learned about Quine's ontological relativity is that it is a *reductio ad absurdum* on the premise that meanings don't exist—difficult to individuate or not.

Here is just one example of the many errors and mistakes about scientific realism found among contemporary philosophers (especially among "Quineans", but also surprisingly in Sartre too). (Incidentally, I don't identify scientific realism with naive realism, even though they share something—viz., common-sense knowledge. But, of course, the triumphs of scientific realism are the achievements—first, historically, in engineering (building, metallurgy, medicine) and later in the hard sciences of the modern and contemporary eras—which exceed commonsense knowledge and belief though remaining built upon it.)

It is a mistake to think that realism requires a transcendental, noumenal world to exist and be (somehow) the cause of our knowledge and/or what it is about, while at the same time being in itself cognitively inaccessible. This preposterous (though widespread) view of realism is the result of manipulations of terminology by skeptics (often Anglo-American empiricists, but Sartre too), as well as by Kant's counterproductive "foundations" for ethics. Rather, realism is the view that nearly all knowable (thus *not* cognitively inaccessible) reality is in itself absolutely independent of its being known by anyone at any time or necessarily being part or an aspect of a knower or the achievement of knowledge. (The exceptions are facts like that I know I exist, which is not just the fact that I do exist.) In sum, reality is knowable (in part at least), but it is not caused or determined to exist by the knowing of it (even if perplexing problems arise like the uncertainty principle in physics). So, within a scientific realism, the transcendental, noumenal view of reality and knowledge is just not possible. Kant, long ago, demonstrated that. As far as I can see, it is Kant's unfortunate choice of the term "idealism" in his phrase "critical idealism", together with his entirely separate (and dubious) moral theory, which led interpreters away from the correct understanding of his philosophy—that wherein its realism shines through. For an objective reality exists in Kant's view, one which makes our beliefs true or false. It is simply a mistake (even incoherent) to set that reality up as permanently unknowable—even though its independence cannot be doubted. It's a mistake because we have some knowledge to start with (and we know we do)—e.g., of the facts that $5 + 7 = 12$, that I exist, that sugar is sweet, etc.

The complaint will be that such Kantian-styled knowledge is not of an independent-to-knowers (or of their judgments-beliefs-or-cognitive achievements) reality. Rather, what is known is *within* the knower's judgments, minds, thoughts, or experience (so idealism or Putnamian "internal" realism, or Sartrean phenomenalism is true). No, again. For we do know something about our knowing. Concerning my knowledge that 5 + 7 = 12, I know that it is not a truth just about components of myself, my conceptual system, or cognitive achievements. There are, of course, cognition-dependent facts like that; e.g., that I have distinct powers of imagination and memory and that I can often tell/know which thought is an imagination and which a memory. But the fact that 5 + 7 = 12 is not that kind of fact. Rather, the nature of numbers and mathematical relations that I did not create, invent, or cause to exist make my belief that 5 + 7 = 12 true. (And I did not create the law of addition. It's a valid rule of mathematical manipulation which I learned—or found myself correctly following, if my knowledge of it was innate.) And I know this. That is part of realizing facts about what I know of this sort—viz., among these facts are the fact that I myself am not the source for the truth of 5 + 7 = 12. Something beyond me is, and that is the nature of knowing on the realist's view. (Of course, we didn't first postulate realism and then look for evidence for it. It's the other way around. Instances of knowledge, when reflected on motivate the recognition of realism.)

Someone will object that I don't really have these second-level cognitions about my knowledge (about first level facts such as that 5 + 7 = 12). Well, such second level knowledge certainly does not seem to be required to have the first level knowledge (the KK thesis is false). However, I also know that sometimes it is possible to know that I am knowing something. Why is this possible? Because it's actual for some cases—e.g., my knowledge that I know that 5 + 7 = 12. What makes the fact known independent of me and my knowledge of it is not a magical transcendent leap, but rather that *mistakes occur* (and maybe are always possible—weak fallibilism). The occurrence of mistakes and errors (false beliefs, and ignorance where knowledge was believed to exist) very strongly suggests that known fact is independent of knower and event (or fact) of knowing. "Thinking makes it so" is almost universally false! We know that. The only cases where it's not false are Cartesian; e.g., that I am thinking that now I am thinking. Aha, that particular thinking does make what I'm thinking (that I am thinking) true. But that is the "exception that proves the rule"; i.e., as a rule, thinking is not like that. As a rule, thinking something is so (believing it, hoping, fearing, claiming to know it, etc.) over and over proves not to make it (the proposition thought) true. We know this! It is the primary evidence for the independence of many facts that we know (or believe to be true)—their mere existence not depending on our having any cognitive sort of relation to them.

Knowledge of the "external world" (i.e., not of my thoughts or sensations, or of mathematical facts) is the same way. A bridge from non-independent-to-knowing facts (such as that I am thinking now about something, or am suffering a headache) to knowledge of facts of the "external world" is most usefully built via cases of *a priori* knowledge like those just mentioned. For some *a priori* knowledge is about independent-from-us realities. We know that via knowing we have made past mistakes, as well as by realizing that acquisition of such knowledge is often difficult and complex—e.g., in demonstrations. Many will object here that pure Platonism is the only way to defend such realism on *a priori* knowledge and pure Platonism must be rejected. My answer is: "Oh? Is this "Quine" again? We should *start* with metaphysical rejections? Essentially on aesthetic grounds perhaps, like Quine's [10] desert landscape preference? Why not eschew preconditions to metaphysical and epistemological inquiry and see where the attempts to explain actual knowledge (including *a priori* knowledge and scientific knowledge, such as "Einstein"—not to mention aesthetic judgments) lead us? Wouldn't that attitude towards inquiry have more in common with actual scientists than the empiricisms of "Quineans" like Quine?"

The independence of the (often contingent) facts of the "external world" is supported not only by the sort of evidence that supports realism for *a priori* knowledge (i.e., past mistakes and ignorances, plus knowledge of our own lack of total contribution to cognitive achievements), but by new evidence—viz., that science exists. Science (which we know exists) is an accumulation of knowledge. We have learned that space and time are ultimately relative. Einstein discovered it and explained it to us. So, space and time (though still mysterious in many of their/its other aspects) aren't quite what we had thought (e.g., permanently independent-of-each-other dimensions of perceptually experienceable reality). Rather, real space and time (it is now known) are inherently interrelated. Foundational facts about very specific, absolute spatial-temporal properties are not to be found (and don't exist). But different basic facts do. Unexpectedly, c is constant across inertial systems. We did not create that fact (nor did any scientist). (There are lots of absolute negative facts too—e.g., that c is not equal to or larger than 200,000 miles/sec; our moon has no moon itself; that no temperature falls below -273 Celsius). It is knowledge and knower-independent (though knowable). It's a fact of physical nature (which, perhaps, is all the reality there is; cf.[9] Peterson-Chomsky on the "physical").

IV

There are two realist philosophers in the 20th century which we should look to as models for how to "overcome" "Quine" and re-evaluate "Einstein" and "Magritte"—viz., Bertrand Russell and Noam Chomsky. Space and time [sic] prevent me from any explanation of this. So, I will just provide some quotations which illustrate what I see, realism-wise, in Russell and Chomsky. (All the inserts in square brackets, "[...]", are my comments.)

Recall, from early writings, Russell saying: "The first truism ... is that the world contains facts, which are what they are whatever we may choose to think about them, and that there are also beliefs, which have reference to facts, and by reference to facts are either true or false." [14, p. 182] "It is important to observe that facts belong to the objective world. ...the outer world ...is not completely described by a lot of 'particulars', but ...you must also take account of these things that I call facts, ...these, just as much as particular chairs and tables, are part of the real world ... When we speak falsely it is an objective fact that makes what we say false, and it is an objective fact which makes what we say true when we speak truly." [14, p. 183] Slightly later, Russell says. "Man is a part of nature, not something contrasted with nature. His thoughts and his bodily movements follow the same laws that describe the motions of stars and atoms." [15, p. 48] "The philosophy of nature must not be unduly terrestrial; for it, the earth is merely one of the smaller stars of the Milky Way. It would be ridiculous to warp the philosophy of nature to idealism, for example in order to bring out results that are pleasing to the tiny parasites of this insignificant planet." [15, p. 55] Similarly, he later says: "To scientific common sense which I accept) it is plain that only an infinitesimal part of the universe is known, that there were countless ages during which there was no knowledge, and that there probably will be countless ages without knowledge in the future. Cosmically and causally, knowledge is an unimportant feature of the universe; a science which omitted to mention its occurrence might, from an impersonal point of view, suffer only from a very trivial imperfection." [16, p. xi] And, as R. Marsh says: "It is Russell's hope that as philosophy comes closer to a scientific point of view, the right of the philosophers to alter his opinions in the light of his later thoughts will be accepted as a matter of course." [17, p. 283]

Noam Chomsky says: "...our mental constitution permits us to arrive at knowledge of the world insofar as our innate capacity to create theories happens to match some aspect of the structure of the world." [1, p. 21]

And, "If, as Russell frequently expressed it, man's 'true life' consists 'in art and thought and love, in the creation and contemplation of beauty and in the scientific understanding of the world,' if this is 'the true glory of man', then it is the intrinsic principles of mind that should be the object of our awe and, if possible, our inquiry." [1, p. 50] Also: "[Steven Weinberg] points out that 'it is remarkable that physics in this Galilean style should work ... I recall a statement by J.B.S. Haldane in which he said, in effect, that the universe is not only a good deal queerer than we know; it is a good deal queerer than we can know. Sometimes I believe that is true. But suppose it is not. Suppose that by pursuing physic in the Galilean style we ultimately come to an understanding of the laws of nature, of the roots of the chains of explanation of the natural world. That would truly be the queerest thing of all.' These remarks seem to me quite apt. It might be that physics with its various offshoots, developed through inquiry in the Galilean style, is a remarkable historical accident resulting from chance convergence of biological properties of the human mind with some aspect of the real world. Whether this proves true or not, we have no present alternative to pursuing the 'Galilean style' in the natural sciences at least....It is not unlikely, for example, that literature will forever give far deeper insight into what is sometimes called 'the full human person' than any mode of scientific inquiry can hope to do. But I am interested here in a different question: To what extent and in what ways can inquiry in something like 'the Galilean style' yield insight and understanding of the roots of human nature in the cognitive domain?" [2, pp. 8-9] Finally: "Suppose now that someone were to approach this astronomer with the following contention: 'True, you have presented a theory that explains the available evidence, but how do you know that the constructions of your theory have physical reality—in short, how do you know that your theory is true?' ...One might argue that the evidence is inconclusive or that the theory is objectionable on some physical (or conceivably, methodological) grounds. But it is senseless to ask for some other kind of justification for attributing physical reality to the constructions of the theory, apart from consideration of their adequacy in explaining the evidence and their conformity to the body of natural science as currently understood. There can be no other grounds for attributing physical reality to the scientist's constructions." [2, pp. 189-190]

Finally, I have frequently alluded to Nietzsche. But shouldn't he be classed as an early "Quinean" (as would most of his contemporary (mis)interpreters among the postmodernists)? No! Science is at the forefront of his positive doctrine, certainly to be among what epistemological and moral values arise (or are found or created) after we overcome human politics and religions, Western World theologically-ridden philosophies and cultures, and associated systems of epistemological, as well as, moral and aesthetic evaluation. (Nietzsche's so-called "perspectivism" has been grossly overemphasized lately, not to mention misunderstood.) I read his remarks as *transparently* realistic (as I have introduced realism above), as well as suitably scientific. For example: "On the whole, the procedures of science are at least as important a product of inquiry as any other outcome: for the scientific spirit rests upon an insight into the procedures, and if these were lost all the other products of science together would not suffice to prevent a restoration of superstition and folly. There are people of intelligence who ... lack the spirit of science: they have not that instinctive mistrust of devious thinking which ... has put its roots down in the soul of every scientific man ... It is for this reason that everyone now should have acquired a thorough knowledge of at least one science: then he would know what is meant by method and procedure and how vital it is to exercise the greatest circumspection. ... Insofar as genius of every kind maintains the fire of convictions and awakens distrust of the modesty and circumspection of science, it is an enemy of truth, no matter how much it may believe itself to be truth's suitor." [5, section 635, pp. 201-202] And, of course: "*Long live physics!* How many people know how to observe something? Of the few who do, how many observe themselves." [4, section 335, p. 263] " ... we must become *physicists* in order to be able to be *creators* in this sense—while hitherto all valuations and ideals have been based on *ignorance* of physics or were constructed so as to contradict it. Therefore: long live physics! And even more so that which compels us to turn to physics—our honesty." [4 p. 266] (Walter Kaufmann's views support my attribution of scientific realism to Nietzsche. For example, in note 67 to section 335 of *The Gay Science* quoted in part above, Kaufmann says: "Even as one has to know physics to build airplanes and make flying [them] possible—*apparently* defying the laws of gravity—one also has to know physics [= science] to become autonomous: that is Nietzsche's point." [4, p. 267] And Kaufmann's quotes from Nietzsche's letter to Overbeck (22 December 1984): "I do not know how much I would give if only I could bring it about, *ex post facto,* that such a glorious, great nature [viz., Emerson], rich in soul and spirit, might have gone through some small discipline, a really *scientific education.* As it is, in Emerson we have *lost a philosopher.*" [4, p. 7] And in his book *Nietzsche* [3], Kaufmann says: ".[Nietzsche] insisted that we must elevate philosophy to a science—and this programmatic declaration in the preface of his first book holds equally for all his later works....Nietzsche did not want philosophy to be less scientific than this but rather more so; only he had in mind the "gay science" of fearless experiment and the good will to accept new evidence and to abandon previous positions, if necessary." (p. 86) And: "Those who take refuge in irrationality, dogma, or systems based on unquestioned premises, seemed slack and weak to him. 'The priest knows only one great danger; that is science.' ([Twilight of the Idols] A49) By 'science' Nietzsche means the willingness to question, to submit one's opinions to experiments, and to revise one's belief in the light of new evidence." (p.232))

AFFILIATION

Philip L. Peterson
Department of Philosophy
Syracuse University
United States

REFERENCES

[1] **Chomsky, N.**, *Problems of Knowledge and Freedom: The Russell Lectures*, Vintage Books, 1971.

[2] **Chomsky, N.**, *Rules and Representations*, Columbia U. Press, 1980.

[3] **Kaufmann, W.**, *Nietzsche*, Princeton U. Press, 1974.

[4] **Nietzsche, F.**, *The Gay Science*, (W. Kaufmann, trans.) Vintage, 1974.

[5] **Nietzsche, F.**, *Human, All Too Human*, (R.J. Hollingdale, trans.), Cambridge U. Press, 1986.

[6] **Peterson, P.L.**, "Philosophy of Language", *Social Research*, Winter, **47**(4), 1980, pp. 749-774.

[7] **Peterson, P.L.**, "Semantic Indeterminacy and Scientific Underdetermination", *Philosophy of Science*, **51**, 1984, pp. 464-487.

[8] **Peterson, P.L.**, "Revealing Designators and Acquaintance with Universals", *Nous*, **20**(3), 1986, pp. 291-312.

[9] **Peterson, P.L. & Chomsky, N.**, "Correspondence 29 Juin-8 Novembre 1993", *Revue Philosophique de la France et de l'Etranger*, **185**, 1995, pp. 83-96.

[10] **Quine, W.V.**, *From a Logical Point of View*, Harvard University Press, 1953.

[11] **Quine, W.V.**, *Word and Object*, MIT Press, 1960.

[12] **Quine, W.V.**, *Ontological Relativity*, Columbia U. Press, 1969.

[13] **Quine, W.V.**, "On the Reasons for Indeterminacy of Translation", *Journal of Philosophy*, **67**, 1970, pp. 178-183.

[14] **Russell, B.**, "The Philosophy of Logical Atomism (1918)", in: Russell, 1956, pp. 175-281.

[15] **Russell, B.**, "What I Believe (1925)", in his *Why I Am Not a Christian*, (P. Edwards, ed.), Simon and Schuster, 1957.

[16] **Russell, B.**, *Human Knowledge*, Simon and Schuster, 1948.

[17] **Russell, B.**, *Logic and Knowledge: Essays 1901-1950*, (R.C. Marsh, ed.) George Allen and Unwin, 1956.

APPENDIX: QUINE'S FAILURE (VIA QUOTATIONS FROM PETERSON)

Einstein's relativity can certainly be regarded as a species ontological relativity—since it is a new relativity of space *vis à vis* time (mutual dependence as versus supposed independence), and space and time are certainly ontologically fundamental (if anything is). Quine's relativity is not, however, a scientifically motivated kind of ontological relativity. It results from a particular kind of account of the semantics of the languages we use (or could use—artificial as well as natural). Quine's semantics is not motivated by scientific observations, anomalies, or challenges, but by nominalism and skepticism—both philosophical, not scientific, concerns. Scientists should doubtless adhere to Ockham's razor and also be continually skeptical about unusual data and surprising hypotheses. But neither of these activities requires them to become metaphysical nominalists or epistemological skeptics. (Many empiricists, since they admire scientific methods and results, have presumed that anything which they similarly admire, such as nominalism, must be—or is a good candidate for being—scientific in some appropriate way; fallacious reasoning, of course.)

Quine pursues *nominalism* (as well as anti-mentalism, so he can't be a Berkeleyan nominalist) by defending an extreme form of meaning nominalism:

> [But his] pursuit of extreme meaning nominalism ...has not proved classical nominalism—or even supported it very well. [For] Quine has assumed that there are no meanings, no meanings to be identified with universals, or with ideas, or with anything else [10, Chapters 1 and 2]. Then Quine has pursued *inter alia* the explanation of language possession and use with a new relationship of so-called synonymy (not sharing the same meaning, but a replacement for it) and a new property of significance (not a relationship of a word to a meaning, but a replacement for it). Quine argues (validly, I think)

> that this leads to the destruction of the analytic/synthetic distinction, the indeterminism of radical translation [11], an analogous indeterminism of meaning 'at home' in one's own language [12], and ontological relativity (a surprising nihilism in which there are simply no ultimate metaphysical facts with respect to basic entities or categories of reality). [Even] disregarding ontological relativity, these other consequences appear to be a *reductio* on the assumption that there are no entities called meanings. [6, p. 765]

There is, however, an appropriate reply to the conclusion that Quine has supplied a *reductio* proof that meanings exist—from Quine's skepticism:

> ...try to get in a [skeptical] Quinean frame of mind in the following way. Consider all the psychologists and/or others throughout the world who are investigating ESP. Now what of the skeptic who says there are incoherencies in the very idea of ESP? He might say that the term(s) for the alleged phenomena—"extra-sensory perception"—are incoherent, because (say) "perception" includes the idea of a mechanism and process for yielding its typical achievement, so referring to nonsensory perception is contradictory. It is thinking of perception as both including some mechanisms and also not including it. (Although this is a superficial objection to ESP, I think there are deeper problems of a similar nature that do vitiate the whole enterprise so that so-called scientific investigations of "Psi-phenomena" aren't investigations of anything at all—or, at least, aren't investigations of what they seem to be.) ESP experiments, demonstrations, discussions, etc., may go on and on. No matter what we do about ESP, if it simply doesn't exist, then we are not investigating it (but, at best, something else). One reason it might not exist is that we don't know what we are talking about. That would be the case if that skeptic about ESP turned out to be right, that the very idea was incoherent at bottom. 'No experiment can straighten out a confusion of thought' as Geach said (1957, p. 190) in another connection. (Someone else might be analogously skeptical about IQ testing and the concept of intelligence allegedly involved in contemporary psychological research.) So, of course you might reply to the skeptic (like Quine) who thinks the proposal [that meanings exist and are among the objects of study in scientific linguistics] is very suspect ... by saying that we have linguistics—a real ongoing productive contemporary science. But a skeptic (like Quine) can reply that we also have ESP research (supposedly scientific investigations of 'Psi-phenomena'). (And another skeptic—some of these, by the way, are

found among the linguists—might reply that we also have IQ testing and real ongoing productive contemporary investigations of human intelligence.) Yet, if ESP is, at bottom, incoherent—and analogously, one single unequivocal concept of human intelligence as a topic for scientific investigations is similarly incoherent—then no matter what we 'have' (ESP research and/or IQ research) the studies are undermined. They are about nothing—or, at best, not at all about they seemed to be. [7, p. 472]

What shall we do [since other methods alleged to save linguistics and/or meanings from such Quinean skepticism simply fail; cf. [7, pp. 473-6]? The only alternative to Quinean semantic indeterminacy ... that I see is to admit that linguistics is irrevocably entangled with philosophical entities. In particular (and only in one of possibly many alternative terminologies), the philosophical entity in question with respect to Quinean criticism (leading to semantic indeterminacy) is the concept of concept itself—or (when properly understood) the idea of idea and/or meaning. Quine is right, linguistics is irremediably involved with references to meanings. There is no way out of this except to admit that linguistics is irremediably entangled in (or is it just as a simple part of ?) epistemology—and perhaps metaphysics. (Truer-believers-in-Quine-than-I will,˙ of course, hold that this cure for semantic indeterminacy is no better than the disease—and probably worse through its great potential for self-deception) For the concept of concept is irremediably epistemological. It is concepts which are (or are among) the crucial 'material' in the content of our knowledge (if we have any) and beliefs (which we certainly do have). And these concepts (or meanings, or ideas, or something of that ilk no matter how labeled) are precisely what we must posit to be the determinants of linguistic significance and (cognitive) synonymy. Quine himself rejects any such posits in 'On What There Is' ([10, p. 11]). It is clear he holds that positing any such entities constitutes positing vacuous potentialities (p. 48). Up against semantic indeterminacy, however, some of us might be willing to grasp at even these suspect straws. If there is anything to them, it could 'save' linguistics—though it wouldn't save if from philosophy, [but] only make it make more obviously philosophical.

...here is the argument that [we] should [make]...against Quine's semantic indeterminacy thesis [thereby undermining Quinean ontological relativity]. Since the semantic indeterminacy thesis is parallel to the uncertainty relations in the received interpretation of quantum mechanics (namely, Bohr's), any alternative to semantic indeterminacy must be capable of explaining everything that semantic indeterminacy explains, just as any alternative interpretation of quantum mechanics must explain all that it can explain (not to mention solve some problems or anomalies that the received interpretation cannot handle). Hanson (1967, pp. 41-49) lists ten such points (drawn from Van Vleck) about quantum mechanics, running from '...explain the spectral frequencies of such atoms as hydrogen and ionized helium' to 'Make available at least the rudiments of a quantum electrodynamics...'. So, any new interpretation must account for all these ten point as well as Bohr's interpretation does. (To be an interesting alternative, of course, it must do even more than that.) In an analogous way, alternatives to semantic indeterminacy (e.g., the Chomskyan approach to scientific linguistics, or the speech-act approach of Searle, or of Austin) must explain all of what the Quinean approach (containing IRT) can explain. But here is where the trouble with Quine's approach occurs. Quine's approach (with semantic indeterminacy at its core) does not explain much of anything. It is a retreat compared to other theories, not an advance over them. Many alternatives to Quine's theory (before him and now) already have richer explanatory power. So the parallel of semantic indeterminacy to the uncertainty relations of quantum mechanics (both of which might be characterized as holding that beyond a certain degree of refinement there is simply 'no fact of the matter', such as about meanings, or about position-and-momentum and particles-vs.-waves) does not in the end aid Quine's cause, but hinders it. For it is very easy to propose alternatives to Quine's view which explain all that his view encompasses and even much more, since his view doesn't in general explain very much at all (particularly in linguistics), but rather prevents many explanations (descriptive accounts, calculations, etc.) that most live alternatives to it permit... [7, pp. 476-7]

H.-H. VON BORZESZKOWSKI & R. WAHSNER

NOT EVEN CLASSICAL MECHANICS IS MECHANISTIC

1. INTRODUCTION

Contemporary natural science or, put another way, the rationality of the classical sciences variously have been held responsible for the disconcerting, if not desperate, state of human affairs and future prospects of humanity. An unbridgeable gap is seen between the "two cultures" or two types of rationality, the one being that of the "hard and cold" natural sciences which indulge in measurement and calculation as their basic methods, and the other the moral standards of human rationality. Usually, the natural sciences, due to their methodology and state, are seen as turned against human culture, and a way out of the depressive state of human culture is hoped for by transforming the very nature of the sciences. This transformation is to entail first and foremost the "historification" and "humanization" of the sciences so as to allow for the incorporation of the thinking and acting subject into theory. However, as shown elsewhere [1; 2]., both the criticism leveled against the sciences and the expectations as to their future transformation rest on a misapprehension of the characteristic features of their epistemology.

The expectations arising from a science-to-be-transformed build on the unquestioned assumption that scientific propositions are (or could be or even should be) statements about the real state of nature, about concrete objects in their mutual connectedness. (Thus it is argued that it is necessary to change the conception of natural sciences because only artificial products can be deterministic and reversible whereas nature contains accidentalness and irreversibility as essential moments [cf. 3, 19].) This identification, however, of theory and reality is not an inherent component of the ("mechanistic") sciences, but is rather prompted by a philosophical mechanistic world view. Neither are the natural sciences in general mechanistic nor is this the case of classical mechanics [4]. Given the non-mechanistic character of mechanics, consequences for modern physics are warranted. After all, also modern physics is based on the conception of the sciences developed by Copernicus, Kepler, Galileo, and Newton. Though Newton's mechanics, including his theory of gravitation, certainly have been superseded in part by quantum mechanics and Einstein's theory of relativity, Newton's mechanics still has to be regarded as the prototype of physical theory. This does not refer to the range of validity of physical theories but to the structure and character of their concepts: all physical theories are constructed according to the rules of classical mechanics. It might also be said that unlike natural philosophy, all physical theories share basically the *epistemological* status of classical mechanics. This is even true of physical theories which do not relate to mechanical phenomena and of theories in respect of which Newton's theory has been shown to be no more than an approximation in terms of validity range. More importantly, Newton's mechanics not only serves as model for all physical theories, it also plays a central role in the modern structuring of physics.

This is because it is an essential part of the theory of measurement which is basic to all physics. The legitimacy of calling modern physics mechanistic depends on the legitimacy of doing so for classical mechanics. We however advocate the non-mechanistic character of classical mechanics, and in what follows we shall argue towards this end. Hence, we maintain that the notion of an inherently mechanistic nature of classical mechanics results from confusing this physical discipline with its inadequate interpretation by a mechanistic philosophy.

In what follows, we first outline the world view of mechanicism and show that the mechanistic features usually associated with Newtonian mechanics really spring from this philosophy. Second, by way of comparing Newtonian mechanics (inclusive of Newton's own arguments) with this mechanistic outlook we provide evidence showing that this mechanistic feature is foreign to classical mechanics. This paves the way to the more general question of the differences between physics and philosophy which will be considered in greater detail in the final section.

2. ON THE PHILOSOPHICAL WORLD VIEW OF MECHANICISM

After its foundation, classical mechanics remained for a long period of time the only fully elaborated field among the sciences and stood paradigmatically for what genuine science was to become. This is also why philosophical interpretation built largely on the experience of classical mechanics or why the development of the sciences resulted in a mechanistic world view. (For the basic features of the philosophical mechanistic world view see [4, 13-18].) Above all, the whole of the Enlightenment was permeated by this belief. Here there was a science that finally demonstrated the possibility of man's rational mastery over nature and society. It was thought that classical mechanics allowed to grasp the whole of the world on a rational scientific basis and in particular that the law of gravity was the principle to which all explanations answered.

Such a world law was conceived of as a universal mechanism to which all happenings in the world are subjected.

This way, classical mechanics and its ideological application became an important element in the intellectual preparation of the French Revolution. Thus it served as a weapon against the clergy, and in this way classical mechanics helped in founding the exclusive claim of reason. This, again, reinforced the urge to demonstrate that it was possible to govern rationally the whole of nature and society, and to show how this was done; that is, by reason based on knowledge of the inevitable lawfulness of the world. All this demanded that mechanics be considered a totally flawless representation of reality. Indeed, whenever this theory would display gaps and limitations, then these were gaps and limits of reason.

This philosophy was mechanistic, not because it built on mechanics, but because it considered mechanics (or physics, which was at the time the same thing) *the only possible* way to comprehend nature. Thus mechanics rose to an all-embracing natural philosophy which moreover was seen as superior to any other philosophy. Ultimately, the claim was that only those things qualified as existing which could be explained in terms of the laws and notions of mechanics (physics). The outcome was an identification of theory and reality, or theory was made the arbiter of existence. By abstracting from the epistemological and axiomatic assumptions of classical mechanics (physics), the mechanistic inference was made that mechanics itself was the world's natural condition or that the world itself was governed by eternal laws—so that man could rule nature and subject it to his needs like a machine.

3. CLASSICAL MECHANICS AS A PHYSICAL THEORY AND THE PHILOSOPHICAL MECHANISTIC WORLD VIEW

For discussing the relation between classical mechanics and mechanicism it is useful first to note what Newton understood by mechanics (physics), how he determined its subject-matter, its presumptions, and its limits.

Such an analysis has been performed elsewhere [4], and the conclusions repeated here are as follows: In contrast to the mechanistic advocators of mechanics to come, Newton was keenly aware of the fact that mechanics (physics) builds on epistemological, axiomatic, and of course experimental assumptions and that the validity range of his physics displays limits. He considered mathematics and the experimental method as means of physical cognition characterizing and determining the range of its validity, and he has developed these tools.

The way he presented his mechanics in the *Principia*, the *Regulae philosophandi* added to the 2nd edition of this work, the passage *Scholium generale*, and the 31st of the Queries added to his *Opticks*, all this attests to Newton's awareness of the indispensability of epistemological assumptions prior to and underivable from the body of physics. For him the reduction of the essentials of human knowledge and reality to the laws of mechanics such that no room left for anything else in the world was inconceivable.

With his mechanics, Newton laid the foundations of physics as a discipline the specific feature of which was to comprehend the laws of nature as mathematical relations between physical quantities which are, in principle, accessible to measurement. Thus physics can neither exist without its empirical basis nor without its mathematical language. Mathematics and the experimental method were in Newton's understanding the physicist's indispensable tools.

Newton's views and the mechanistic identification of physics and mathematics are worlds apart. That he considered mathematics as a method is obvious already from his Foreword to the 1st edition of the *Principia*, from the 3rd rule of the *Regulae philosophandi*, and so on.

To see the empirical orientation of Newton's mechanics (physics) there is not even any need to turn to individual passages of his text. It suffices to point to the structuring of the *Principia*: in the beginning, there are definitions that ascribe a physical meaning to the theory by relating it to reality. This is accomplished by defining the physical quantities in terms of a rule about how to measure them. The physical laws by which these quantities are connected then are introduced as statements of experience.

The definitions (explanations) and laws of motion in the 1st book of the *Principia* are followed by mathematical propositions. In the 1st and 2nd books mathematical conclusions are drawn and, finally, in the 3rd book Newton considered the special case of gravitational forces and showed how "the motions of the planets, comets, the moon, and the seas" can be derived therefrom.

The mechanistic extrapolation of classical mechanics and especially the idea of explaining the whole world by one single "law of attraction" emerged notably from the misinterpretation of the 3rd book.

Already Newton opposed this misinterpretation, mainly by the following three arguments:

(i) The mechanistic interpretation took the force term by itself for the principle of motion. It overlooked the fact that the basic laws are given by three axioms of mechanics and that the gravitational law is only one special case of this general theory. Neither was it realized by this view that there should be found various force terms which may be inserted in the right-hand side of the second axiom. Thus there is no such thing as *the* law.

(ii) A solution of an equation of motion always requires assumptions about supplementary (initial and boundary) conditions. These follow from the experimental arrangement or from the dynamics and structure of the system of which the system under consideration is a subsystem. The universe, however, is neither subject to manipulation by experimental physicists nor is it a subsystem of anything. The universe, by definition, has no supplementary conditions outside of itself and thus cannot be handled like a usual physical system.

(iii) It is impossible to derive a physically interpretable model for homogeneously and isotropically distributed matter (Newton's or Seeliger's paradox).

4. ON THE DIFFERENCE BETWEEN PHYSICS AND PHILOSOPHY OR THE CONFIRMATION OF THE CLASSICAL CONCEPT OF SCIENCES

The idea to extend or even to overcome the boundaries of mechanics, and this not by an ideological over- or misinterpretation but by developing the science *mechanics* further, is not new. Already in 1755, with his *Allgemeine Naturgeschichte und Theorie des Himmels*, Kant intended to extend Newtonian mechanics to a natural-scientific theory of evolution. This means, he wanted to solve one of the tasks that today is required of natural science to solve for the purpose of its renewal. But as can be shown, Kant did not achieve this aim [5, 66-90]. His paper is not consistent and runs into contradiction with its own basic presuppositions. Kant himself arrived at this insight. He came to it at that moment at which, animated by David Hume, he became conscious of the problematic involved in the human cognition of laws. This insight can be grasped in the thesis: It is not possible to construct physics only on active, dynamical principles; an empirical theory requires presupposed or so-called passive principles (i.e., principles which the theory itself does not explain, but takes as a basis for its explanations). As a consequence of this insight, Kant grounded his space-time apriorism. This apriorism has been and still is criticized—and rightly so if one understands it as something the origin of which cannot be inquired into, if one understands it as the thesis that space and time *in themselves* lie ready in the human spirit. However, as has been shown [6; 7; 8; 9], it is quite valid with respect to the process of physical cognition and it implies the proposition that the epistemological starting points of natural science and of philosophy are necessarily different in principle.

This difference (difference but not separation!) ascertained by Kant has not yet been superseded. And there seems to be no reason why it should be superseded. Rather it can be seen that the current discussions regarding the status of science are permeated with a confusion which is due to the fact that, on the one hand, the status of the individual sciences is identified with that of philosophy and, on the other hand, scientific experience and everyday experience are identified (i.e., only the connection but not the essential difference between scientific and everyday experience is perceived). This means, however, precisely the following: Instead of superseding the shortcomings of Kant's insight, this insight is liquidated or ignored. Current discussions in philosophy of science show this clearly. Here, those views are best received which claim to have founded a physics of evolution and thus to have achieved what Kant failed to accomplish. There is no consideration of Kant's reasons for refraining from the expansion of mechanics to a theory of evolution, which also moved him to construct his critical philosophy postulating the necessity of different epistemological starting points for natural science and philosophy. No inquiry is made as to whether these reasons are of a principle nature and so are still valid today. It is thought that a different physical theory than mechanics can achieve what Kant recognized as doomed to failure. It is not asked whether the reason for this failure is specific to the physical theory "mechanics" or specific to physics as such [10; 11].

That physics neither can be turned into a theory of development or can describe the world in its totality results from the fact that it has to determine and presuppose etalons in order to become a measuring science. Naturally, the etalons must not act on physical bodies (or: they may only act in a well-defined manner, under *one* welldefined aspect) and they must not be effected by them (or: they are only influenced by this *one* action). On the other hand, physical bodies are considered as mutually effecting. Of course, such a situation does not exist in nature and thus it has to be constructed in a experimentally and theoretically "artificial" way. Thus it is an error to think like Hegel that only philosophy—unlike other sciences—has not the advantage to admit their *subjects* immediately from the idea and to presuppose the *method* of recognizing for the beginning and continuation as already given. ("Die Philosophie entbehrt des Vorteils, der den anderen Wissenschaften zu Gute kommt, ihre *Gegenstände* als unmittelbar von der Vorstellung zugegeben, sowie die *Methode* des Erkennens für Anfang und Fortgang, als bereits angenommen, voraussetzen zu können" [12, 41].) Therefore, one is not dealing with concrete bodies but with abstract ones to which a specific property or behaviour is attached.

Thus a measuring science will never be able to conceive the world in its totality; it can only be valid to the extent to which the abstraction made reflects an essential connection in a good approximation. It does not describe real but possible modes of behaviour of real objects. This specific character of physics results in the fact that physical theory cannot grasp historicity immediately. It was constructed just in such a manner that closed systems can be established that make is possible to define the measuring body "space" or the etalon of motion. The definition of all physical quantities is made via "conservation laws" that is, via basic dynamic equations or specific conservation laws; all physical natural laws are formulated as equations. To violate basic equations means to violate the theory. A historical component, like it is necessary in cosmogony, is introduced into physics by initial and boundary conditions which, as mentioned above, must always be determined when specific physical situations are to be calculated by means of general laws. So-called evolutionary models of the cosmos usually take evolution to refer to mere expansion of the cosmos which does not differ in status from simple spatial motion. And whenever such models seem to tend towards some description of the history of the universe, this description is contained rather in the supplementary assumptions than being a product of the physical laws of such a model. Physical laws themselves do not contain this component; but they are compatible with it. This is the difference between a measuring and calculating science and philosophical wisdom [13; 14]. By misunderstanding this basic situation of physics, by attempting, for example, to conceive the world in its totality on the basis of classical mechanics, the idea of historicity degenerates into the idea of a universal linear causality or into the Laplace demon.

The entire forming of concepts and theories in physics is essentially determined by measurement requirements [2, 239-285]. Thus it is not possible to discuss about problems of the forming of physical concepts or theories or about physical truth by abstracting from measurement and its requirements. Without measurement, the subject of cognition cannot be mediated with the object of cognition. Thus its possibility must be presupposed in theory. This possibility of presupposition is not identical with theoretical advances on which the respective theory is based. Certainly, all theories are based on preceding theoretical findings. But, in general, they do not mediate the subject and object of cognition. To see measurement only as the reading of positions of pointers, as the comparison with given standards or as observation of natural connections of action in a calculable manner cannot result in understanding its outstanding position in the process of physical cognition. The fact has to be taken into account that measurement always presupposes quantities, to which belong also rules saying how to measure these quantities, how to compare the measurements made at different space-time points and how to think the realization of etalons theoretical-physically. Thus a physical theory must always be formed in such a manner that these demands are satisfied. Physics (by necessity mathematically formulated) differs from mathematics by that it works with measurement quantities and the consequences arising from this. To objectify *(vergegenständlichen)* the equalities grasped by the respective quantities the real measurement requires the artificial production of ideal situations; it requires objects that were prepared by real idealization, that is, by the purposeful exclusion of certain influences and effects, in such a manner that they can be used as objectified measures (e.g., as clock or measurement rod). In experiments, one operates with ideal and not with concrete bodies under ideal and not under concrete conditions. Both the bodies and the conditions must be produced, but one also must be able to think them. The experiment is a method to realize (purely) equalities and their relations. Only through this mutual approaching of theory and reality, measurement can become mediation. The fact that measurement-theoretical determinations must be both presupposed and attached to dynamics was called by us as dualism.

The fact that physics excludes historicity in the manner explained above does not mean at all that physics is placed besides philosophy without mediation, but, on the contrary, it indicates the necessity of philosophy for physics or, more generally, for the empirical, mathematized natural science as a whole. This is true because there is no natural science that is able to explain per se its epistemological basis. This was known already by Proclus who wrote: "There is no science demonstrating and discussing its own principles, but it is certain about them. It is more clear about principles than about derivations. The former it recognizes in their own light, the latter, however, by the principles" [15, 75, 6 ff.].

Looking back on the above-mentioned and analysing the epistemological problems of modern physical theories like thermodynamics, quantum mechanics, and theory of relativity, we may state: Hitherto, there has been no revision nor is there any need for a revision of the classic concept of science philosophically grounded by Kant.

Often, however, things are argued the other way around, and the usual argument runs like this: The classical (Galilean) concept of science regarded the world as an "object" and tried to describe the physical world as if it were an object to be described from outside, as a thing which does not include ourselves. This attitude was most successful in the past. Now, however, we meet the limits of this Galilean concept. To make further progress, we must better understand our position, the standpoint from which we describe the physical world. This does not mean that we have to return to a subjectivistic concept of science. But to a certain extent we have to bring physical cognition in relation to the characteristic features of life [cf., e.g., 16, 15].

The background of this reasoning is: Living objects are objects far removed from equilibrium. And because we can now describe them physically, we can and must incorporate them into physics. Identifying now also "subject", "cognitive subject", and "living being" one believes that one is able to grasp the subject natural-scientifically. But this completely misses the point of the philosophical concept *subject.*

If we regard the cognitive subject as a mere object among other objects of nature we do not really study the cognitive subject, but rather take the human being and consciousness as the object of recognition. This approach is of course legitimate for science, and there is no doubt that progress in physics and other sciences leads to further scientific knowledge of the human being and his consciousness. However, by this approach one does not learn anything about the subject as a subject. By this, one does not understand how the subject is given to itself and how the cognitive faculties of the human species are constituted. The answer to this question is an essential task of philosophy. Physics as a natural science is epistemologically so grounded that it cannot do this. Let us attempt to clarify this a bit with the following remarks: For natural science, objectivity conceived in contradistinction to subjectivity is precisely the characteristic of true cognition. However, considering the world exclusively under the form of the object excludes the subject as such. The subject remains here explicitly only as the external observer; it reflects on itself implicitly in the agreement to recognize an objective, real external world, to presuppose it as given [17, 94-95, 155; 18, 33, 56, 62-63, 72].

This essential withdrawal of the cognitive subject is by no means a defect of natural sciences which has to be removed. It is, without a doubt, as Schrödinger said, a trick, but this trick is necessary. The need for it stems from the same grounds as the necessity of science as such: science is required because the phenomenal form and the essence of things do not immediately coincide. One must therefore search for a method to bring the essence to appearance, to reveal it as something objective, constant, and reproducible. This aim can in principle only be attained via experimental comparison. For this purpose, the cognitive subject must also determine itself as a comparable object. The essence which is brought to appearance as constant and re-producible must have these properties at each time and place and also for each subject (each individual, each class, each generation). The cognitive subject makes itself a comparable object by changing into a specially prepared natural body and in the presumption that an external observer and a real external world exist.

The natural-scientific standpoint does not simply negate the subject; it rather does not conceive of different subjects as possible and cannot grasp their characteristics. From this standpoint the subject as a subject cannot be investigated. Therefore, from the purely natural-scientific point of view the epistemological basis of natural science cannot be investigated; natural science as natural science does not reflect on its own epistemological position. This position, however, must somehow be reflected upon. And this is the task of philosophy, whose concern is the subject (i.e., not the individual, but the species *man*). It must explain its own standpoint as well as all the other possible (and necessary) standpoints including their conditions. It posits as given in its existence that part of reality which is the subject itself and then investigates its properties, in order afterwards, taking account of the cognition of nature (or, better, of the cognition of the object), to conceive the unity of subject and object or, in Hegel's words, in order to supersede the dichotomy which is the source of the need for philosophy [19, 20-25].

Against this differentiation between the epistemological standpoints of natural science and of philosophy it is sometimes argued that such a difference cannot exist in principle because there is no fundamental boundary between man and nature. Of course, one must admit that there is no sharp boundary. Man is also and essentially a natural being, and one may not even abstract from this, if one wants to understand his social essence. However, there is without doubt a difference in principle between man and nature in the difference between subject and object. The possibility of conceiving the subject (more precisely, the concrete thing representing the subject) as an object as well does not obviate the necessity of conceiving the subject as a subject. And in doing the latter, one is dealing with philosophy, not with physics.

The epistemological basis of physics outlined here has not been revised hitherto not even by modern physics. If one wanted to follow the proposals as mentioned at the beginning and revolutionize the Galilean-Kantian concept of science, then one would have to revise the foundation on which the achievements of physics are based.

The progress made by modern physics has without doubt great physical significance. The widespread philosophical interpretation discussed here rests, however, on an antiquated ontologization. To echo Husserl, what is in fact a method is taken for real being. This procedure is the normal way of mechanicism: It abstracts from the presuppositions of the physical theory and overinterprets the results. In this manner the genuine achievements of physics are overlooked and those alien to it are ascribed to it. Physics then appears only negatively, as not-yet-philosophy, and philosophy appears as not-yet-physics. *In the last analysis* such a procedure satisfies neither physics nor philosophy.

Attempts to introduce the dimensions of history and morality into the sciences by mechanistically smoothing down the differences between philosophy and science are bound to turn against one's own intentions by undermining the real unity between these two different kinds of rationality.

AFFILIATION

Horst-Heino von Borzeszkowski and Renate Wahsner
Max Planck Institute for the History of Science
Berlin
Germany

REFERENCES

[1] **v. Borzeszkowski, H.-H., and Wahsner, R.**, *Physikalischer Dualismus und dialektischer Widerspruch. Studien zum physikalischen Bewegungsbegriff*, Wissenschaftliche Buchgesellschaft, Darmstadt, 1989.

[2] **Wahsner, R., and v. Borzeszkowski, H.-H.**, *Die Wirklichkeit der Physik. Studien zu Idealität und Realität in einer messenden Wissenschaft*, Verlag Peter Lang, Frankfurt a. M./ Berlin/ Bern/ New York/ Paris/ Wien, 1992.

[3] **Prigogine, I.**, "Natur, Wissenschaft und neue Rationalität", in: **Sandkuhler, H.J., and Holz, H.H.** (eds.), *DIALEKTIK 12. Die Dialektik und die Wissenschaften*, Pahl-Rugenstein, Köln, 1986.

[4] **v. Borzeszkowski, H.-H., and Wahsner, R.**, *Newton und Voltaire. Zur Begrundung und Interpretation der klassischen Mechanik*, Akademie-Verlag, Berlin, 1980.

[5] **Wahsner, R.**, *Das Aktive und das Passive. Zur erkenntnistheoretischen Begrundung der Physik durch den Atomismus—dargestellt an Newton und Kant*, Akademie-Verlag, Berlin, 1981.

[6] **v. Borzeszkowski, H.-H., and Wahsner, R.**, "Erkenntnistheoretischer Apriorismus und Einsteins Theorie. Einstein in seiner Beziehung zu Newton und Kant", *Deutsche Zeitschrift fur Philosophic*, **27**, 1979, p. 213.

[7] **v. Borzeszkowski, H.-H., and Wahsner, R.**, "Zur Beziehung von experimenteller Methode und Raumbegriff", *Deutsche Zeitschrift für Philosophie*, **28**, 1980, p. 685.

[8] **Wahsner, R.**, "Erkenntnistheoretischer Apriorismus und die neuzeitliche Physik", *Deutsche Zeitschrift für Philosophie*, **40**, 1992, p. 24.

[9] **v. Borzeszkowski, H.-H.**, "Raumbegriff und Einsteins Theorie", *Deutsche Zeitschrift fur Philosophie*, **40**, 1992, p. 36.

[10] **v. Borzeszkowski, H.-H., and Wahsner, R.**, "Did Nonlinear Irreversible Thermodynamics Revolutionize the Classical Time Conception of Physics?", *Foundations of Physics*, **14**, 1984, p. 653.

[11] **v. Borzeszkowski, H.-H., and Wahsner, R.**, "Evolutionism as Modern Form of Mechanicism", *Science in Context*, **2**, 1988, p. 287.

[12] **Hegel, G. W. F.**, *Enzyklopädie der philosophischen Wissenschaften im Grundrisse. Erster Teil. Die Wissenschaft der Logik*, Werke, Bd. **8**, Suhrkamp Verlag, Frankfurt a. M., 1986.

[13] **v. Borzeszkowski, H.-H., and Wahsner, R.**, "Über die Notwendigkeit der Philosophie für die Naturwissenschaft", **Heidtmann, B.** (ed.), *DIALEKTIK I. Orientierungen der Philosophie*, Pahl-Rugenstein, Köln 1980.

[14] **v. Borzeszkowski, H.-H., and Wahsner, R.**, "Noch einmal über das Bedürfnis der Naturwissenschaft nach Philosophie", in: **Bayertz, K., Heidtmann, B., and Rheinberger, H.-J.**, (eds.), *DIALEKTIK 5. Darwin und die Evolutionstheorie*, Pahl-Rugenstein, Köln, 1982.

[15] **Proclus, D.**, *Euclidis Elementorum librum commentaria*, in: **Friedlein, G.** (ed.), Lipsiae, 1875.

[16] **Prigogine, I.**, *From Being to Becoming. Time and Complexity in the Physical Sciences*, W.H. Freeman and Company, San Francisco, 1980.

[17] **Schrödinger, E.**, *Die Natur und die Griechen*, Zsolnay, Wien, 1955.

[18] **Schrödinger, E.**, *Was ist ein Naturgesetz?*, Oldenbourg, München/Wien, 1962.

[19] **Hegel, G.W.F.**, *Differenz des Fichteschen und Schellingschen Systems der Philosophie*, Werke, Bd. **2**, Suhrkamp Verlag, Frankfurt a.M., 1986.

HELENA KNYAZEVA

THE SYNERGETIC WORLD VIEW AND ITS SYNTHETIC VALUE

INTERDISCIPLINARY SYNTHESIS OF KNOWLEDGE

Synergetics is a new interdisciplinary movement in modern science signifying a process of becoming in the world. Synergetics is a new dialogue of humans with nature, a new synthesis of human knowledge and wisdom.

Synergetics can serve as a basis for an interdisciplinary synthesis of knowledge. The theory endeavors in revealing common patterns of evolution and self-organization of complex systems of any kind, regardless of the concrete nature of their elements or subsystems. The concepts of self-organization, chaos and order, nonlinearity and complexity are being widely used both in natural sciences and in the humanities. Different scientific schools all over the world are working in the field. It's worth to mention the names of such scholars as H. Haken [2, 3], I. Prigogine [13, 14], S.P. Kurdyumov [1, 7], E. Laszlo [8], K. Mainzer [9], B. Mandelbrot [10], E. Morin [12], F. Varela [11].

A non-traditional synergetic world view is in the process of formation now. It means a radical shift of paradigm from being to becoming, from stability to sustainability, from images of order to chaos generating new ordered evolving structures, from self-maintaining systems to fast evolution through a nonlinear positive feedback, from evolution to coevolution, reciprocal evolution of different complex systems.

The theory of self-organization might be considered as an initial foundation for cross-disciplinary, cross-professional and cross-cultural communication. The possession of synergetic knowledge or, at least, of synergetic style of thinking could be a background of an open creative dialogue between scholars, thinkers, artists with different creative intentions, research positions, world views, etc.

HOLISTIC CHARACTER OF SYNERGETIC KNOWLEDGE

The synergetic world view is nonlinear and holistic. The paper is based on generalizing and philosophical comprehension of the recent results obtained by the Moscow synergetic school at the Keldysh Institute of Applied Mathematics of the Russian Academy of Sciences. Synergetics is viewed here as a theory of non-stationary, fast evolving dissipative structures. As it's supposed, synergetics investigates the appearance of such structures, their growth, transformation, synthesis and decay, i.e., formulates principles of evolutionary holism [1], [5], [7], [15]. It's worth to add that the investigation of constructive principles of coevolution, that is principles of integration of evolutionary structures into more complex ones, is one of the peculiarities of the Moscow studies in the theory of self-organization in comparison with other scientific schools in the field.

The future states of complex systems escape our control and prediction. The future is open, not unequivocal. But at the same time, the spectra of "purposes" of development are available in open nonlinear systems. If we choose an arbitrary way of evolution, we have to be aware that the way may be not feasible in a given complex system. Only a definite set of evolutionary ways, evolutionary structure-attractors can develop. There are discrete spectra of evolutionary structure-attractors for different complex systems. The spectra are determined exclusively by inner properties of corresponding complex systems.

One of the most essential questions is that of ways of integration: how a complex evolutionary whole is being assembled from integrated parts. Integration of structures is not merely putting together: the regions of structures localization are overlapping with energy defect taking place. In this case a whole is not equal to the sum of its parts. Generally speaking, it is neither more nor less than the sum of parts, it is qualitatively different. It is an evolutionary whole, because it unites structures of "different ages", the structures at different stages of evolution.

There are various, but not arbitrary, ways of integration of simple structures into complex ones. Synergetics formulates some "rules" of integration of structures evolving with different speeds into a whole one according to their "architecture", topology of organization, the rates of evolution. The main principle is as follows: synthesis of relatively simple structures into a more complex one occurs through the establishment of a common tempo of evolution in all unified parts (fragments, substructures). Structures start to co-exist in one and the same 'tempo-world', i.e., they acquire a common rate (tempo) of evolution. Besides, in the case of right topology of integration an arising whole organization can accelerate its tempo of evolution [5],[7].

Thus, the synthetic value of synergetics is determined by the interdisciplinary character of synergetic knowledge. It results from the development of the new evolutionary holism in the frames of synergetics as well.

WAYS OF SYNTHESIS

What could be synthesized on a basis of the synergetic world view. The possible paths towards a kind of synthesis are as follows:
- the Eastern (holistic, synthetic) and the Western (analytical) world views and ways of thinking;
- natural sciences and the humanities (Natur- und Geisteswissenschaften);
- foundations of science (strict models, mathematical theorems, physical mechanisms of phenomena) and its applications (including the most remote applications of synergetics to the social and human systems, its application as a kind of heuristics);
- descriptive and normative aspects of science, knowledge and values, "units" of information and value of the information for human activities, for instance, the description of spectra of structure-attractors as spectra of possible evolutionary ways of complex systems and the opportunity of human choice connected with values (human preferences);
- science (new science of complexity, nonlinearity and chaos) and culture, art, philosophy; synergetic knowledge and cultural images as well as philosophical ideas resonant to synergetics; strict models and means of visual presentation of evolution and self-organization patterns.

As to the latter correlation between synergetic ideas and images from the history of culture, let me give some examples.

The synergetic concept of *"attractor"*, or, as I used to say, "structure-attractor", is close to that of "goal". Attractors look like goals of evolutionary processes. I understand this in a non-anthropomorphic sense as trends in the behavior of nonlinear systems. The concept of attractor can be compared with that of Plato's *eidos*, and with Aristotle's *forms* as well as, in regard to human psychology, with *archetypes* in C.G.Jung's sense.

One can visualize the synergetic notion of *"bifurcation"* (branching in the evolutionary paths) by using one of the most ancient images known to humanity—the image of the World Tree. The image is present in the mythology of practically every civilization, with many different variants: the Tree of Life, the Family Tree, the World Pillar, the Evolutionary Tree, the Tree of Cognition, etc. World trees are simply versions of a cosmic organizational model in which spatial opposites (top - bottom, heaven - subterranean realm) are integrated. Temporal distinctions are removed as well: the past, the present and the future are represented synchronously in, for example, the image of the family relations, or in the organization of the developed scientific school (teacher and generations of his pupils). That is, all possible paths of development display themselves in the present [6].

Another synergetic concept is that of *"fractal"*. Fractals are objects which have property of self-similarity, or scale invariance. This means that the structure of a small fragment of such an object is similar to that of another, larger fragment, or even to the object as a whole [10]. As in the case of bifurcations, we can discover traces of this new synergetic concept in the existing images of culture. Consider, for example, Leibniz's concept of the monad. According to Leibniz, any monad reflects like a mirror all of the properties of the cosmic totality. The ancient Orient was aware of the same principle: "One thing in everything and everything in one thing". Thus, the old Chinese proverbs are: "When one speck of dust rises it contains the whole Earth", "When one flower blossoms the whole world opens".

SYNERGETICS AS A BRIDGE BETWEEN THE EAST AND THE WEST

It's rather surprising that synergetics being a product mostly of the Western science turns out to be very close to the Eastern way of the understanding of the world (presented, first of all, in Buddhism, Taoism, Yoga). The latter is closely connected with the hieroglyphic writing and can be characterized by integrity, in particular, by a non-dual principle of mental perception of opposites.

The synthetic way of thinking and some holistic picture of the world are peculiar for the East, while the Western style of thinking is used to be considered as analytical, logical, verbal and conceptual one. The Eastern people comprehend the world mostly through images, while the Western people understand it through words, concepts, numbers. Roughly speaking, the Eastern thinking is "right-hemispheric", whereas the Western one is a kind of "left-hemispheric" thinking. One of the main Eastern symbols is TAO, whereas the central Western symbol is LOGOS.

Synergetics discovers laws of new evolutionary holism; the synergetic way of thinking is holistic. This is one of the reasons of some astonishing closeness of the synergetic and the Eastern world views. I venture to believe that synergetics could serve as a bridge between the East and the West. Synergetics allows to synthesize some elements of the Eastern and the Western ways of thinking and outlooks on the world.

From the Orient synergetics perceives and develops the idea of integrity (everything in everything), the idea of recurrence and the common Law, i.e., the common path which the world as a whole and a human in it follow.

From the West synergetics takes positive aspects of the analytical methods, such as experimental basis, general significance and readability of scientific conclusions, their ability to be transmitted (from one scientific school to another, from science to culture) through scientific texts, specific mathematical means, and even through computer diskettes.

Besides, synergetic notions correlate with some ideas of the Western philosophical teachings, namely: Plato's notion of *eidos* as potential forms-patterns of processes and his idea of "perfect bodies of nature", symmetrical polyhedra; Leibniz monadology (pre-established harmony, coherence and mutual correlation of all parts of the universe, the ideas of a pre-formed and potentially pre-given), F. Nietzsche's ideas of eternal recurrence and pre-determination of the present; A. Bergson's notions of creative evolution, its moving force in a form of *elan vital*; P. Valery's ideas of chaotic regularity of the world, relative pre-determination of a new, paradoxical creativity of the world and of a human in it; C.G. Jung's archetypes of the collective unconscious; P. Teilhard de Chardin's ideas of direction of evolution, convergence and an evolutionary purpose, the Omega point; N. Hartmann's ontology of an hierarchically organized world and structures of teleological thinking.

Probably, one might speak about a special integrating mission of synergetics in culture. Synergetics appears to be a fruitful new formation into "the body of culture" which implicitly contains significant elements of both the Eastern and the Western way of understanding of the world. The "treads" of cross-cultural interaction of the East and the West could be stretched exactly trough this "knot".

ORIENTAL IDEAS REOPENED IN SYNERGETICS

Synergetics forces us to re-discover and comprehend in an unusual way some principles of the Eastern wisdom. Let me draw your attention here to some basic oriental ideas getting concrete re-interpretations in synergetics.

First of all, this is an idea of unity and harmony of the world. According to Buddhist and Taoist teachings, there is a unique all-penetrating connection of everything with everything. Each smallest particle of our universe is a special world inspired with its own life as well as with a general unique life of the universe.

Scientists are interested in a metaphorical sense of the idea. Indeed, synergetics gropes for some inner connection of world elements which can occur through small influences, fluctuations. Synergetics reveals the limits of truth of the idea. Small influences actually can determine a macro-picture of events, but only in particular cases. It may happen in certain types of complex systems (media) which are able to amplify the small influences with a nonlinear positive feedback. Complex systems become to be sensible to fluctuations (chaos on microlevel) in states of instability, i.e., near a bifurcation or blow-up point.

Another astonishing oriental idea is the idea of the potential and the unrevealed. It was presented in teachings of ancient Greeks too. An observable harmonious structure of the universe, Cosmos, arose from primordial Chaos, Being - from Non-Being, observable phenomena sprang up from an abyss unlimited in its potentials. Non-Being is a potential and an unrevealed, something without a certain form. It is an eternal, all-begetting and all devouring, basis of things. On the contrary, Being is something actual, revealed and already formed; it is a temporal and transient manifestation of Non-Being.

The potential and the unrevealed is, from the synergetic point of view, the availability of evolutionary structure-attractors. A result of fundamental importance has been obtained by the Moscow synergetic school: a continuous open nonlinear medium (complex system) potentially contains in itself different kinds of localization processes (different kinds of structures). The medium is a carrier of different forms of future organization, i.e., of a discrete spectrum of structure-attractors, a set of potential final states.

The idea of cyclic, eternal recurrence, the Yin-Yang rhythms is one more oriental idea consonant with synergetics.

The evolution of complex structures undergoes the periodical change of states, of various regimes of process development. There can not be sharp growth of structures without a threat of their fall and destruction. There are some universal laws of rhythms. They are peculiar to the living being as well as to the inanimate nature. There are some cyclic changes of states: upsurge - slump - stagnation - upsurge - slump etc. Only obeying these "life rhythms", oscillatory modes, the complex systems can maintain their integrity and develop dynamically. The mechanism of the "self-movement", auto-oscillations is profoundly analogical to the oriental image of Yin-Yang. Yin is complete potentiality and aspiration: the subconscious, unverbalized and unrevealed. Yang is realized: the verbalized, revealed.

According to synergetic models used here, changing of the two different, complementary regimes, HS- and LS-modes with peaking, is taking place in open media (systems) with strong nonlinearity. HS-mode is a mode of "infinitely running out wave" when there is no localization, all structures, heterogeneities are being washed away. LS-mode with peaking is a mode of "converging wave of burning", a mode of localization and intensive growth of processes in a more and more narrow area near the maximum. The switching over these two complementary regimes could be considered as a mathematical analogy of Yin-Yang alternation, i.e., changing of two basic world elements.

Finally, it's worth to mention the oriental idea of non-violence towards nature, human being and social organization. Synergetics develops the idea by investigating of the role of small, rightly organized influences upon complex systems. It proves that for efficient control over complicated systems (open nonlinear media) a right topological configuration, an "architecture" of driving mechanism rather than a power of perturbation itself is the most important. Weak, but rightly organized—so called resonant—influences upon complex systems are extremely effective.

This feature of a complex organization had been guessed thousands years ago by the father of Taoism Lao-Tsu. It was expressed as the weak defeats the strong, the soft defeats the hard, the low defeats the loud. Considered from the modern point of view, complex systems turn to possess a property of selective topological sensitivity. They demonstrate an unexpected strong reply to excitations which are relevant to their inner structural organizations, i.e., to resonant excitations.

TOWARDS A SYNTHESIS OF NATURAL SCIENCES AND THE HUMANITIES

Two main discoveries force us to revise the usual world view and constitute a basis of a new theory, the theory of self-organization. These are the discovery of strange attractors and the discovery of blow-up regimes. Both of them are of great philosophical importance. They open a possibility to build a bridge between synergetics originating mainly from the natural sciences and the humanities (cognitive sciences, epistemology, cultural anthropology, sociology, demography, economics, etc.).

The availability of strange, or chaotic, attractors is one of fundamental facts in the theory of self-organization. The strange attractors have been discovered by now almost everywhere: in various fields of natural and human world, starting with meteorology and ending with neurophysiology, investigations of the human brain activities. Strange attractors show us the limits of predictability of evolutionary processes and the availability of areas of principal unpredictability of some phenomena. This means the understanding of probabilistic, chaotic behavior of dynamic systems which is determined not by the limitation of our investigative instruments, but by the very nature of the systems.

The other fundamental fact is the law of population growth. We are used to believe that the processes of rapid growth, such as the population growth, "economic miracle" or information explosion, occur according to exponential law. As a matter of fact, it is one of the myths of classical science. Most of the processes of the avalanche-like growth occur much more rapidly, in blow-up regime when the characteristic values are increasing infinitely during a finite period of time.

The blow up regimes are under investigation in plasma physics, meteorology, ecology (growth and dying out of biological populations), neurophysiology (modeling of signal propagation along neural nets). There are blow-up regimes in social and cultural development as well. The growth of information and the population on the Earth obey the hyperbolic law, i.e., the processes occur in blow-up regimes [4].

Due to the interdisciplinary character of the fundamental discoveries, synergetics makes steps towards a new dialogue between specialists in natural sciences and in the humanities.

AFFILIATION

Helena Knyazeva
Institute for Theoretical Physics
University of Stuttgart
Germany

REFERENCES

[1] **Achromeeva, T.S., Kurdyumov, S.P., Malinetskii, G.G. and Samarskii, A.A.**, "Non-stationary Dissipative Structures and Diffusion-Induced Chaos in Nonlinear Media", *Physical Reports*, vol. **176**, 1989, pp. 186-372.

[2] **Haken, H.**, *Synergetics*, Berlin, Springer-Verlag, 1978.

[3] **Haken, H.**, *Principles of Brain Functioning. A Synergetic Approach to Brain Activity, Behavior and Cognition*, Berlin, Springer-Verlag, 1996.

[4] **Kapitza, S.P.**, "Population: Past and Future. A Mathematical Model of the World Population System", *Science Spectra*, No. **4**, 1996, pp. 60-65.

[5] **Knyazeva, E.N. and Kurdyumov, S.P.**, *Evolution and Self-organization Laws in Complex Systems*, Moscow, Nauka, 1994 (in Russian).

[6] **Knyazeva, E.N. and Kurdyumov, S.P.**, "Synergetics in Cultural Context. Contribution to Resolution of Current Civilizational Crisis", *Dialectic, Cosmos, and Society*, Chicago, No. **8**, 1995, pp. 3-8.

[7] **Kurdyumov, S.P.**, "Evolution and Self-organization Laws in Complex Systems", *International Journal of Modern Physics C.*, vol. **1**, 1990, pp. 299-327.

[8] **Laszlo, E.**, *The Interconnected Universe*, New York, World Scientific, 1995.

[9] **Mainzer, K.**, *Thinking in Complexity. The Complex Dynamics of Matter, Mind, and Humankind*, 3rd ed. Berlin, Springer-Verlag, 1997.

[10] **Mandelbrot, B.**, *The Fractal Geometry of Nature*, San Francisco, Freeman, 1982.

[11 **Maturana, H.R. and Varela, F.**, *The Tree of Knowledge. The Biological Roots of Human Understanding*, Boston London, New Science Library, 1988.

[12] **Morin, E.**, *Method. Towards a Study of Humankind. Vol. 1. The Nature of Nature*, New York, Peter Lang, 1992.

[13] **Prigogine, I.**, From Being to Becoming, San Francisco, Freeman, 1980

[14] **Prigogine, I. and Stengers, I.**, *Order out of Chaos. Man's New Dialogue with Nature*, New York, Bantam Books, 1984.

[15] **Samarskii, A.A., Galaktionov, V.A., Kurdyumov, S.P., and Mikhailov, A.P.**, *Blow-up in Problems for Quasilinear Parabolic Equations*, Berlin, Walter de Gruyter-Verlag, 1995.

DINO KARABEG

PROSPECTUS FOR POLYSCOPIC MODELING

1. INTRODUCTION

It is best to introduce polyscopic modeling as an *attitude* towards information, rather than as a collection of techniques for managing and representing information. If polyscopic modeling is identified with its current implementation, then its development may be hindered. When, on the other hand, the underlying attitude is understood, those techniques that have so far been developed easily follow.

In order to motivate the attitude which defines polyscopic modeling, we apply its First Principle: (speaking figuratively) we take a few steps back from the picture we are looking at or creating and, like a painter looking at his work, we look at the broad and essential contours of the whole, at its *composition*. The whole, in the present case, is the contemporary Western society and culture. We are interested in understanding the role which information has in it.

Different views are possible. According to one of them, we live in a time of prosperity and great opportunities. Not long ago, the human race freed itself from the irrational beliefs and mores of tradition. No longer a traditional society, we are free to use *reason* to find a way towards a true betterment of the human condition. Science and technology carry us rapidly in the chosen direction. Information plays a central role in our new rational culture, being the very material of which rational decisions are made. We are therefore fortunate to have information technology which provides easy access to information world round. The information technology turns all the world's cultures into a single world culture.

The above *optimistic* view is by its basic emotional hues more in tune with the temperament of the last turn of the century than with the present. Today one is more likely to hear the *pessimistic* view, according to which we are notoriously unable to direct our actions prudently. Science and technology take us rapidly towards some final disaster, be it ecological, nuclear or another unanticipated one. Information technology gives us far more information than we are able to digest, leaving us utterly congested and confused. We try to make sense by superspecializing (we cannot possibly follow *all*). Superspecializations turn a single culture into a myriad of tiny subcultures which do not communicate with one another. At the same time the basic questions which determine the direction of human effort (called the *high-level questions*) remain unaddressed, being too broad for each particular area of specialization.

It is, of course, possible to combine the above two views into a single one, according to which the optimistic scenario is a possibility and an orientation worth pursuing, and the pessimistic scenario is probable reality unless the prevailing attitudes change. That combined view motivates polyscopic modeling.

An objection is heard at once that this motivation is too ambitious, too vague and too general: "In science, which is our chosen path towards truth, we proceed slowly and carefully, studying within established disciplines, with caution and rigor, only those questions which can be properly defined and treated with the trusted techniques of those disciplines. After many small steps we construct a picture of reality which can be relied upon."

While this objection carries a useful message about the need for precision and rigor, the attitude which it represents has a serious drawback—it *breeds* the pessimistic scenario. Polyscopic modeling inherits the modeling-related insights and techniques from science as the best ones available, but it adapts them for its purpose and combines them with an orientation which is not usual in science—it gives a higher priority to higher-level questions. Polyscopic modeling aims to supplement and to complement science, not to replace it. Science, etymologically related to 'schism' and 'scissors', dissects and isolates in order to understand. Polyscopic modeling is an art of putting disparate pieces of information together into a coherent whole.

The objection that high-level questions tend to be too complex to be properly understood in simple terms is answered by the First Principle of polyscopic modeling which extends into the following advice: From the basic composition of the whole choose one or several concepts which characterize it well. When the selected concepts are considered together with the corresponding context, a simple *view* of the complex situation results which reflects its essence and which can be easily grasped. When this principle is applied to modern culture, *information* emerges as a key concept.

2. INFORMATION AND CULTURE

The Second Principle of polyscopic modeling connotes that words should not be assumed to have unambiguous links with reality but that they should rather be *defined*. In what follows, the word *information* is used in a limited sense to denote "the record of experience". Information which is detailed and technical is said to be *low-level*, while the *high-level* information covers basic issues in non-technical terms. Typical low-level information is technical know-how. The high-level information decides what will be produced and bought and what political agendas are worth pursuing. We use the word *culture* to denote both information and various activities oriented towards its cultivation, such as education, art and science. *Ideognomy* is what value judgements are based upon. It is the way of determining what is to be considered as true, good, prudent, useful or just. Since it decides what information is worth cultivating, preserving and relying upon, an ideognomy is a foundation of culture.

Since prehistoric times generations of people have contributed to our basic repertory of skills. It has often been said that the ability to learn from the experience of others, especially through speech and writing, has been the deciding advantage of our species over other species in nature. On the other hand, both our condition and the condition of our environment have gone far from the original natural ones in which we were able to survive guided by instincts alone. We *depend* upon information.

Likewise, we depend upon a functioning culture. The question "Can a culture fail to function?" is therefore a subject of choice for a polyscopic modeling study. One might immediately respond: "Of course not! Culture itself is a time-tested part of our heritage which has *evolved* together with the species and with civilization. How could it possibly stop functioning?" Yet there are sensitive periods in the evolution of culture where substantial structural changes take place during a short period of time. In such times one can not rely upon the cultural status quo as being time-tested and stable. We are now living in a time of profound structural changes in culture.

Until only a generation or two ago, the mechanism of culture was predominantly based upon the customs and rituals of tradition. Information was recorded and passed on through a system of beliefs, habits and obligations which were followed without necessarily being rationally understood. Science, from physics to psychology, has repeatedly challenged the tenets of tradition. From science we have learned not to rely on the authorities from distant past. We have learned to rebel, to question rationally.

During the recent decades the living conditions have changed so much that the traditional ways no longer apply. Our beliefs have changed. We no longer comply to the old ways. The mechanism of traditional culture has been damaged beyond repair. The culture is going through a *change of ideognomy*, a change from a predominantly *traditional* ideognomy to another one which has not yet been formed. The current culture is characterized by a predominant reliance upon rational reasoning and argumentation.

A change of ideognomy is to a culture what a change of foundations is to a house. When a house is being moved to a new foundation it is necessary to secure that the house can be safely carried over and that the new foundation is large and solid enough to carry it.

We must ask whether we are able to recreate a viable culture on the existing rational foundation. It seems inevitable that much of information will be lost during the 'move', because we are neither always able to distinguish those parts of the traditional heritage which are obsolete from those which are essential, nor are we necessarily capable of replacing the dysfunctional mechanisms of tradition by new functioning ones. On the other hand, the change of foundations may be a highly positive development leading to spectacular possibilities. We may be able to create a foundation that will support the development of a culture profoundly superior to the traditional ones. Our handling of information will determine which of the existing hazards and possibilities will materialize.

Information is central not only in culture but also in politics and in economy. The basic political issue is whether the existing social mechanism is *just*, i.e., whether it secures that people can freely benefit from their work and from the resources in their natural and socio-cultural environment. History exhibits an evolution of political justice beginning with the abolition of slavery and ending with modern democracy. Now that people are able to *choose* their government and their political system, has the basic political issue been resolved once and for all? Not necessarily. We may be free to choose provided that we *know* how to distinguish our own from other competing interests, provided that we are informed and not misinformed. The circumstances in the economy are similar.

When we look at the basic contours of information in its socio-cultural context, we see on the one side a high vulnerability of information which is caused by the culture going through a change of ideognomy, and on the other side we see high pressure towards abuse, which is always there when political and economic power is at stake. A critical look at the way in which information is handled is therefore called for.

3. HOW INFORMATION IS HANDLED

We notice at once the curious fact that the most sensitive part of information, its high-level part, is hardly covered by any regulation or organized effort. Cultivation of high-level information seems to have died with the traditional culture. An exception is of course the advertizing industry, which has developed an arsenal of methods for instilling ideas, tastes and values. Advertizing has replaced the Church in telling us about the purpose of our earthly existence.

We do not usually consider the information that has been developed within the world's traditions as something worth preserving. We tend to dismiss it with a sense of superiority, even with a touch of arrogance—we have *science*, we like to believe, which is providing us with knowledge of a superior quality. We tend to overlook the fact that science is not at all preoccupied with high-level issues.

We try to patch the resulting holes in high-level information by using the scientific information on the high level or by drawing high-level conclusions from the scientific information. That gives rise to a number of problems which now mark the entire culture (cf. Appendices I and II).

When the thinking style which is usual in science—isolating small parts from the whole in order to understand them—is applied to high level issues, another bulk of problems result: Through long chains of cause-and-effect relationships running across many areas of specialization, our actions produce harmful side effects which last much longer than their benefits. With time deep problems develop which can only be understood and treated through holistic approaches.

Our culture puts us into a peculiar situation: We are given a right and an obligation to make choices on a multitude of vitally important issues. But our high-level information is such that we are unable to make those choices in a rational way.

The *low-level* information is the best part of our culture, as the power of our technical civilization clearly demonstrates. But without a meaningful direction, the value of such power is questionable.

There is a clear trend towards low-level information becoming more and more low-level and more and more fragmented and disconnected. Disconnected pieces of low-level information are difficult to make sense of and difficult to use. A scientist working in such surroundings is forced to turn the informational chaos into relative order by specializing. A specialization that is thus inspired will tend to take him further away from the higher-level issues which originally motivated his line of work. Ultimately, this process separates the best qualified people from the important issues of their time. While excessive amounts of low-level information continue to be produced, large gaps in high-level information remain unnoticed.

We notice also enormous quantities of information competing for our attention. Like so many other commodities, information is treated according to the rule 'the more—the better'. But when the limited ability of humans to assimilate and digest information is taken into account, it immediately becomes clear that more information does not necessarily mean more knowledge or more benefit of any kind. Even accurate and in principle useful information can be misinforming if its style of presentation and its quantity make it difficult or impossible to digest, or if it eliminates in competition for attention some information which is of a greater value.

The representation method of choice for the most relevant information—the printed text—has remained essentially unchanged in its style and structure for centuries, since the time when there were only one or a handful of books to be read and contemplated. In the meantime the number of books has increased by many orders of magnitude. Text is usually intended to be read sequentially and any statement in it may depend on any preceding statement. Accessing such information is a slow process. We may be coming close to the point beyond which more textual information effectively means *less* knowledge, since that which already exists is made less accessible.

It is interesting to speculate about how the character of information might have influenced the character and mentality of a contemporary educated person. Among a number of possibilities we mention *the loss of perspective* (the present usage of the term is explained below).

4. POLYSCOPIC MODELING ATTITUDE AND PRINCIPLES

The goal of polyscopic modeling is information which *can* fulfill its pivotal cultural and social role. Polyscopic modeling is characterized by what may be called an 'environmentalist' attitude towards information—it endeavors to reduce pollution, to preserve, to protect, to rebuild and to cultivate. As an approach to information representation polyscopic modeling aims to represent information in such a way that it can be understood.

The two main principles of polyscopic modeling correspond to the two words in its name. The principles facilitate practical action based on the above-described attitude.

The First Principle, associated with the word *polyscopic*, states how information is selected and represented. It connotes that the *whole* is represented by multiple simple, related models called *views*. The views are both *macro-* and *microscopic*, both high-level and low-level.

Each of the views is sufficiently concise so that it can easily be *grasped.* Simplicity is not gained by ignoring all information which lies outside of some area of specialization or which disagrees with some theory or metaphysics or tradition, but by organizing information into multiple simple views.

The views are organized in such a way that they provide a *perspective* on the represented subject of study, which clearly delineates the large from the small, the crucial from the less important. This is done by placing that which is large and crucial into high-level views, and that which is small and less important into low-level views. The role of the perspective in information representation is further explained in Appendix III.

The polyscopic representation provides for an efficient access to information, in a similar way as searching for information in a telephone book is more efficient than searching in an unordered list of names. The higher-level views provide the basic composition of the whole, the lower-level views provide the details and the justification for high-level claims. The high-level models makes it easy to select only those low-level that are of interest. If the representation medium allows, a reader is given the possibility to select the *scope* (the terms or aspects of the subject that are of immediate interest) and to obtain in response the corresponding *view* consisting of only that information which is most relevant for that scope.

The polyscopic representation also makes it possible to bind together and present heterogeneous pieces of information or points of view on the same subject. The individual pieces are represented as lower-level models while the statements and ideas which bind them together form the higher-level models.

The Second Principle of polyscopic modeling is associated with the word *modeling*. This principle determines the *ideognomy* of polyscopic modeling (the word in this context replaces the more usual and more specialized word *epistemology*). It connotes that polyscopic modeling views information as no more and no less than a record or a *model* of experience (the word 'experience' is understood in a broad sense).

Usually, epistemology stems from some stated or unstated metaphysics which determines the nature of reality and thereby also decides what in experience should be considered as real or possible. The current metaphysical position that prevails in our culture is the mechanistic metaphysics of classical science, a position which modern science has proven wrong and abandoned [4]. That metaphysical position is a well-known obstacle to cultural integration and to a smooth ideognomy change.

Polyscopic modeling assumes no metaphysical position whatsoever. The construction of a reliable metaphysics is simply regarded as a much harder, more error-prone and less important task than the modeling of experience. The practice of treating the latter task as dependent on the former is therefore considered as not rational.

Freedom from metaphysics means freedom to choose the manner and the terms in which experience is represented which are best suited for its understanding and communication. It also means freedom to open any form and record of experience for rational consideration. Reliability is not gained by limiting the information taken into account to that which agrees with some chosen theory or metaphysics or tradition or data collection methodology, but on the contrary by taking as much as possible all relevant experience into account. That which is common or consistent in various forms of experience is raised to a higher level as both more reliable and more general. The goal of polyscopic modeling is information which is easy to understand and use, and which agrees with or embodies or *explains* much of what is relevant in experience.

The Second Principle connotes also that polyscopic modeling does not search for a single *true* model of reality, but that it considers different models to be more or less successful *approximations* of experience. It follows that different models of a single phenomenon not only *can* be valid, but may also be necessary. Metaphysical beliefs and theories are of interest as much as they reflect and point back to experience.

In modeling *representation* is preferred to mere description or opinion. Modeling aims towards precision and rigor. The word 'modeling' also implies the use of *analogy* for understanding that which is unfamiliar or unknown in terms of that which is familiar or known.

As explained in Appendix IV, both polyscopic modeling principles agree with basic insights and trends in modern science.

5. POLYSCOPIC MODELING TECHNIQUES

Ideas and techniques from science, art and other traditions are combined and adapted for developing polyscopic models. The predominantly theoretical or technical techniques are illustrated in Appendix V. The predominantly artistic or intuitive techniques are shown in a separate article in this volume [5].

Polyscopic modeling *languages* are theoretical tools for defining terms in which models and their relationships can be stated precisely. They combine aspects of mathematical theories and computer programming languages. Polyscopic modeling *environments* are computer tools for reading and writing polyscopic models. They are similar to computer programming environments.

In Appendix V a rudimentary relational polyscopic modeling language called **ab** is illustrated by an example. The approach taken is similar as in mathematics: basic modeling terms and relationships are defined by using formal objects and studied theoretically. The resulting theory can be thought of as an abstract *model* of the real-world modeling. The terms and insights from the theory can be applied in practice by analogy.

In Appendix V a modeling environment based upon the **ab** modeling language is also briefly described. It is shown how this approach can be used for representing information in such a way that it can be accessed efficiently.

The example given in Appendix V emphasizes the use of *patterns* or abstract models. Patterns play a central role in polyscopic modeling theory and practice. Intuitively, patterns represent 'kinds of relationships' that may be formed by two or more entities.

In addition to theoretical and technical tools, intuitive modeling techniques such as *ideograms, metaphors* and *dialogues* are used in polyscopic modeling. Such intuitive techniques are necessary for representing high-level information in an accessible way. Ideograms condense the essence of some complex relationship or situation into a picture which can easily be grasped. When used for representing *patterns*, ideograms have a similar role in practical polyscopic modeling as mathematical symbols in physics. Metaphors provide an alternative way for defining patterns. Metaphors serve for applying insights from a situation which is familiar in a situation which is less familiar. Dialogues are used for presenting and harmonizing multiple points of view. The use of ideograms, metaphors and dialogs is illustrated in [5].

6. TOWARDS DEMOCRATIZED INFORMATION

By representing high-level information so that it can be understood and verified by non-experts, polyscopic modeling democratizes information. The procedure of model reading and verification resembles a jury trial, in which a reader acts as a juror. A high-level claim which is provided within a high-level view is analogous to an accusation in court. Lower-level models contain reports by witnesses and experts and other forms of evidence which the readers can consider as needed.

The polyscopic structure makes it possible to separate the general, high-level statements from technical details of justification whose understanding may require special training. The model is made available to public. When and if sufficiently many readers have been convinced that the claim has been proven beyond reasonable doubt, the public opinion that they have jointly formed corresponds to a positive jury verdict.

Jury trials are a standard device used in jurisdiction world wide in order to prevent justice from slipping out of the hands of the populace. It is the common people and not legal experts who decide on innocence and guilt in court. That makes it less likely in practice that the court could be controlled by state or private interests.

The benefits expected from democratizing information are similar as the benefits of jury trials and the hazards are smaller—unlike court verdicts, the rulings in polyscopic modeling are never irreversible. Polyscopic models change whenever relevant new information becomes available. A reading of a model can be repeated at any time.

AFFILIATION

Dino Karabeg
Institute for Informatics
University of Oslo
Norway

REFERENCES

[1] **Bohm, D. and Peat, F.D.**, Science, Order and Creativity, Bantam, 1987.

[2] **Einstein, A.**, *Autobiographical Notes*, Open Court, 1949.

[3] **Einstein, A.**, *Ideas and Opinions*, Random House, 1954.

[4] **Heisenberg, W.**, *Physics and Philosophy*, Harper, 1962.

[5] **Karabeg, D.**, "Polyscopic study of a basic cultural pattern", in: **Aerts, D., Van Belle, H. & Van der Veken, J.**, (eds.), *Worldviews and the Problems of Synthesis; The Yellow book of Einstein meets Magritte*, Kluwer Academic Publishers, 1998.

[6] **Oppenheimer, J.R.**, *Atom and Void*, Princeton Univ. Press, 1989.

[7] **Schrödinger, E.**, *What is Life?* with *Mind and Matter*, Cambridge University Press, 1992.

APPENDIX I

On Using Low-level Information as High-level Information

An example that illustrates the typical compound of relationships which result when scientific information is used on the high level is the biochemical-anatomical model of the human: Thousands of chemical compounds and tiny and large anatomical details! That model is so complex that no medical doctor and no scientist can understand all of it.

Doctors specialize: one understands the eyes, another one the ears. If one has problems with a certain organ, his specialist is unlikely to determine if the problem with the organ has origin in another organ. He is even less likely to realize, as the case may be, that the actual cause of the problem is in the patient's unhealthy lifestyle or environment.

In spite of its complexity, the scientific model of the human fails to explain some of the most basic questions such as why humans hug and shake hands, not to speak about the biochemical mechanisms that are involved in reading poetry or falling in love. (One may also find more profound scientific reasons why the scientific model of the human is incomplete [7].)

Since the biochemical-anatomical model is our culture's idea of the human, it should not be surprising if those sides of culture whose existence is not justified by the model tend to depreciate in value and fade from daily use. Poetry now tends to play a less visible role than before, while love tends to be reduced to its anatomical aspects.

How can one use a model which is too complex even for a specialist, and yet incomplete, to decide about his daily affairs? He is bound to conclude that the world is just too complex to be understood. He must rely on experts who presumably *do* understand it. When he listen to advertising, he has little choice but to trust that the experts who made all those good things know better than he does what is good for him. If those good things don't do him good and his health deteriorates, he has little choice but to rely on some medical expert to fix it.

Why is our picture of reality composed of models that are clearly too technical, too complex and too incomplete to be practical? The answer is clear: we believe that those models *correspond* to reality, which itself is a huge and tremendously complex mechanism. This metaphysical belief has been with us since the last turn of the century. It is the metaphysical belief that science held at that time when scientific metaphysics replaced the traditional metaphysics in the minds of the intellectual elite. Subsequently, scientists have found out that the reality is *not* a mechanism [4].

APPENDIX II

On Creating High-level Information from Low-level Information

In the following metaphor the hazards involved in using scientific information to construct high-level information are illustrated. The numbers in the metaphor play the role of information.

```
x = 31947332560784434998990756462873458698686990
y =   -9184345987920794757285496668278738883 20256
```

a

$$x = 3.19 \times 10^{43}$$
$$y = -9.78 \times 10^{41}$$

b

```
x =    4 332     84    998           2 73
y =    18 34     9    94    2        6 2       8
```

c

Three representations of numbers x and y are shown. Representation *a* corresponds to low-level information. That representation is completely accurate, but too detailed and too verbose to be suitable for grasping the numbers. Representation (b) corresponds to high-level information. It is less precise but it provides an immediate intuitive idea of the numbers. Representation (b) is for most purposes better suited for human use than representation (a). In representation (c) some but not all of the digits of the two numbers are specified.

Science aims at a precise low-level picture of reality corresponding to representation (a), but the picture it operates with at any time is inevitably incomplete, resembling representation (c) rather than (a). It is easy to see why it so often happens that (b)-like high-level views are constructed from information of type (c), in spite of the fact that the outcome of such a procedure must be grossly inadequate. Such high-level views *seem* highly accurate because the *existing* information is obtained with large reliability and precision, while the *missing* information is not there and is therefore easily discounted. Indeed, we are often so impressed by high technical elaboration and 'precision' of such procedures that we tends to disregard the high-level information which has been produced in other ways as *non-scientific.*

APPENDIX III

Perspective in Information Representation

We naturally understand something by mentally or visually alternating between global and local scopes. When looking from an airplane, one sees only the broadest and, largest features of the landscape: a mountain, a lake and perhaps a village by the lake (a). When looking at the village from a boat on the lake, one sees individual houses and a segment of the shore (b). By looking into the water, one sees the lake weeds and fishes. Every time one changes the context to a more detailed one, one remains aware of the larger picture to which the details belong.

In the same way, information representation is clearest when all the views, from global to detailed, are provided separately. One can then focus on the more global and the more detailed views at will without losing the *perspective*—i.e., the distinction between the large and the small, the most relevant and the less relevant, and the way the details all fit together into a whole. When only a single scope (b) or (c) is provided, it becomes difficult or impossible to reconstruct the global the view (a). When elements from global and local views are brought together within the same frame, confusion (d) results.

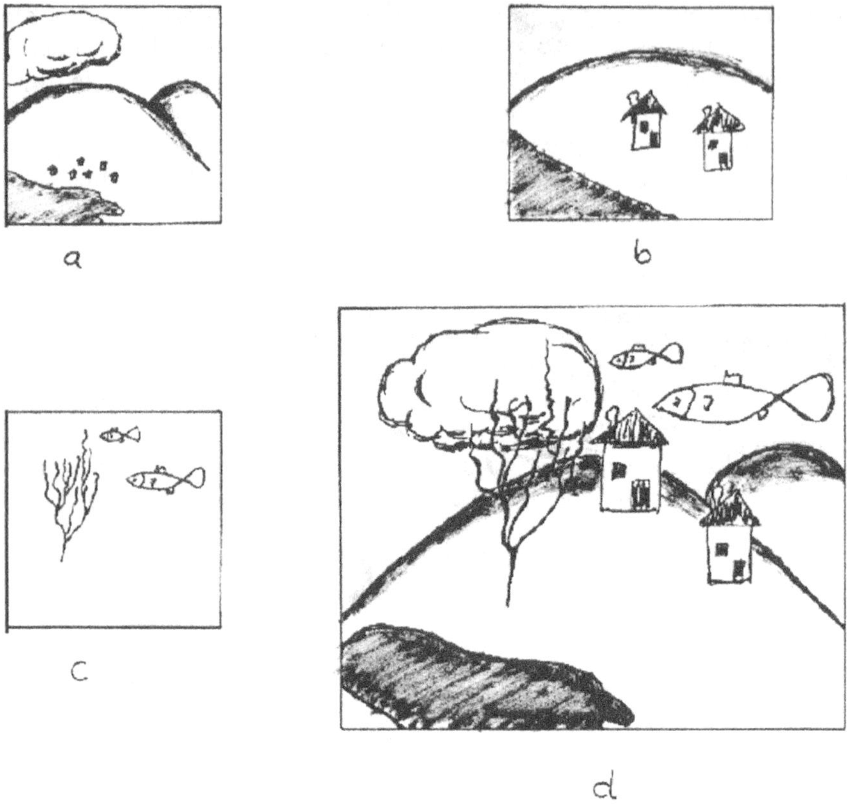

APPENDIX IV

Polyscopic Modeling is Scientific

If we should interpret the word 'scientific' as 'whatever agrees with the prevailing routines in science', then polyscopic modeling can clearly not be honored by that attribute. But if the word 'scientific' is interpreted as 'whatever agrees with the essential insights from science and with its vital trends', then the conclusion is different.

As it has usually been the case in the history of science, modern physics is in the forefront of the methodological trends. The principles of polyscopic modeling are there easily recognized.

It is easy to find examples which show that physics has become polyscopic. Although Newtonian mechanics has been proven wrong, it is still taught at schools and used in practice as a suitable high-level *approximation* of the more accurate relativistic mechanics. The much simpler Newtonian model is clearly precise enough for most practical purposes.

The wave/particle duality of matter is an example of polyscopicity on a single level. Experimental studies have shown that small chunks of matter behave sometimes like waves and sometimes like particles. Is an electron a wave or a particle? It is clearly neither the one nor the other. But it is useful to *understand* an electron in terms of waves and particles which are objects from common experience. On the low level the behavior of electrons is accurately described by using mathematics. That representation is, however, more difficult to talk about and to visualize than waves and particles.

High-level modeling in terms of simple analogies facilitates understanding. However, multiple simple models are often needed to tell the whole truth, and the temptation to oversimplify by considering one of them as the 'real' one needs to be resisted.

The Second Principle of polyscopic modeling agrees with the epistemological position that prevails in modern physics. Einstein wrote in his autobiography [2]:

I shall not hesitate to state here in a few sentences my epistemological credo[...] I see on the one side the totality of sense experiences and, on the other, the totality of the concepts and propositions that are laid down in books. [...] All concepts, even those closest to experience, are from the point of view of logic freely chosen posits, just as is the concept of causality[...]

Einstein defined science as follows [3]:

Science is the century-old endeavor to bring together by means of systematic thought the perceptible phenomena of this world into as thorough-going an association as possible.

In polyscopic modeling a wider range of experience is modeled than in science. The scope of science is restricted for historical reasons: initially, the less 'objective' realms of experience were under the auspices of the Church and the tradition.

The methodological and epistemological advances in physics that have taken place during this century have amounted to a large paradigm change which is now spreading into other sciences and into the culture at large. Our world views are still largely a product of the world view of classical science. Some of the leading physicists have written about the errors in those world views and about profound changes that are expected to result when the insights from modern physics are digested and assimilated [4] [6] [1]. The changes in the way science understands itself and the world give us an opportunity to not only correct our world views according to existing insights, but also to extend the whole project of science much beyond its present scope.

APPENDIX V

Polyscopic Modeling Theory and Computer Environments

The role of the modeling languages and the associated theories in polyscopic modeling is similar to the role of mathematics in physics. Polyscopic modeling environments are similar to computer programming environments. They employ information technology for reading and writing polyscopic models.

We introduce the polyscopic modeling theory and environments by an example. The example involves a prototypical situation which has been called *the voter's dilemma*: During an electoral campaign a voter hears a presidential candidate support his agenda by claiming that *tax cuts remedy recessions* (statement A), because they stimulate spending and investments. While preparing to cast his ballot in support of that candidate, the voter hears another candidate claim that *tax cuts cause recessions* (statement B), because they damage the infrastructure.

The voter is perplexed. He is determined to fulfill his democratic duty conscientiously by making a well qualified decision. However, several obstacles stand in his way.

- Common logic rules out the possibility of both statements being true. Yet the voter feels that in some ways both are true and both false. He finds it unacceptable that politicians can come up with such contradictory statements. But when he looks at any one of them, he feels that there is some important truth in it. He observes that one can 'prove' almost any claim about a complex system such as the economy by conveniently selecting some of the parameters and leaving out the others. He concludes that logic is of little use in dealing with such things as politics.
- He would like to use an economic textbook to decide the issue. But he finds out that none of the available textbooks has a conclusive concise explanation of the particular relationship he is interested in. Of course, the information he needs is probably scattered around

different chapters and texts and in principle available provided that he is able to put it together. But if he is to do that, he must read and understand large parts of economic texts. Even if he were to begin such an undertaking he can not properly conclude it, since no book will ever tell him whether the information he has assembled is sufficient. Perhaps another crucial aspect of the situation is described on another page or in another book?

The voter's dilemma is a *never-ending* dilemma.

THE **AB** POLYSCOPIC MODELING LANGUAGE

The notion of a *modeling language* is central in the theory of polyscopic modeling. A modeling language consists of formal objects called *models* and of functions and relations defined on models which constitute a *modeling logic*. By using the functions (operations) it is possible to construct new models out of existing ones. By using the relations it is possible to give names to properties of models and to define relationships that are formed by models.

In a construction of a polyscopic modeling language the choice of models and modeling logic is tailored to suit the purpose of the language. A simple relational polyscopic modeling language called **ab** is used for defining the basic polyscopic modeling terms.

An **ab** model consists of *concepts* and of binary *relationships* among concepts. Each concept and each relationship has a *quality* which can be either *positive*, *negative* or *unspecified*. In addition, each relationship has a *time scale*, which can be *long-term*, *short-term* or *unspecified*.

Models can be represented by graphs in which nodes correspond to concepts and arrows to relationships. The values *positive* and *negative* are represented by light and dark shading respectively, the *unspecified* values are marked by absence of shading. The shading of the head of an arrow denotes its time scale: a dark head shading denotes short-term, while light shading denotes long-term relationships.

In model M_1 in Figure 1 recessions are represented as a negative concept. Tax cuts are represented as having a negative influence on (i.e., reducing) recessions.

If a model can be changed into another model by renaming the concepts, we say that the two models are *isomorphic*. Mutually isomorphic models constitute a *pattern*. Intuitively, patterns are generic relationships or kinds of relationships among multiple concepts. In Figure 1 the "solution" pattern is represented by its *generic model*. The concepts in a generic model ("problem" and "remedy" in the figure) are said to be the *roles* of the pattern.

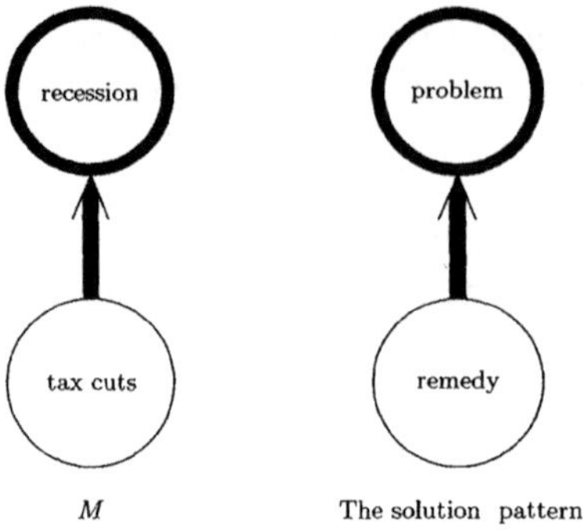

Figure 1. A simple **ab** model (M) and the corresponding pattern.

Generic models are a convenient tool for giving names to roles which objects may perform within a compound relationship and for defining kinds of relationships. In the example, the terms "problem", "remedy" and "solution" are defined by the corresponding generic model. By exhibiting an isomorphism between model M_1 and the generic model of the remedy pattern, the statement "Tax cuts are a remedy for recessions" is shown to be an *interpretation* of model M_1.

Once statement A is represented by a model, it can be formally *justified.* A model is justified by showing that it can be derived from previously justified models using modeling logic (an alternative form of justification is by referring to experience). In Figure 2 model M is justified by *simplifying* model M_2, which is assumed to be already justified.

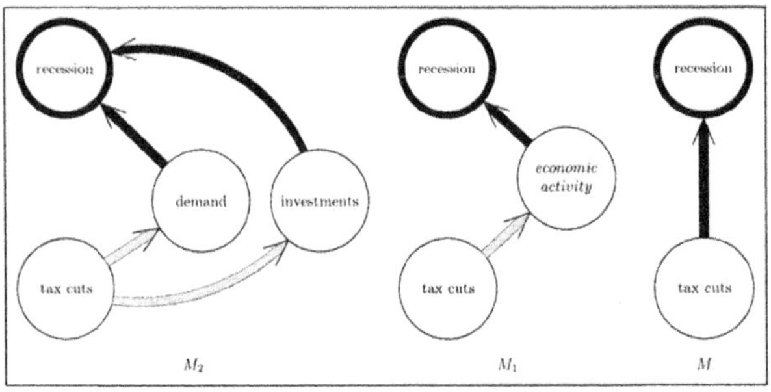

Figure 2. Model M is justified by simplifying model M_2: model M_1 follows from M_2 by abstraction; model M follows from M_1 by concept elimination.

The simplification is done in two steps: first concepts *investments* and *spending* are joined together by *abstraction* into a concept called *economic activity* yielding model M_1 (abstract concepts have those relationships which are common to all contained concepts). Model M is then showed to follow from M_1

by *concept elimination.* In concept elimination, a concept and is removed from a model along with the relationships it forms and the remaining relationships are adjusted as follows: each pair of relationships such that the first leads to and the second from the eliminated concept is replaced by a single relationship, their *composition.* The composition of two relationships is positive if and only if the relationships are either both positive or both negative. The composition of two relationships is long-term if and only if at least one of the relationships is long-term.

In figure 3 model N corresponding to statement B is shown. The model is derived from model N_1 by eliminating "poor infrastructure".

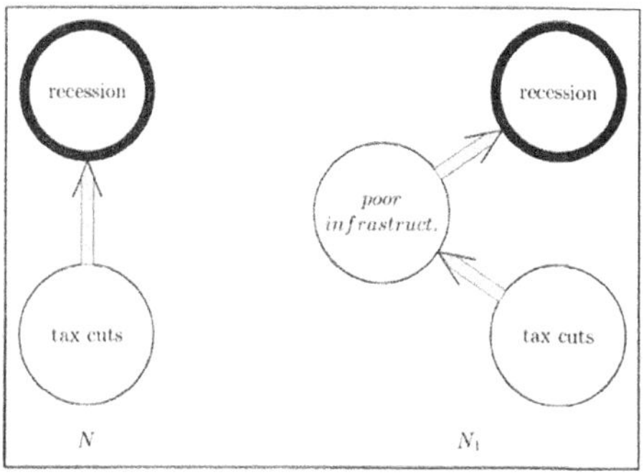

Figure 3. Model N follows from model N_1 by concept elimination.

Since models M and N contain contradictory information, they are said to be *incompatible.* The emergence of an incompatible model pair invalidates the justifications of both models of the pair. A new *coherent* model (i.e., not containing immediate contradictions) needs to be constructed which is in agreement with *all* available information.

Incompatible models M and N originated by *restriction*, i.e., by *leaving out* information from a larger model or from experience without making the corresponding adjustments to the information that is preserved. Since restriction may lead to incompatible views, it is not permitted in model justification.

In order to reconstruct a coherent model that includes the information of models M and N, it is necessary to *undo* the restriction somewhat and go back to more detailed models. That can be done by enlarging the *scope.* A scope includes concepts and relationship qualities and time scales (intuitively corresponding to the concepts and kinds of relationships that are of immediate interest).

Models M' and N' in Figure 4 are obtained by adding time-scale information to models M and N. Models M' and N' are compatible. Their *model union* is a coherent model P, which belongs to the *temporary solution* pattern. The statement *tax cuts are a temporary solution to recessions* is thereby justified.

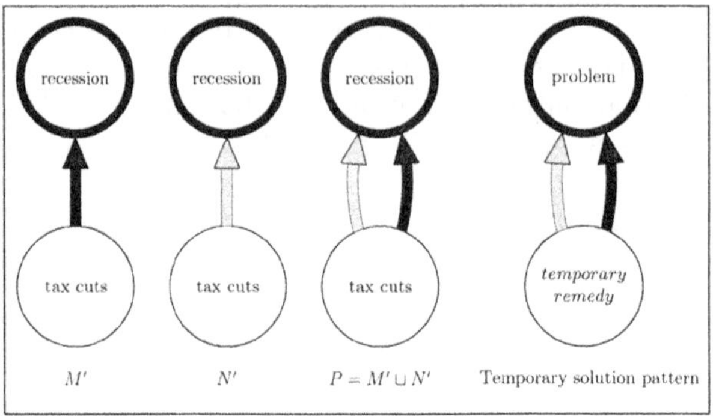

Figure 4. By specifying the time-scale information in models M and N, the result is a pair of compatible models, M' and N'. Their model union P is a coherent model which belongs to the temporary solution pattern.

In practice model union serves for joining pieces of information into a single model. Other model operations and relations such as model intersection, inclusion and closure are defined in a similar way as model union.

Model P_1 on Figure 5 is the union of models M_1 and N_1. Model P_1 is coherent. *Views* of model P_1, such as model P_2, can be computed by simplifying model P_1. Model P_2 belongs to the *symptomatic solution* pattern. Thus the statement *tax cuts are a symptomatic remedy for recessions* is also justified. However, since neither P nor P_2 can be simplified to M or N, neither the statement A nor the statement B can be justified.

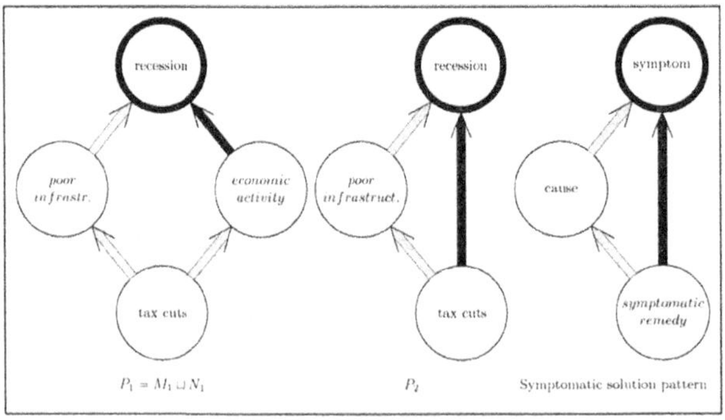

Figure 5. The union P_1 of models M_1 and N_1 is coherent. Model P_2 is a view of model P_1 which belongs to the symptomatic solution pattern

Simplifications of a model are said to be the *views* of the model. A view can be defined by a model and a scope. The possibility of defining views by selecting scopes is essential for polyscopic modeling.

Views can be used in practical modeling for joining together models by extracting information which they have in common (common views) and by using it as higher-level models (and connecting elements) in a model hierarchy. When it is shown that two or more models can be simplified to models which belong to the same pattern, the pattern emerges as their common abstract view. That approach has been used for intuitive modeling in [5].

The same reasoning which is used for discussing the relationships among models naturally applies to relationships among patterns. From the fact that neither the *symptomatic solution* pattern nor the *temporary solution* pattern can be simplified to the *solution* pattern it follows that neither symptomatic solutions nor temporary solutions are solutions.

POLYSCOPIC MODELING ENVIRONMENTS

Information technology has produced powerful alternatives to printed paper. The fact that an increasing proportion of information now exists on-line presents an opportunity to search for more efficient alternatives to linear text. Polyscopic modeling environments combine the advantages of polyscopic models and modern information technology. They provide facilities for on-line model reading, writing and updating.

In Figure 6 a screen from the **ab** modeling environment is shown. By clicking on the *economic activity* concept a concept menu is activated, which offers various concept-related operations. If the option *expand* is selected, the concept (presumably a result of abstraction) is expanded, and a model similar to M_2 in Figure 2 is displayed. Alternatively, by selecting the option *reduce*, a model similar to model P_2 is shown. By clicking on a relationship arrow, an analogous relationship menu is shown.

By selecting a menu item *text*, textual information can be displayed, along with formulas, figures, historical data etc. Thus a polyscopic model subsumes text. A polyscopic model enhances the access efficiency of text by effectively freeing the text of its one-dimensionality.

The described representation makes it possible to select the scope of interest (including, say, tax cuts and recessions) and obtain a minimal view as a result, consisting of exactly those concepts and relationships that are necessary for understanding the concepts selected within the scope. The text corresponding to the view can then be read. In that way a solution to the voter's dilemma can be provided.

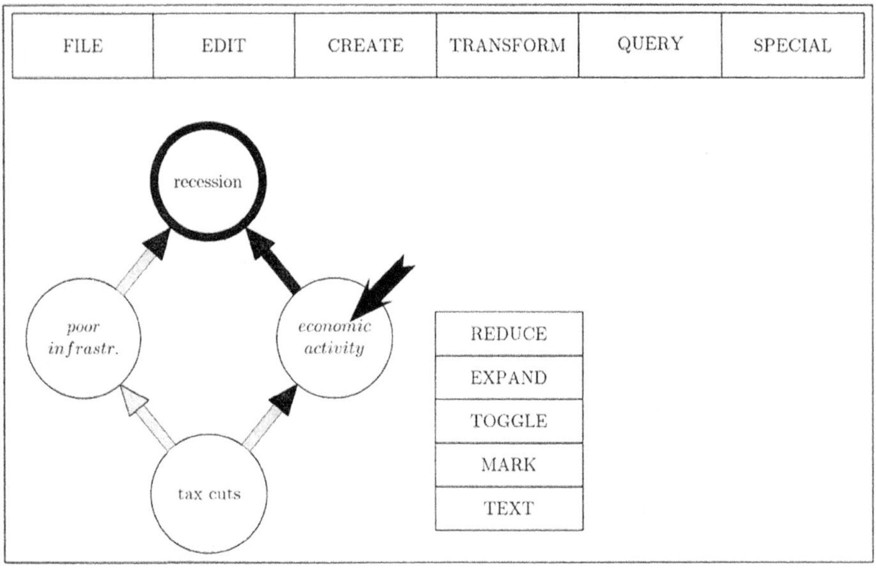

Figure 6. A tentative screen layout during an interactive session within the **ab** modeling environment.

DINO KARABEG

POLYSCOPIC STUDY OF A BASIC CULTURAL PATTERN

1. INTRODUCTION

In another paper in this volume [7] the motivation and the main ideas behind polyscopic modeling are explained. In an appendix to [7] some polyscopic modeling techniques rooted in mathematics and informatics are outlined. The present article illustrates applied polyscopic modeling and intuitive polyscopic modeling techniques. Since one of the main objectives of polyscopic modeling is to create *high-level information* (covering basic issues in non-technical terms) intuitive representation techniques such as ideograms, metaphors and dialogs play a central role.

Polyscopic modeling treats the reader's attention as a precious resource. Verbose descriptions and unnecessary details are avoided as a form of misinformation. The saying that a picture is "worth one thousand words" is about information content of pictures. But one thousand words take a lot more space and are understood a lot slower than a picture. It is for that reason that the ideograms are used.

Metaphors are a means of defining types of relationships (called *patterns* in polyscopic modeling) by referring to their familiar instances. Metaphors extend insights from a familiar context to a less familiar one. Dialogs serve for illustrating and harmonizing multiple points of view on a single issue. They can appear anywhere in the text as needed.

The choice of subject and of the terms in which the subject is discussed is another key aspect of high-level polyscopic modeling, since those choices influence the reader's perspective. Contrary to the usual practice, an author is responsible not only for what has been written, but also for that which has *not* been written. As explained in [7], polyscopic modeling considers giving a distorted perspective by leaving out the essentials as a form of misinformation. The present study of culture focuses on the issue of *ideognomy* (the term is explained below).

Polyscopic modeling aims to organize and unify information by joining disparate and heterogeneous pieces of information into a single structure. The present study uses a hierarchically structured polyscopic model for representing the heterogeneous pieces of information as lower-level models and the essential information which they have in common as higher-level models.

Polyscopic modeling uses essential ideas from science for creating reliable high-level information. In a polyscopic study justifiable or 'provable' claims are preferred to mere description or opinion. In the present study a hierarchical model structure is used for representing the claims (as higher-level models) and the justifications (as the lower-level models).

2. IDEOGNOMY AND CULTURE

It is recommended practice in polyscopic modeling to define the main terms. We use small capitals when we want to emphasize that the intended meaning of a word is as defined in the text.

The word INFORMATION denotes a record of experience. Accumulated, refined and systematized information is an essential part of the heritage of a civilized human. CULTURE denotes information and various activities oriented towards cultivating information such as education, art and science. IDEOGNOMY is the way of deciding what is to be considered as true, good, prudent, desirable or just. Since it determines what information is worth cultivating and preserving, an ideognomy is a foundation of a culture. This study distinguishes between three types of ideognomy: the NAIVE ideognomy, characterized by reliance on uncultivated sense impressions, the TRADITIONAL ideognomy, which relies on customs and mores of a tradition, and the RATIONAL ideognomy which relies on rational reasoning and argumentation.

In every culture all three forms of ideognomy are used for different purposes, but their relative importance varies. A recent major change in our culture has been a dissolution of the traditional ideognomy. The rational ideognomy, represented by science, has ousted the traditional ideognomy, but has not assumed its role. The reason for this is simple: science has little to do with high-level information which covers the basic issues of human life. In traditional culture the high-level issues used to be regulated by tradition. As the role of tradition diminished, its high-level tenets have largely been replaced by the NAIVE ones.

At the first glance, the naive ideognomy resembles the scientific judgement, since it recognizes the primacy of sense experience: what *feels* good, pleasant, easy or desirable is automatically considered as such. On that foundation rational reasoning is easily applied towards securing as much as possible of the 'good' things and towards eliminating those 'not-so-good' ones. Thus the overall ideognomy seems rational.

With the rise of science and technology and fall of the traditional culture, *convenience* has been elevated as the prime value [11]. The desire to live easily and pleasantly now directs the development and use of technology and even rules the countries through democratic mechanisms.

But whether something should be considered as pleasant or unpleasant, easy or difficult, that is decided NAIVELY. It is therefore of a central importance to ask to what degree can the naive judgement be considered as a meaningful criterion for convenience?

3. TOWARDS A POLYSCOPIC MODELING RESULT

We define the CONVENIENCE PARADOX as a contradiction to the naive judgement on what should be considered as convenient (easy or pleasant).

In the rest of the article a justification of the following claim in the form of a polyscopic model is outlined (definitions and detailed justifications are provided in [6]):

The Convenience Paradox is a Characteristic Pattern.

Roughly, the meaning of the claim is that the convenience paradox occurs so commonly that it is a characteristic aspect of the human situation.

The convenience paradox is introduced and represented by an ideogram. The rendition of the paradox serves as the highest-level view of the model. On the second level a high-level view of the human is given which reflects and explains the convenience paradox. The view consists of two related patterns called EMANATION and EMANCIPATION. On the third level a large number of models—instances of those two patterns—are shown to emerge in the experience of a variety of schools of HYGIENE and THERAPY including

- Western and Oriental medicine.
- Psychoanalysis
- Work of Western pioneers of body and mind therapy (F.M. Alexander, W.H. Bates, M. Feldenkrais, I. Rolf).
- Several Oriental schools of therapy (Shiatsu, Jin Shin Do, Yumeiho, Macrobiotics).
- Qigong, traditional palm healing, martial arts, Yoga, Buddhism, Sufism and Taoism.

We use the terms HYGIENE and THERAPY to refer to activities oriented towards the cultivation of wellbeing or VITALITY (as defined below) of healthy and ailing people respectively. The model justification is based on the observation that although the HYGIENE and THERAPY schools represent a variety of cultural traditions, cosmologies and world views, their tenets reflect the *experience* of working with large numbers of people during long time periods. Polyscopic modeling aims to restore what is essential in that experience by eliminating idiosyncratic details of cosmology and metaphysics as modeling artifacts and by searching for that which is common in experience. In the present model it is shown that the experiences accumulated within the HYGIENE and THERAPY schools from a variety of traditions reflect in different ways a single idea of the human. The convenience paradox appears in all of them as a fundamental insight.

Because of lack of space, the third level of the model is only briefly sketched here. A small sample of the fourth level which contains first-hand anecdotal material, opinions of well-known experts and witnesses and other documents justifying the third-level models is given in the appendices. The incomplete model presented here will provide a rough illustration of the polyscopic modeling approach to high-level information, but will not be sufficient for establishing the result. The complete model will be given in [6].

FIRST LEVEL: THE CONVENIENCE PARADOX

On the CONVENIENCE PARADOX ideogram, a figurine of a human is shown, about to choose which way to go. The way going downwards appears to be easier and more comfortable but it ends up being the more difficult one.

The Chinese yin-yang symbol at the base of the figurine represents the idea that the opposites tend to change into one another incessantly. Only the natural laws which underlie and regulate change tend to be stable. Only by knowing the laws by which the phenomena change and basing one's actions on them can one avoid illusion and govern one's actions prudently. The quotation in the lower-right corner of the ideogram is from a twenty-five centuries old Chinese classic. It suggests that in avoiding the illusions one can fortunately relie on culture—on the accumulated and refined experience and insights of others.

Science has taught us that the naive ideognomy is not a reliable criterion of *truth*. When studying the objective reality we have learned from science to replace the naive judgement by carefully developed procedures. But in deciding what is to be considered as *easy, comfortable* and *pleasant*, it appears that we can and, indeed *must* rely on our senses.

Figure 1. The CONVENIENCE PARADOX ideogram

What else? After all, it is the senses which tell us what is easy, what is comfortable and what is pleasant.

The source of the convenience paradox is a common source of paradoxes: We fail to take into account ourselves, i.e., the bias introduced by the inner condition of the observer. It may well be the case that our own condition makes us experience a multitude of random daily events as happy moments or mishaps and ultimately makes the difference between a 'wonderful' and a 'horrid' day. And our inner condition may be a result of factors well beyond our awareness. Like distorting eyeglasses, our inner condition may determine our view of reality without being perceived as a part of that reality. Unable to experience our inner condition directly or to understand its origin, we may view the symptoms which make it manifest as its actual *causes*, and to attribute to our feelings of comfort and discomfort an *objective* quality. The appearance of paradoxes at the illusive borderline between us and the external world has been observed by many people in the past (cf. Appendix I).

It is easy to see how an alternative which appears as easier or more pleasant can turn out to be more difficult or unpleasant in the long run by affecting our inner condition adversely. We define VITALITY to be that aspect of the inner condition of the human which influences the sensations of ease, comfort and pleasure.

Thomas: *This is all just sheer platitude! Everyone knows that what is comfortable in the short run differs from that which is comfortable in the long run. Don't we all go to school (which is more difficult than playing) in order to have it easier over-all? Don't we exercise physically (which is less comfortable than lying on a sofa) in order to enjoy the comfort of good health? There is no need to make a theory out of that.*

Michael: *The things are not so simple. When the reason contradicts the feelings most of us will go along with the feelings.*

Nancy: *In fact, for most people sense impressions* are *objective reality. There can be no conflict between sense impressions and reason, because there is no way to question the sense impressions rationally. All thinking takes place later, on sense impressions as a foundation. As I understand it, the main issue here is a style of thinking which can penetrate beyond that.*

Hermann: *I don't agree with you, Thomas, that the convenience paradox is something which is obvious to all. Far from that. What is here called the convenience paradox has been known in various traditions under different names: the Indians called the world as perceived by the senses 'Maya' or semi-illusion, the Chinese called the senses 'the five robbers', the Sufis talk about the veils that hide the reality... All believed that the way towards happiness is hidden behind the illusion created by the senses. But such insights have always been accessible to a small minority of people. And in our tradition, that minority seems to be particularly small.*

Hanne: *You mean, Hermann, that very few* men *are aware of the paradox. You guys take your ideas too seriously. But as soon as you relax your minds a bit and just look at the world, you'll find an abundance of paradoxes which are both amusing and inspiring.*

Thomas: *I still think that one can easily see the paradoxes with a help of some simple thinking, provided one is ready to question the things which everyone considers as obvious. Take, for example, the common idea that the pleasure in eating comes from eating. That's obvious, isn't it? One gets more pleasure by eating more. The common image of a guy who enjoys food is a kind of a glutton.*

Michael: *So, what's wrong with that?*

Thomas: *Well, we enjoy food the most when we are hungry. A guy who eats all the time probably just has an eating disorder. He can never really enjoy food because he is never really hungry. If one should* cultivate *gastronomical pleasure, one must do it by cultivating desire, perhaps even by fasting, rather than by eating excessively. The Epicureans used to do exactly that. But if you shall satisfy your desires, then there go your desires! And with them the possibility for satisfaction.*

n: *Let me ask you something else, folks: I wonder why is the conclusion that we should not trust our senses worth a penny? Aren't we already trusting our senses too little? Aren't we thinking when we should feel, calculating when we should relate? We are being told here that we should trust our senses even less than we do, that we should plunge even deeper into alienation.*

nne: *No, Ann, I don't think so. I am afraid that* you *are now thinking naively. Action guided by naive feelings might* feel *more related, but really lead into deeper alienation.*

rmann: *Precisely right. In a natural state of affairs our feelings should indeed point to right action. But we are very far from being natural, and so is our environment. We are caught up in a vicious cycle where our unnatural feelings lead us to make both ourselves and our environment ever less natural. Developing the natural unconditioned state of awareness is the central subject in spiritual traditions. It is the inner state attained by the spiritually advanced after years of work. But that type of work is very different from following naive impulses and desires.*

no: *There may well be instances of the convenience paradox that are either too obvious or too profound to write about. But we shall not try to cover all of them, not at all. We mainly want to see if we can use the cultural experiences of various traditions that are now available to us to create a different, rational perspective on our culture and our life style.*

SECOND LEVEL: EMANATION AND EMANCIPATION

The high-level view of the human which reflects VITALITY consists of two related patterns: EMANATION and EMANCIPATION. Each pattern is defined by an ideogram and a metaphor. As suggested by the ideograms, the two patterns are often two different views of the same situation.

The EMANATION PATTERN reflects an essential aspect of living: development. Life always begins as a possibility—a seed, an egg or an embryo, from which a mature organism slowly develops provided that the conditions for its development are favorable. VITALITY develops in a similar manner.

Figure 2. The EMANATION ideogram

The EMANATION ideogram shows a human figure emanating from the earth towards the heaven. The earth provides (and *represents* in the ideogram) the SUSTENANCE for the human. The heaven represents VITALITY as a consequence of development or of one's *growing up*, or of one's upright posture which affords maximal motility.

The earth (and thus also SUSTENANCE) is what is felt directly and always. The heaven (and thus also VITALITY) is a distant and long-term goal. The convenience paradox arises when SUSTENANCE stands in opposition to VITALITY, i.e., when that what is experienced as an immediate need or goal differs from that which is vital in the long run. In the ideogram, the human grows *away* from the earth, against its gravitational attraction. The convenience paradox is reflected in the upright posture of the human.

The earth is of an inert, material nature. The ideogram suggests that life, whose manifestations are growth and motility, arises when vitality wins over the inert tendencies of the matter. The materials of the earth then become endowed by life and they rise (within the human body) towards the heaven.

A metaphor which illustrates and explains the paradoxical aspect of emanation is the emanation of a plant from a seed. One removes from the soil the wild-growing plants, tills the soil, plants a seed and waters it. At first there are no visible effects. But after some days of watering a small plant emerges from the soil. If the plant is continually cared for, after several years its branches are sufficiently strong to bear savory apples. One's need for apples is at that point perhaps taken care of for

Figure 3. The EMANCIPATION ideogram

ever.

The procedure of planting and watering the seeds is hardly something a hungry person who is craving for an apple would think of. Some exceptionally insightful historical humans were able to devise the procedure for us. The rest of us would be more likely to eat an apple and throw away the seeds than to plant them. Hunger and its satisfaction—the experiences at the level of subsistence—provide no visible clues that planting the seeds is a prudent direction to follow.

Since in cultivation of plants we fundamentally depend on the experience of others we say that it is a *cultural* phenomenon. Is the cultivation of humans, and in particular the cultivation of VITALITY, also a cultural phenomenon? The situation there is similar because there also our senses are tuned in such a way that the change happens too slowly to be perceived (cf. Appendix II). Already the commonsensical observation that the faculties that are exercised develop and the ones that are not exercised decay suggests that the direction of maximum vitality will not be the same as the direction of maximum comfort.

The EMANCIPATION PATTERN arises when some condition called a VITALITY BLOCK inhibits the development or expression of vitality. The ideogram of the pattern is a bent human figure contained within a circular crust representing the VITALITY BLOCK. A VITALITY BLOCK is depicted as something which inhibits growth, uprightness and motility i.e., VITALITY.

The convenience paradox arises when the person having a VITALITY BLOCK is unaware of the block and perhaps even perceives the blocked condition as favorable and resists change. Alternatively, some actions and conditions that are experienced as comforting and pleasant may be undetected causes of VITALITY BLOCKS

It may at first seem implausible that major vitality blocks could remain undetected. Would we then not feel discomfort, or that we are not functioning as well as we should? What would prevent us from spontaneously making corrections? The explanation of the paradox is found in the existence of *habits*. Habits make any condition which persists a sufficiently long time *feel normal.* Indeed, subsequent to any deviation that our minds or bodies might suffer, the rest of us adjusts in such a way that an optimal functioning *with the deviation* is secured. Following such an adjustment, a condition without the deviation would feel abnormal and probably even *be* less functional than the one with the deviation until a new adjustment is completed.

A metaphor illustrating the paradoxical character of the vitality block pattern is a spring. A spring normally absorbs shocks by oscillating around some equilibrium position. But when the spring is pulled or twisted by an exceedingly large force, it becomes distorted and continues to operate around a new, less functional equilibrium. Once the spring has been deformed, a large force and a carefully organized action is required if it is to be brought into its original condition. In physics, the large force which takes the spring out of equilibrium is called *stress.* By analogy, stress is a common name for causes of vitality blocks.

A SKETCH OF THE THIRD-LEVEL MODELS

On the third level a number of models are shown which reflect the EMANATION and the EMANCIPATION. We give here only a rough sketch of the models.

The *Posture in the Gravitation Field* model reflects both second-level patterns. Vitality in the model is reflected by the ease of movement which results from a correct upright posture. When a posture is incorrect, the body is supported by permanent tension in muscles and tendons. That tension acts as a VITALITY BLOCK in a number of ways. A waste of energy and a loss of motility are the most obvious among them.

Permanent tension is known to be a common cause of discomfort, whose origin in the incorrect posture is usually not understood. Instead, the ensuing loss of vitality is manifested as a general sense of fatigue, lack of inspiration, bad mood or dissatisfaction with objects, situations and people. Further details showing the paradoxical relationship between posture and vitality, and how that relationship is exploited in some therapy schools, are given in Appendix III.

The view of the human which is common in Oriental medicine, in most oriental therapy schools and in traditional healing around the world is the one which is based upon the concept called 'Qi'. ('Qi' has in polyscopic modeling the same status as all other concepts, namely, it is considered as a modeling construct, as a means for organizing experience. The question whether Qi *really* exists is of little interest.) Roughly, the model identifies good health with a free flow of Qi (also called Prana or Chi or bioenergy). The blockages which prevent the free flow of Qi, often manifested as tension in the tissues, are considered as root causes of disease.

Each in its own way, the HYGIENE and THERAPY schools aim at restoring the free flow of Qi by eliminating the blockages, or to restore the balance of Qi throughout the organism which should normally result from an uninhibited flow. Qi is considered to be like water—when it flows it gradually clears and deepens its own channels. Qigong is an ancient Chinese healing practice based upon stimulating one's own Qi. In traditional palm healing a healer runs her own Qi through the body of the patient. In Hatha Yoga one stretches the tense tissues. In Yogic Pranayama the Qi is cultivated through breathing. In Shiatsu and some other physical therapy schools the acupuncture channels and points and the tense tissues are pressed and massaged.

It is natural to ask about the relationship between the above two models: Does the tension resulting from an incorrect posture also affect the flow of Qi? Have the two kinds of VITALITY BLOCKS a similar nature? A scientific study performed at UCLA Medical School involving Rolfing therapists and Rosalyn Bryere, a well-known aura reader, has indicated that they do [2].

A simple way to stimulate both motility and Qi is movement. Indeed, movement is known to stimulate vitality. Simple forms of movement combined with self palm healing have been used successfully to cure even such problems as near-sightedness [5].

The view that equates movement with effort and inactivity with comfort is clearly naive—when unused, our bodies and our abilities quickly deteriorate to the point where even inactivity presents discomfort. But movement also brings fatigue and causes stiff muscles. It may seem then that our choices are confined to either decay or effort, but that is fortunately not true. It is known that fatigue is predominantly a product of tension which results from *wrong* use of the body, and that correct use of the body is pleasant and effortless [4]. In the martial arts tradition various techniques for learning effortless movement have been developed.

The art of effortless movement involves tensing up only those muscles which are needed for the movement while keeping *all others relaxed.* That is, of course, possible only if one is reasonably free of physical VITALITY BLOCKS.

We can summarize what has been said so far by saying that motility—of our bodies and of our Qi—reflects our physical vitality. Physical vitality blocks manifest themselves as tense tissues which block the free movement of our bodies and of our Qi. They keep us moving in fixed, mechanical and inefficient patterns, and standing or sitting in difficult, uncomfortable postures. They cause permanent discomfort and limit our freedom without us being aware of them. They also cause damage to our health.

Complexes are the main object of study and treatment in psychoanalysis. They can be thought of as stiff and sore spots in our psyche which are caused by some painful experience. Complexes keep us emotionally tied to past experiences and make us unable to relate correctly and freely to events and people. They prevent us from seeing the relationships and situations as they are, and make us see the projections of our past painful relationships and situations instead. Psychologically, the complexes prevent us from perceiving, thinking and relating freely and hold us in not-so-good psychological states. We are then justified in calling them *psychological vitality blocks.*

Psychoanalysis endeavors to free us from complexes. It is a basic insight in psychoanalytic tradition that a patient will resist healing, as it involves reexperiencing *more* pain initially, when the repressed painful experiences are called back to memory (in psychoanalytic tradition this is phenomenon is called *resistance*). Furthermore, a person is usually unaware of her complexes. The complexes modify and create our experience of the world, but they themselves are not a part of that experience. The CONVENIENCE PARADOX can then be identified as a central aspect of the psychoanalytic view of the human.

Psychoanalysis focuses on major painful experiences which eventually cause deep crises. It is often a crisis which brings a person to an analyst. But what about *small* complexes? Could the same psychological mechanism be causing a multitude of small negative experiences to remain recorded in our unconscious mind and distort our vision and determine the character of our emotions? Since they do not have a single identifiable cause, such small complexes would not be as easily detected as the large ones. But being more common and more numerous, the small complexes might nevertheless have an even greater impact on the lives on most people than the psychoanalytic complexes. In Buddhist psychology which, as a school of mental and spiritual hygiene, works on improving and preserving the wellbeing of 'normal' and 'healthy' people, such small complexes do indeed play a central role. They are called 'knots' or 'internal formations'.

Regarding the prevention and healing of psychological vitality blocks, both psychoanalysis and the spiritual traditions provide valuable insights. On the one hand, an attitude of understanding and forgiveness is known to release the blocks, while anger and remorse tend to tie them faster. This insight points towards a possible link between VITALITY and ethics. In psychoanalysis, free associations and a corresponding spontaneous playful attitude are the foundation of the healing technique. This insight points towards a possible link between psychological VITALITY and art on the one, and meditation on the other side.

Freud has shown that the unconscious acts as a repository of repressed material which inhibits normal functioning of the rational mind and hinders the development of the ego. C. Jung has shown that in the unconscious also vital elements of the psyche are found which are called archetypes. Archetypes are usually accessed through corresponding symbols, such as truth, art, beauty, justice, God, mother, father or hero.

As Jung and his followers have shown, for persons who are psychologically centered in the ego, and identified with the ego, the ego and the rational mind act as vitality blocks. A connection with an archetype, experienced as a connection with something *larger than oneself* is then feared and resisted. The experience of making a connection has been compared to physical death and rebirth. But once the connection is established, *inspiration* comes—a surge of energy which revitalizes the person and her every action. The archetypes are known to be rich sources of vital energy and motivation. Art, religion, science, parenting, statesmanship and other cultural and social functions depend on functioning relationships with archetypes.

We summarize by saying that along with physical vitality blocks, psychological vitality blocks also exist which have a similar effect on our wellbeing and which exhibit a similarly paradoxical pattern. Is there a relationship between the physical and the psychological vitality blocks?

Although we usually consider the psyche as distinct from the body, the idea that the two kinds of vitality blocks are closely related is not implausible. It is common knowledge that a psychological condition of a person—fear, pain or pleasure—is reflected in the bodily posture, as a specific pattern of muscular tensions. Also, stress is usually thought of as being a *mental* cause of physical problems.

Wilhelm Reich developed these observation further into a psychotherapeutic method [8]. He showed that the patterns of tension in the body can be used for diagnosing psychological problems and that one can work on psychological problems by working with the body. Towards the end of his life Reich rediscovered Qi and he called it 'orgone energy'. All of Reich's latter work was based on a bioenergetic model. Reich's results point towards a close relationship between physical and mental EMANATION and between physical and mental EMANCIPATION. Reich also observed a common tendency to *resist* the free flow of Qi in the body, i.e., to resist healing. Many of Reich's written works have to do with that basic insight and its psychological and cultural consequences.

The possibility of working with the mind through the body is an attractive one, because the body is far more easily accessible than the mind. An instance of that approach is Hatha Yoga, where one works towards spiritual realization through bodily postures. But the reverse approach—working on the body through mind—is no less common and no less necessary. A central insight in both the Alexander and the Feldenkrais techniques is that a certain nonassertive attitude of the mind, characterized by a focus on the action itself rather than on the goal of the action, is essential for the correct and effortless functioning of the body. A more developed form of the same attitude is the foundation of Karma Yoga (the yoga of action).

Although this brief sketch only scratches the surface of a wonderfully rich subject, it allows us to observe that different traditions have understood different aspects of a mechanism which governs VITALITY. When combined together, the insights of those traditions give a clear and coherent picture. The convenience paradox emerges as a major aspect of that picture.

The experience of a variety of THERAPY and HYGIENE schools has shown that the vitality is a continuum. On the opposite side from dying and disease are abilities developed by martial artists, Qigong masters and Yogis. The blissful inner states of the Sufis and the Zen adepts suggest that suffering may indeed be rooted in ignorance (of the convenience paradox) and that the range of possibilities open to humans is by far larger and more pleasant than what is currently made available to us by our culture.

4. CONCLUSIONS

Given the CONVENIENCE PARADOX and the fact that the profound changes in the living conditions made possible by the Industrial Revolution have been directed by the NAIVE IDEOGNOMY, it should not be surprising if those changes have very often been misdirected. Indeed, already from our brief sketch it can be seen that it is possible to formulate some simple principles which govern the cultivation of vitality, and that in modern living those principles are violated with a remarkable consistency. It is common knowledge that modern living is stressful, but there is little awareness of what that means. While rushing after naively understood comfort we squander our vitality and thereby miss the opportunity to use the scientific and technological progress to our advantage. Technology whose development is directed naively either directly harms our VITALITY or allows us to survive with less vitality.

Throughout history, the CONVENIENCE PARADOX has been rediscovered by many people in many cultures. They expressed their insight in whatever way was available at the time. But in the long run, for the majority of people, the NAIVE IDEOGNOMY always prevailed:

My words are easy to understand,
And easy to put into practice.
Yet no one in the world can understand them,
And no one can put them into practice.

Lao Tzu

The challenge of explaining the CONVENIENCE PARADOX in such a way that it can be understood by most people as a very basic aspect of the human condition still continues to be open.

AFFILIATION

Dino Karabeg
Insitute for Informatics
University of Oslo
Norway

REFERENCES

[1] **Batenson, G. and Rieber, R.W.**, "Mind and Body: a dialogue". In: **Rieber, R.W.** (ed.), *The individual, Communication, and Society. Essays in Memory of Gregory Bateson*, Cambridge Univ. Press, 1989.

[2] **Bryere, R.**, *Wheels of light*, Simon and Schuster, 1989.

[3] **Alexander, F.M.**, *The Alexander Technique*, (Edited by E. Maisle), Thames and Hudson, 1990.

[4] **Feldenkrais, M.**, *The potent Self*, Harper, 1992.

[5] **Huxley, A.**, *The Art of Seeing*, Harper, 1994.

[6] **Karabeg, D.**, *Polyscopic Modeling*, Book in preparation.

[7] **Karabeg, D.**, "Prospectus for polyscopic modeling", in: **Aerts, D., Van Belle, H. & Van der Veken, J.** *World Views and the Prob-*

lem of Synthesis; The Yellow book of "Einstein Meets Magritte", Kluwer Academic Publishers, Dordrecht, 1999.

[8] **Reich, W.**, *Character Analysis*, Farrar, Straus and Giroux, 1994.

[9] **Rolf, I.**, *Rolfing Your Way to a Better Self*, Harper, 1994

[10] **Schrödinger, E.**, *What is life?* with *Mind and Matter*, Cambridge University Press, 1992.

[11] **Tierney, T.F.**, *The value of Convenience*, SUNY Press, 1993.

APPENDIX I

A Meeting Point

Many well-known people in the past have realised that paradoxes arise at the illusive borderline between us and the external world, between the subjective and the objective, between our images and reality. Some of them, like René Magritte, saw the poetic or the *artistic* side of the paradoxes, others their rational or philosophical or *scientific* side:

During philosophy's childhood it was rather generally believed that it is possible to find everything which can be known by means of mere reflection. [...] Someone,indeed, might even raise the question whether, without something of this illusion, anything really great can be achieved in the realm of philosophic thought—but we do not wish to ask this question.

This more aristocratic illusion concerning the unlimited penetrative power of thought has as its counterpart the more plebeian illusion of naive realism, according to which things "are" as they are perceived by us through our senses. This illusion dominates the daily life of men and animals; it is also the point of departure in all of the sciences, especially of the natural sciences.

(Albert Einstein: Remarks on Bertrand Russell's Theory of Knowledge. Tudor, 1951.)

APPENDIX II

Perception and Change

It is easy to see that perception is related to change: we perceive only hat which changes in time or in space. Perception is best when the *rate of change* is within a certain interval which is determined by the sensitivity of our senses. When the rate of change is outside of that interval, either much faster or much slower, perception becomes impossible: slower changes are eliminated by habit, while the faster ones are filtered out by the coarseness and the inertia of our sensory organs.

The existence of the rate-of-change interval of perception is essential for the functioning of our senses for the following reason: while on a very small scale (both in time and in space) physical phenomena exhibit highly irregular behavior (called 'random', 'Brownian' or 'thermic'), in the scale where our perception is best the irregularities cancel out to reveal far more regular 'average' aspects of physical reality.

In his essay "What is Life?" [10] Erwin Schrödinger explain this phenomenon on the example of the movement of fog in a glass jar. While each individual fog particle bounces up and down randomly, together (as *fog*) the particles are seen as sinking gradually downward. Schrödinger concludes:

This example shows what funny and disorderly experience we should have if our senses were susceptible to the impact of a few molecules only. There are bacteria and other organisms so small that they are strongly affected by this phenomenon. Their movements are determined by the thermic whims of the surrounding medium; they have no choice.

Because of coarseness and slowness of our eyes we are able to perceive regularity in the midst of irregularity:

If it were not so, if we were organisms so sensitive that a single atom, or even a few atoms, could make a perceptible impression on our senses—Heavens, what would life be like! To stress one point: an organism of that kind would most certainly not be capable of developing the kind of orderly thought which, after passing through a long sequence of earlier stages, ultimately results in forming, among many other ideas, the idea of an atom.

A physicist has indeed good reasons to appreciate how finely our senses are tuned to perceiving the regularity in physical nature. The rate-of-change interval of perception there works for our benefit. But that may not always the case. It is easy to see that there are phenomena which affect us profoundly but whose rates of change lie outside the interval.

We don't see a tree grow. All living beings including ourselves grow at a similarly slow rate. Our bodies, our abilities, our health, our wellbeing, our characters and our happiness all grow or decay slowly. We can neither see them grow or decay, nor can we see directly what causes them to grow or decay and orient our actions accordingly.

It is therefore conceivable that we too, like bacteria, may be adversely affected by the existing rate-of-change interval of our senses. We may be tossed up and down by random daily events, unable to perceive slow regular trends concealed behind them. By reacting to the comfort and discomfort caused by the senses we may fail to become aware of a greater freedom of choice which may be inherent in the human situation.

Perhaps if we could ignore the momentary ups and downs and orient our actions towards slow and stable change, inconceivable possibilities would open up for us. What has been called *wisdom* may indeed be not much more than such a predisposition of the mind. Perhaps the so-called *supernatural* abilities of martial artists and yogis and countless others are accessible to all humans who are able to make a simple attitude change and *cultivate* their VITALITY.

Possibilities may be hidden from us by the limited rate-of-change interval of our perception, but so may dangers. Those dangers did not matter much during the long history of humankind because nature had so much more power than the humans. Nature was able to correct our mistakes. But now our mistakes *may* matter. All sorts of degenerative trends might be allowed to develop which one would perhaps immediately notice if one could fall asleep and wake up a century later. But given the limitations of our perception, the quality of our natural and socio-cultural environment as well as our own wellbeing may deteriorate at a slow rate without us taking notice.

A social scientist with enough wisdom to see those trends would have much less reason to be pleased by our senses than a physicist. As Gregory Bateson [1] put it,

it's like a frog in a saucepan of cold water. If its temperature is raised so slowly that at no moment is there a sufficient doc/dt [temperature change per second] to activate his neurons, he does not jump. There is no time marker to tell him when *to jump. He gets boiled. He just wasn't that much alive. I hate to tell you what this reflects upon our society.*

APPENDIX III

Partial Justification of the 'Posture in the Gravitation Field' Model

Ida Rolf and F.M. Alexander are progenitors of highly successful and now widely known and globally practised therapeutic methods. Both methods focus on correcting the human posture, Rolfing via a rather painful deep-tissue massage, the Alexander technique by teaching the correct use of body in posture and movement. Thus Ida Rolf and F.M. Alexander are well suited as witnesses for our model. Two claims need justification:

- Human posture in gravitation field is relevant for VITALITY as we defined it.
- Humans are usually unaware of the influence their posture has on their sense of wellbeing (the CONVENIENCE PARADOX).

Regarding the first claim, we quote Ida Rolf from her book "Rolfing your way to a better self" [9]:

In order for a living body to be at ease in its spatial environment on earth gravity must be able to deal positively with it. We can't change the gravitational field; we haven't gotten that big yet. What we can do is to change the way the parts of the body fit together into a whole which can transmit the gravitational field through that body in such a way that it enhances its energy field.[...] This is the basic concept of Rolfing. We are using gravity as our tool, not chemistry. When you come to look at it, this is quite an idea.

The impact of posture is clearly not limited to physical comfort and discomfort:

Rolfing is not primarily a psychotherapeutic approach to the problems of humans, but the effect it has had on the human psyche has been so noteworthy that many people insist on so regarding it. [...] The amazing psychological changes that appeared in Rolfed individuals[...] inevitably suggest that behavior on any level reflects directly the physical energy level of the initiating physical structure. [...] It would seem appropriate, at this point in time, to state that following Rolfing a man's greater awareness suggests to him that his energy has been increased. In fact, Rolfing has simply freed his energy, made it more available. His greater structural competence makes it possible to utilize his energy more efficiently.

We add the following quotation of F.M. Alexander [3] in support of the second claim:

Perhaps the most striking and at the same time the most pathetic instance of human delusion is to be found in the human creature's attitude towards his own psycho-physical defects [...] in connexion with unreliable sensory appreciation and with perverted ideas or conceptions of what is "right" or "wrong," where the human creature's uses of his own mechanisms are concerned.

In the following passage Alexander describes how, after years of work on correcting his own functional problem, he made a breakthrough and discovered the principle which now forms the foundation of his technique. The insight which made the breakthrough possible was an instance of the CONVENIENCE PARADOX, namely, that the habitual wrong posture feels right, and that the corrected posture will therefore feel wrong.

I saw that if I was ever to succeed in making the changes in use I desired, I must subject the processes directing my use to a new experience, the experience, that is, of being dominated by reasoning instead of by feeling, particularly at the critical moment when the giving of directions merged into 'doing' for the gaining of the end I had decided upon. This meant that I must be prepared to carry on with any procedure I had reasoned out as best for my purpose, even though that procedure might feel wrong. *In other words, my trust in my reasoning process to bring me safely to my 'end' must be a genuine trust, not a half-trust needing the assurance of* feeling right *as well. I must at all costs work out some plan by which to obtain concrete proof that my instinctive reaction to the stimulus to gain my end* remained inhibited, *while I projected in their sequence the directions for the employment of the new use at the critical moment of gaining that end.*

The last witness is John Dewey, the great American philosopher, who was taught the Alexander technique by Alexander himself [3]:

And so I verified in personal experience all that Mr. Alexander says about the unity of the physical and psychical in the psycho-physical; about our habitually wrong use of ourselves and the part this wrong use plays in generating all kinds of unnecessary

Figure 4. Misdirection in common designs

tensions and wastes of energy; about the vitiation of our sensory appreciations which form the material of our judgments of ourselves; about the unconditional necessity of inhibition of customary acts, and the tremendous mental difficulty found in not 'doing' something as soon as an habitual act is suggested, together with the great change in moral and mental attitude that takes place as proper co-ordinations are established. In re-affirming my conviction as to the scientific character of Mr. Alexander's discoveries and technique, I do so then not as one who has experienced a 'cure,' but as one who has brought whatever intellectual capacity he has to the study of a problem.

APPENDIX IV

Misdirection

An expectable consequence of the CONVENIENCE PARADOX is the misdirected development and use of technology. The misdirection present itself in most banal objects and circumstances.

The misdirection pattern is illustrated metaphorically by the ideogram in the figure. From what has been told it can easily be seen that sitting in a chair and reading or writing text which is lying flat on a desk leads to incorrect posture and VITALITY BLOCKS. Yet that is the posture in which we undergo most of our education. Is it true that in the process of becoming civilized we sacrifice a major part of our vitality?

DIRK SIEFKES

TURING MEETS THOREAU
AN ECOLOGICAL APPROACH TO COMPUTER SCIENCE

1. INTRODUCTION

To the naturalist Thoreau "in the wild there is the liberation of mankind". To the mathematician Turing human communication can be replaced by computation. To the anthropologist Bateson evolution and thinking follow the same principles, thus mind and nature are a necessary unity. I try to make Turing and Thoreau meet: On Bateson's ground I build a model of mind where thoughts and concepts, feelings and values evolve in interaction like living beings and their genetic representations. Communication then happens locally, in stories and pictures; even when we work with formalisms we have to allow for free interplay with our imagination. The model is fruitful in teaching and research, and helps me to see the development of science as social history. Even as a computer scientist with this view I might help to keep mind and nature in balance, and thus both of them alive.

2. ECOLOGICAL ORIENTATION IN COMPUTER SCIENCE

As a computer scientist, trained in mathematics, I work with formalization all day, in teaching and research. As a naturalist I am involved with nature "all night", in gardening, stalking and reading. There are people like Thoreau who think that "in the wild there is the liberation of mankind". In [19] (ca. 1850) he states as a goal: "Let us spend one day as deliberately as nature". Then there are people like Turing who think that computing, or all of human reasoning, can be reduced to symbol manipulation, and thus to machine operation. In [20] (1950) he proposes that by the end of the century there will be machines who beat people in simulating human behaviour. How do I resolve the contrast between these two orientations?

For some years now I have been working on the concept of "small systems", through which we can keep our environments on a human scale, even when working with computers and formalisms. (See [13]; some of the papers are available in English, see the references there and in other papers.) On this ground I joined a group of computer scientists searching for a theory of our discipline which rests not only on mathematics, but on other fields like psychology, sociology, and education as well.

In the volume [5] on Computer Science perspectives edited by the group I published my first paper on what I now call ecological models for computer science. Following the anthropologist Gregory Bateson I set up a precise analogy between biological evolution and mental development of human individuals. In contrast to Bateson, however, who regards everything—even a technical apparatus—as a "mind" if it satisfies the six criteria in [2], my model includes feelings and values as central parts. Thus, although my theory yields a rather detailed model of developmental processes in mind and culture, including technology, I do not reduce human development to biology. Rather I use the model for a better understanding of how human cultures develop, and how they could and should fit in with nature; this is why by now I call it 'ecological' rather than just 'evolutionary'.

In 1992 FIFF (the German equivalent to "Computer Scientists for Peace and Social Responsibility") started a movement "Information technology for a world worth to live in—Visions for the 21th century". This initiative sparked a working group "Ecological Orientation in Computer Science" where my ecological models fit in; see [15, 16]. How can we as computer scientists help to keep nature as something worth to live in, and what does this mean for Computer Science? This is a quest as well as a question, a point of departure as well as a goal, a place where ethics and theory meet.

3. THE BIOLOGICAL MODEL

Biological evolution can be depicted as a threefold stochastic process, resting on the relation between individual and species, between living being and its genetic representation, and between living being and environment.

In our language individual and species are not distinguished. The gourmet and the fisherman are after the individual salmon whereas the scientist is concerned with the salmon in general; both types of predator, however, use the same word, 'salmon'. Still the species is not just the collection of all individual salmons; or else, how could it evolve? The individual salmon develops and changes through the "genetic instruction" it carries in each cell and through the random influences from the environment. The changes, however, vanish with the individual salmon whether he be eaten up or died naturally; they become permanent only on the level of the species through natural selection. Thus biological evolution needs reproduction of the individuals. And reproduction, sexual or otherwise, works through genetic representation, and thus is subject to more random influences like mutation and recombination. Formerly one would have said that this all happens in an environment, and that natural selection has the salmon adopt better and better to it. Actually in this process the environment changes as much as the salmon. Sure enough, man provides breeding grounds for the salmon because he wants to eat more. But might you not say as well that salmon provides himself with more breeding grounds by tasting so good to man? Every environment consists mainly of living beings anyway, and every species is environment for numerous other species. To sum up: Living beings evolve in a threefold interplay between individual and species, and between living process and both, its genetic representation and its environment. The genes and the environment act in an analogous way as internal and external "memory" for the species and the individual. To stress the social aspect one speaks of co-evolution. These more modern views and concepts of evolution theory are treated for example in [8] [21].

Still the picture is too crude, of salmons swimming about and reproducing in a suitable environment, with both environment and species evolving thereby. Actually the whole business happens neither individually nor globally, but locally. Fish swim in schools, wolves range in packs, people live in families and gangs of many kinds, and work in teams and committees and groups of still other kinds. Ecologists speak of (ecological) niches, I prefer the neutral term *center*. Life is centered around these social nuclei "where the things happen"; all development radiates from there to the individuals as well as to the species, or the culture, or the society, only to be reflected back into the centers as social condition, or species characteristics, or individual character, which together constitute the life there.

4. THE MENTAL MODEL

Along with genetic representation and social interaction animals have evolved mind as a third means to guide their way in the world. Minds connect with other minds through communication, and through action and perception with the physical or social world. To understand these connections as evolutionary processes, we first have to see mind itself as an ecosystem evolving on a much faster pace. This is Bateson's thesis [1] [2], that individual learning and biological evolution follow the same principles. Thus from the above biological model I construe a model of mind [6, 14, 16].

In the cove of my mind, thoughts and feelings float like so many fish and weed, constituting species of knowledge and emotional settings. This conforms well with the notion from philosophy and psychology that we abstract knowledge from many individual experiences and guide experience therefrom—which applies easily to emotions as well, see below. So we have individual and species; but where are the genes? Here pedagogical experience comes in: To understand something we have to do both, to use our imagination to get a vivid picture ("anschaulich machen"), and to work out the ideas conceptually. Thus concepts are the genes of our minds, whereas images and ideas are the live part. To see this you have to loosen the grip of our cultural heritage, by which genes are "canned inside" living beings like berries in a pale, and thoughts are "built up" from concepts like houses from bricks. The biological perspective is more appropriate for living phenomena than the engineer's terminology. Bateson uses the pair of *process and form* here, abandoning the traditional pair of form and content. Processes happen: We eat and sleep, we work and dance, we are born and bear and die soon enough. Form is everything we need or produce to have all this happen: food and eating habits, tools and body posture, rituals and beliefs. Processes run in given forms, and produce new forms. In this sense forms represent processes, and allow them to happen; processes create forms, and allow them to be. In this sense, images and ideas "harden out" into bodies of concepts which in turn represent them. So far, representations—that is, both forms and processes—are in the mind; we will se in a moment how they connect to the world.

The same picture works for emotions, with values and beliefs as genes. Thus emotions are the processes that allow us to call things and events good or bad, and values—not value–free physical experiences—cause emotions. I find it tempting to associate thinking and feeling with animal and plant life respectively in the biological analogy. In any case, one is the environment of the other, as we very well know: Bitter or sweet feelings can give different meanings to the same statement; and cool reasoning has cut apart enough affairs of love or hate.

5. COMMUNICATION

In the engineer's view communication is an exchange of messages through a channel. Translated into ecological terms: I send my word fish, neatly swimming in line, through a channel from my cove to yours where they grow and multiply; then you send some of your fish in return; and so on. In this way we both get our fill. But wait. Words are labels for concepts, or so one says. And concepts are genes in the ecological model.

Is it genes we exchange? How do I excise them out of my fish, and how do you implant them back? And through what channel could we pass them? Something is wrong here.

How do we communicate? I talk, you listen. So why are you always on the verge of interrupting me? And why do I watch your reactions so carefully? You don't just sit there patiently receiving my words in your mind. Rather more or less openly you formulate objections and approvals, you fill in and erase details, you want to take over the story and have it your way. You talk while you listen, formulating silently the questions to which my statements yield the answers. If you did not, you would hear nothing and fall asleep. In the same vain, while I talk I listen to the silent messages that I read from your facial expressions and your twisting body, or extract from your half-begun interjections. If I did not, how could I find answers to your emerging questions? [13, 16]

This fits to how ecosystems "communicate". They do not pass fish through some neutral channel. Rather they form a joint part without giving up their own integrity. And if their gene pools are compatible enough, then the new part grows into an ecosystem itself; which evolves, so that now life forms appear and spread into the separate parts. Thus the ecosystems exchange "experiences"—bits and pieces of their evolutionary history—well enough, but not in the form of genetic material; indeed, not even in the form of a school of fish with some seaweed for decoration. Rather what they contribute are small life ecosystems. The fish and the seaweed come in when we try to analyze the "communication"; and the genes which represent the "experience" as described above, are part of the evolutionary "mechanism". (This corrects my ideas on communication in [14].)

So any communication process is an evolving ecosystem. What then are the ecological niches, the centers of development which connect the micro-level of individual life and the macro-level of the slowly changing ecosystem? In language we call them stories. Stories are the nuclei of conversation "where things happen". We speak in stories and we hear stories. Nothing bigger nor smaller makes sense in itself. We use words and sentences to set up a story; there would be none if we could not understand them. But only through the story we know what is actually meant at that point. We use an awful lot of language to set up a story. But it is through stories that we change language slowly.

We use words as labels for concepts. Both, however, are not singletons, but distributed throughout our knowledge and language respectively, as genes are distributed throughout the gene pool. A word does not have a fixed meaning, or a fixed number of meanings. Rather, like anything living, a word covers a certain range of meanings which is narrowed down differently by different contexts, but never completely (which accounts for misunderstandings and other creative activities in listening). Eleanor Rosch [12] found that concepts are not categories in the logical sense, but have an inner structure and open boundaries. The ecological model tells you why. And it also shows how we use words with certainty nevertheless.

6. VERBAL AND NONVERBAL COMMUNICATION

To understand this better let us look at emotions more closely. Together with thoughts feelings of love and hatred, of anguish and joy populate the cove of our mind. They nurture our thoughts and are moved by them, just as fish eat seaweed and move it about. Like thoughts feelings come and go and form our emotional world, consisting of the emotions we know, we once had, and might have again. Like thoughts feelings are lively, but "harden out" into values, which are persistent, and make the feelings come again and again.

We communicate feelings through words. "I am sad. You look happy." But more profoundly we communicate our feelings through our actions, our body movements, our facial expressions. We look happy or sad, and we act that way. We communicate emotions visually, or rather through all our senses. This makes emotional communication so much richer than talking, and it is obvious that we do not just exchange messages thereby. I understand your feelings by getting into them, or into other moods, that is by developing feelings myself. Like two communicating ecosystems we set up a joint emotional center for a while, which influences our own feelings one way or the other. We do not exchange feelings one by one, but share an emotional setup as we would share a landscape through which we walk together. Only that we change the emotional landscape by walking.

With verbal communication it is harder to accept. We are too thoroughly trained to take words as pointers to something else, and thus sentences as series of pointers interspersed with constructors. Of course it is tempting to compare the linear structure of our speech, and of our writing in particular, to the linear structure of genes in a chromosome. But the comparison works on neither side. In modern genetics the importance of the linear ordering of genes is highly questioned if not given up; even the concept of 'gene' as a basic part of a chromosome has become uncertain [3]. Similarly in linguistics one knows that the structure of a sentence can be changed in many ways without changing the meaning too much; words influence each other across sentences or even whole paragraphs by back and forth reference; some people even question the basic character of words [11]. Surely the notion from logic that the meaning of a sentence can be build up according to its structure from the meaning of its words, does not hold for natural languages.

When we speak we attempt to create a joint world, not to set pointers from one world to another. This world can live only if it evolves, and evolution works through genetic reproduction. For communication to be fruitful our systems of concepts and values have to be compatible—similar enough but not identical—to interact. In a story we share thoughts and feelings so that our values and our knowledge may grow and change. When we converse our theory and our ethics grow jointly.

7. SOCIAL DEVELOPMENT

In the biological model the living being evolves as the phenotype interacts with two types of representation, the genetic material and the environment. The salmon evolves genetically through reproduction, and interacts with the environment through perception and action. In the mental model evolution happens analogously: The mind evolves through interaction between imagination and conceptual apparatus and between special and general knowledge; we call these processes "verbalization" and "contextualization". In the social model the two models, of mind and of nature, are interrelated: The individual being knows enough about its environment to perceive it and interact with it, and thus in turn builds up and changes its knowledge. This is most important with humans: We know so much about the world (including ourselves and our self) that we change it much faster and more thoroughly than biological evolution would permit. Our cultural evolution thus is intertwined with mental, not biological evolution. Our cultural "genes" are in the mind, not in the artifacts; thus selection does not work directly on the artifacts, but through the interaction between environment and mind. (This corrects my ideas on social development in [16].) We build and demolish cathedrals and computers first mentally and then physically, although the physical artifacts may change our mind as much as their mental forms.

Like biological life social activity is always centered locally. It happens in and around small units like families, teams of work or play, groups with a common interest or goal or past. As individuals we are formed, though not determined, through the social units we take part in. Conversely, willingly or not, as individuals we form, though do not determine, these units. Thus humans develop in co-evolution with these centers of their social life. Similarly the social centers are formed, though not determined, through the conditions society imposes or indulges on them. And conversely changes in society are induced by changes in social activities that are distributed through group after group until they are generally accepted as a tradition, or turned into a law. Thus there are no direct connections between individual and society except through social groups; they build the centers of cultural co-evolution. (This is my answer to the question of how the micro–level, the individuals, connects to the macro–level, the society; see also below.)

To many people the biological connotations in this model and their apparent consequences are most distressing. To some there is too much freedom: Political, economical, and cultural development seem reduced to random events; there seems to be no ethics, no overall goal, nothing that distinguishes humans from animals. To others there is not enough freedom: Human life seems reduced to the evolutionary machinery. For me neither is the case. As human we are responsible for the atmosphere in our social centers, thus the model calls to action. Take a seminar group as example. We all know how easily the situation becomes fruitless. We ruin the discussion by too much material support (technical or other), by using too strict rules of conduct, by insisting rigidly on a certain language or conceptual system, by imposing our emotions or intentions on the others—thus by overloading on any of these and possible other human bearings. Underweight bearings that are too weak to carry the discussion are as bad. Therefore I call a center *ecological* if all people involved act flexibly in all these respects, sticking to neither extreme of too much or not enough. Note that the golden mean is no better. And to reduce one dimension to others is worst; for example, to try to get a good discussion by organizational means alone or on a purely emotional level.

A center of social activity cannot contain too many people; otherwise subgroups form where the real things happen. With ecological centers the limits become even stricter. I have never experienced a center of more than twelve people remain ecological over an extended time. For this reason formerly I used the term *small system* [13]. With 'small', however, most people connect numerical size, although a numerically small group need not to be ecological according to the above definition. And a system in general usage consists of elements interrelated for a certain purpose. A center, and even more so an ecological one, constitutes the people involved as much as the people constitute the center. 'Self–organization' or 'autopoiesis' in this sense is incompatible with the technical meaning of 'system'. In an ecological center people do not ask or wait for a free space or a niche to develop in; they do not take the freedom for granted either, but try to grant it each other and themselves.

8. FORMALIZATION AND TECHNOLOGY

You might accept the ecological model as depicting cultural evolution. But does it work for science and technology? There is the world of hard facts, not of thoughts growing in emotional settings, and of hard machines which are constructed and used under strict economic conditions and goals. There is no room for soft metaphors. Or is there? How in our culture science and technology push each other forward is rather well documented. We need mathematical formalisms to invent and build machines which in turn have us look for even stronger mathematical tools that help us calculate. How this enterprise is embedded in economical and political development is even better known. But why "in our culture"?

In [4] Wolfgang Coy describes how in the Renaissance the ideas for more and more sophisticated machinery developed together with a more and more technical view at the human body. Similarly in [9] Bettina Heintz describes how in our century a more and more formalized approach in (certain areas of) mathematics arose together with the spread of "formal rationalization" in industry and administration. She claims that this development lead to the invention of computers and thus the introduction of computer science.

A machine is a thing, not a mental construct nor an animal. But in our mind the three realms are connected. In old German a "Thing" is a meeting of the wise old men(!) who agree upon what the matter is. Not that we were free to decide on what there is and how it affects us. The wise old men were bound in their decisions by their tradition, and so are we in our actions and perceptions, even if we are not burdened by wisdom or age or by being male. In this sense even the development of science and technology is culturally embedded, is in fact part of our culture. Through the ecological model I see how [6, 15, 16].

For science this is easy. Science is a way of viewing the world and thinking about the world in a restricted, concise formalized way. We set up rules and procedures how to handle words and perceptions, and in this way get results which are both sharp and narrow. This happens in the humanities as well as in science, and most thoroughly in mathematics. (It happens also in art and religion, though mostly not on the verbal level. This is the exact point where *Einstein meets Magritte.*) In the ecological model of mind formalization is "breeding". If people want their salmon meat more pinkish and less fat, the breeder achieves just that by influencing natural selection. And if people want their knowledge more objective, the scientist achieves just that by putting more math into his science. People concerned about ecology are worried about the mass of chemicals the salmon breeders use, and how they destroy the environment and endanger the species. But pink lean salmon simply tastes great. Things are not bettered by not eating it. Or are they? This is where *Turing meets Thoreau.*

For technology it is harder, just because the effects just described are there to see for everybody. Through the ecological model we see behind the effects. We need formal scientific knowledge and procedures to build and use machines; and science depends more and more directly on better and better machines. But more importantly, the usefulness of machines induces scientists to develop science further, and thus to push technology. Our everyday wishes and dreams harden into scientific facts and materialize as machines much more directly than through political and economical pressure, namely through our orientation towards facts and machines. The computer scientist who sees his teenage sons handle computers so light–handedly (and light–headedly!) likely is influenced much more thoroughly than through any new interface he might develop or purchase. And likely this will influence his scientific work. "The genes of our machines are in our mind" is not just a neat metaphor. Machines and scientific knowledge meet with culture in our minds, and develop together with it.

9. ON CHANGE IN COMPUTER SCIENCE

With traditional machines engineers mimic human actions and natural processes, and change them thereby. Like the live originals good machines are robust, since they have clearance; small changes in the make or the surroundings do not result in disastrous effects. In their mathematics engineers work with continuous realms to the same effect. It is to this type of situations that our mind is well adapted; sweeping back and forth between imagination and conceptualization and between emotions and values, we leave enough free play in our thoughts to arrive at mental models that are robust enough for both, machines and natural processes. The bigger and more precise the machines become, however, the more difficult they become to understand, since too much formalization leaves not enough clearance, and thus stalls the mental models.

This problem becomes much more urgent with computers. With computers we mimic mental processes in a purely formal way. Computers and the corresponding mathematical formalisms work with discrete descriptions and data which are even more detached from living experience than traditional mathematics, and thus are not robust. Neither are our mental models of them, since they consist of representations only, and thus do not give enough free play for building up a reliable imagination.

Discrete models resist visualization; it is the visual, however, where emotions and intellect meet (as we saw above), where our creativity springs up. Therefore even small programs are hard to understand, and large applications are impossibly so. For this reason the development and use of computers evokes both fascination and a deep frustration. Everything seems possible, even "artificial intelligence". And computers are everywhere now, and normally well working. The development of new software (that is, computer programs), however, is risky as ever: too expensive, too slow, too error–prone; the "software crisis" torments the discipline almost from the beginning. Most importantly, the effects of computerization seem not to be foreseeable. Often the bad effects outweigh the good ones; but worse, mostly people do not agree on whether they are good or bad.

In an Interdisciplinary Research Project "Social History of Computer Science" at the TU Berlin we try to understand the dangerous relation of the field to the culture bearing it. We see some patterns slowly emerging: The development from "codes" for computers to more and more abstract "programming languages" came along with corresponding changes in the idea and use of language and in other areas of our culture [10] [15–17]. What changed what, and what might be the long–term effects [16] ? Computer Science as a discipline was institutionalized along with strong pushes towards mathematization and formalization not explainable from technical necessities [6]. Again there are corresponding pushes in mathematics at schools and universities. The software crisis is not just a technical phenomenon ("more complex technology and applications lead to more complex programs"). Rather wild expectations of the computer as "electronic brain" or "personal assistant" have lead to developments that were unfeasible economically or under a human perspective as well as technically [18].

In the ecological model the social centers serve to relate the activities and ideas of the individuals with the conditions and developments of the society. Different orientations towards computers and computer science grow and change and die in these centers. People get and bring their orientations there, not individually in empty space, but constantly on the move between centers of work and leisure, between friends and relations and job partners. In this way orientations spread, slowly or quickly, and many become seemingly universal. And since the mechanisms of keeping them in power have been established thoroughly for so long, what once drifted up from the depths of the social pool seems now a social necessity.

You can take the last sentence as pessimism, and give up on action: There is no way to change computer science, evolution runs its course. Or you my turn it around: What now seems a deep social necessity, once sedimented down from the heights of so many social centers. And since you are a human being, not a salmon, you may swim in many different schools and there is always something you can do about them, even if it does not look like much. When you lecture, you can stop to just pour the material over the students. Try instead to find the story behind the material that makes you think it is worthwhile to teach; then both the students and you will learn something [13–16]. If you develop software, you will need some formal tools and some hacking; but do not rely exclusively on either. You can do prototyping instead, that is, go in cycles of "design–implement–test–discuss", until part after part the whole thing is done. In this way you work in various groups of small size and do "software development as reality construction". (This is the title of [7]; from Christiane Floyd I heard first about prototyping and "evolutionary software design", and introduced both concepts into my teaching and research.)

Thoreau went into the woods for two years and lived there all by himself, to see whether that way he would not learn more about life than in the crowd. He thoroughly enjoyed the time. Some people call him conceited, since he—without stating this in his journal—went to friends for dinner when felt like it. He was fascinated by trains, but objected that the "sleepers" lay under the rails for so many dead Irish workers. He ridiculed modern life and especially technology as superfluous and a serious hindrance for real life. He was engaged against slavery, and wrote a piece "Civil Disobedience".—Turing had friends, too, but seems to have lead a lonely life throughout. When jogging he wore a gas mask to avoid poisonous air. Apparently he died through a poisoned apple from his own hand. He wrote famous papers on computability and intelligent machines; during the war he was influential in helping to crack the German code with his machines. He proposed to compare the intelligence of men and machines through a game where communication is reduced to typing symbols.

Neither of them truly represents his field. Thoreau's message, however, seems the more important to me. He shows us how to tap life, not how to close it off. *It is high time that Turing were to meet Thoreau to learn some.*

AFFILIATION

Dirk Siefkes
Department of Informatics
Technical University of Berlin
Germany

REFERENCES

[1] **Bateson, G.**, *Steps to an Ecology of Mind*, Ballantine 1972. In German: *Ökologie des Geistes*, Suhrkamp, stw **571**, 1985.

[2] **Bateson, G.**, *Mind and Nature—a Necessary Unity*, Bantam 1979. In German: *Geist und Natur—eine notwendige Einheit*, Suhrkamp, 1982.

[3] **Beurton, P.J., et al.**, *Gene Concepts and Evolution*, Workshop Jan. 95, Max Planck Institute for the History of Science, preprint **18**, Berlin, 1995.

[4] **Coy, W.**, *Industrieroboter*, Rotbuch 1985.

[5] **Coy, W., et al. (eds.)**, *Sichtweisen der Informatik*, Vieweg, 1992.

[6] **Eulenhöfer, P., Siefkes, D. and Stach, H.**, *Informatics as Cultural Development*, Techn. Univ. Berlin, FB Informatik, Bericht No. **97-**2, 1997.

[7] **Floyd, C., et al. (eds.)**, *Software Development and Reality Construction*, Springer, 1992.

[8] **Gould, S. J.**, *Ever Since Darwin*, Norton 1977. In German: *Darwin nach Darwin*, Ullstein, 1977.

[9] **Heintz, B.**, *Die Herrschaft der Regel—Zur Grundlagengeschichte des Computers*, Campus, 1993.

[10] **Pflüger, J.**, "Leitbilder der Programmiersprachentwicklung", In **Friedrich, J.** (ed.), *Informatik und Gesellschaft*, Spektrum, 1995, pp. 196-210.

[11] **Pinker, Steven**, *The Language Instinct*, Morrow 1994.

[12] **Rosch, E.**, *Prototype Classification and Logical Classification—The Two Systems.* In **Scholnick, E.** (ed.), *New Trends in Cognitive Representation. Challenges to Piaget's Theory*, Erlbaum, 1983, pp. 73–85.

[13] **Siefkes, D.**, *Formale Methoden und kleine Systeme—Lernen, leben und arbeiten in formalen Umgebungen*, Vieweg, 1992.

[14] **Siefkes, D.**, *Fish in Schools or Fish in Cans—Evolutionary Thinking and Formalization*, International Computer Science Institute Berkeley, TR–92–009, 1992.

[15] **Siefkes, D.**, "Computer Science as Cultural Development—Toward a Broader Theory", In **Freksa, C., Jantzen, M. and Valk, R.** (eds.) *Foundations of Computer Science—Potential, Theory, Cognition*, Springer, 1997, pp. 37-47.

[16] **Siefkes, D.**, "Die Rolle von Gruppenprozessen in der Informatikgeschichte", In **Eulenhöfer, P., et al.** (eds.), *Socialgeschichte der Informatik. Kulturelle Praktiken und Orientierungen*, To appear in Deutscher Universitätsverlag, 1998.

[17] **Stach, H.**, *The Image of Man as a Self-Steering Device as Constituent for the Concept of the Stored-Program Computer*, In **Eulenhöfer, P., et al**, 1997, pp. 3-9.

[18] **Theißing, F.**, *Auf dem Weg in die Softwarekrise?*, Techn. Univ. Berlin, FB Informatik, Bericht No. **95**–14, 1995.

[19] **Thoreau, H. D.**, *Walden (and other writings)*, Bantam 1962 (1854). In German: *Walden oder Leben in den Wäldern*, Detebe **19**, 1971.

[20] **Turing, A.**, *Computing Machinery and Intelligence*, Mind **LIX**, 236, 1950, pp. 433–460. Also in **Hofstadter, D. & Dennett, D.**, *The Mind's I*, Bantam Books, 1981.

[21] **Varela, F., Thompson, E., and Rosch, E.**, *The Embodied Mind —Cognitive Science and Human Experience*, MIT Press, 1991. In German: *Der mittlere Weg der Erkenntnis*, Scherz, 1994.

FRANCISCO CAMPOS

TOWARDS A 'VIRTUAL' RATIONALITY?

INTRODUCTION

Psychology has, for a long time, shown the boundaries of human rationality. The rationalistic model of human behaviour assumed by neoclassical economics has broken down; the idea that companies are guided by managers who are able to weigh all the options and make the most rational or optimum decisions in accordance with maximizing principles, such as 'profits', cannot be held any more.

Another added problem is that reality changes. The human environment is not 'natural' any more. Especially in cities 'the cultural environment'—in a broad sense—is as important as 'the natural environment'. Technical and scientific development is moving so fast that in the very near future it will take us into such surprising surroundings that we will be able to call it 'virtual reality' without exaggeration. Will it also be necessary to start talking about 'virtual rationality'?

1. THE BREAKDOWN OF THE RATIONALITY OF ENLIGHTENMENT

A clear example model of the rationality of Enlightenment can be found in the neoclassical economy. The economic neoclassical model suggests that companies are guided by maximizing principles, e.g., profits, and led by managers who are able to weigh all the options and make optimal decisions which are both rational and 'necessary'.

This has been strongly criticized by H. Simon [1] with his bounded rationality concept. Simon propounded a humbler reality of human beings who are satisfied because they do not have the wit to maximise [2]. Besides, human beings have limited knowledge and a limited capacity to learn, so: "It is impossible for the behaviour of a single, isolated individual to reach any high degree of rationality. The number of alternatives he must explore is so great, the information he would need to evaluate them so vast, that even an approximation to objective rationality is hard to conceive" [3]. It is not only a question of limited information, knowledge or ability, but members of organizations also have desires, drives, passions, which along with their knowledge constitute their frame of reference. To this, one should add that the individuals of an organization are not isolated, the socio-cultural context also plays a role [4]. In short, "we can only speak of rationality relative to a frame of reference" [5] which is a function of ... a culturally embedded cognitive framework and consequent subjective expectations about future states of the world" [6].

It can be said that most of the time the beliefs and behaviour of human beings are neither based on knowledge, nor on rational decision making criterion.

The problem is that human beings are not always rational.

1.1. *Rationality and the Perception of Reality*

It is traditionally believed that irrationality is the result of emotional influences on the ability to reason. However, Cognitive Psychology shows us that there are other aspects of irrationality in the processes of reasoning which are independent of emotional and motivational influences.

1.1.1. *Limitations in Dealing with Information*

The human capacity to deal with information is limited. Miller [7], in his already classic study, revealed that human beings cannot keep more than seven mental units active at the same time in short term memory.

Nor, during the reasoning process, can human beings, including experts, take more than five to ten factors into consideration when making a judgement [8].

Few people know of this limitation in the cognitive process. In addition, the more information available, the more people trust their judgements although the amount of information surpasses the human ability to process it, which reduces the quality of the judgement even more.

1.1.2. *Time and Space*

Time is another important factor in the perception human beings have of the environment. People feel less involved in events that are distant in terms of time and space.

Björkman [9] proposed the "cognitive discount" concept to explain this ever-decreasing involvement with the future.

1.1.3. *Lack of Realism*

Human beings tend to confuse their desires with reality [10]. Positive things are considered to be probable and negative things, improbable. It is evident that a certain lack of realism is necessary for the individual, above all in stressful circumstances, as human beings need to feel secure. However, it is sometimes rather dangerous. For example, one of the problems which arises within road safety is that no one behind the wheel expects an accident to happen to them. Evidently, it is good to feel safe and not constantly afraid of having an accident. However, a belief that accidents 'happen to other people' is especially evident among certain sectors of the young. When an accident does happen to someone they know, even where there are no emotional ties, it has a strong emotional impact.

1.1.4. *Presentation of the Problem*
Another factor which affects judgement is the presentation of the problem. People can judge the same options very differently depending on whether they are perceived as losses or gains.

Attitudes towards losses and gains are asymmetric, losses being much more badly accepted than gains. Losses cause people to take more risks: people are prepared to risk a greater loss in order to avoid a lesser but certain one. It is the triumph of the 'double or nothing' challenge. On the other hand, in the case of gains, people tend to avoid risk (e.g., "A bird in the hand is worth two in the bush"). Human beings prefer a gain that is certain to a mere possibility of a greater one.

1.1.5. *Perception of Probabilities*
This is an amazing problem: the majority of people find it difficult to understand the concept of probability. Human beings normally pay too much attention to factors with a very low, although higher than zero probability.

Another important error in the perception of probability is that of 'scene'. It is considered that the 'whole' is more probable than the probability of its components. It seems completely absurd to believe that the probability of various events happening simultaneously is higher than them happening separately. However, people are very affected by what we could call 'coherent scenes'. These 'scenes' function like informal explanatory theories.

1.1.6. *Too Much Confidence*
People are usually convinced that they know all the answers to every question. This means that they are not very prepared to accept new information or new ways of addressing problems or looking at the world [11].

1.2. *Cognitive Deviation*

Research in Cognitive Psychology has also revealed strong effects of deviation due to exclusively cognitive factors. A review of the subject can be found in Sjöberg [12].

The most understood effects of cognitive deviation are known as heuristics or biases in information processing:

1.2.1. *Heuristics of Representativity*

People frequently base their judgment on a prototype they have. But this comparison to a prototype is done more by subjectively perceived likeness than objective data, e.g., people are really convinced that the clothes make the man.

Another aspect of the bias of representativity consists in giving excessive importance to the details of a given case while importance is not placed on more fundamental aspects.

Slow, but relevant changes cannot be perceived and remain absorbed in the overall context while excessive importance is placed on secondary or anecdotal changes.

1.2.2. *Heuristics of Availability or Priming*

This involves exaggerating the importance of new events, especially those of a spectacular nature or, more precisely, which attract the attention, because they stand out more.

This is well known and exploited by the sensationalist mass media. For example, in a country where the means of communication continually show signs of environmental degradation, the inhabitants could become sensitive to the problem. But, because of the manipulation of availability, the means can cause people not to notice important events and focus instead on totally unimportant facts and circumstances.

The same happens with the spreading of opinions. People pay more attention to what their neighbours say than they do to the means of communication. The bad experience of a friend can have a much greater impact than any statistics that might prove otherwise.

1.2.3. *Anchoring or "Nothing New under the Sun"*

People need to anchor new ideas into a pre-existing system. This involves basing a judgement on preconceived opinion.

This bias in the information processing refers to our difficulty in modifying our beliefs in the light of new information that contradicts those previous beliefs. The initial opinion can have a very strong effect on subsequent judgements, ensuring that they vary little from the initial point.

In more colloquial terms, people tend to say: "there is nothing new under the sun". There are many studies which have determined that people tend to understand new information in terms of what they already know instead of approaching it as additional or different information.

Expectations of change are approximately nil. In general, human beings expect 'repetitions' rather than 'changes'. In general it is difficult for people to imagine scenes of radical change in the short term. Judgements relative to rapid change are very difficult and tend to underestimate the exponential growth rate [13].

A person's initial impression is usually decisive and leads to the attribution of stable characteristics. If new information contradicts the initial impression it tends not to be taken into consideration.

One positive effect of this anchoring is the reinforcing of self-confidence and the action to be taken in the face of conflicts and threats.

1.3. *Provisional Conclusions*

The frequency of cognitive deviations is surprising. The majority of human beings appear to be affected by these deviations or prejudices, including experts.

Cognitive limitations which are serious in themselves in normal situations increase in situations of emotional tension. Emotions also influence understanding. In emotionally tense situations, people tend to have a limited perspective, that is, they bear fewer things in mind.

But 'emotional states' are not the causes of this cognitive bias. As human beings we tend to overestimate our cognitive abilities. Our ability to pick up information, process it and make judgements is not very reliable.

The introduction of completely new ideas and changes is very difficult. People expect 'repetition and stability'.

People need to feel secure at the expense of accepting desires as realities.

This is the dominant human tendency in relation to his or her understanding of the world. We can say that these are the limits of the rational perception of reality. Can a link be established between this rational perception of reality and mental health?

2. ON MENTAL HEALTH IN OUR SOCIETIES

This paper suggests that a link could be established between the rational perception of reality and mental health. In order to do this, we will examine studies of the environment, the human being's perception of it and the possible resulting effects on mental health.

2.1. *Urban Environment vs. Rural Environment and City vs. Suburbs*

One characteristic of big cities is the presence of poor and marginal zones on the outskirts and in the centre, frequently situated near middle and upper-middle class residential areas. The problems derived from overcrowding, social disintegration, malfunctioning of development and delinquency are in evidence here. I will show that this leads to social and psychological instability, both in the individual and in social groups and ultimately to the loss of social and personal health.

At the beginning of the 20th century Simmel [14] studied the predicament of urban life in contrast to rural life. Along the same lines, Thomas and Znaniecki [15] analyzed the effects immigration had on Polish peasant farmers when they moved to industrialized urban areas of the USA. They collected data from period documents and concluded that change from cohesive, rural surroundings to the impersonal environment characteristic of an industrialized city caused behavioural problems and social disorganization in the immigrant.

At the same time the so called School of Chicago, among whose members can be cited R. Park, E. Burguess and R. McKenzie, began to study the geographical distribution of social problems. They were interested in analyzing the effects of the rupture of social networks.

According to their studies the highest levels of social disorder were recorded in areas of Chicago where a progressive industrialization process was taking place [16]. According to these writers the behavioural and social problems of these areas were the result of a disintegration in the previously socially cohesive communities as a result of the progressive industrialization of the area.

In present times it can be observed that it is precisely in the analysis of existing differences between urban and rural environment and between residential zones and suburbia where the majority of research into the variations of social relations has been carried out. Some examples follow. Fisher [17] observed that the composition and structure of social networks were different in urban and rural communities and that rural social networks were more dense and family-based, while in urban zones networks tended to be less dense and composed of fewer family members. In urban areas, Crenson [18] observed that the atmosphere of the neighbourhood in which a person resides can be as important for the development of a sense of belonging to that community as other factors such as social class or race. In addition, research comparing deprived urban areas with residential areas of higher socio-economic status, suggests that in problem areas, social isolation, the loss of a feeling of belonging and conflict are more common [19].

In this sense Gracia *et al.* [20] have recently examined variations in different aspects of community social support in relation to the level of risk of the residential area [21]. The results obtained in this research indicate that the very socio-economic status seems to have a different meaning in relation to the residential area. Thus, levels of integration, satisfaction in the community, participation, association and level of contribution to community organizations were significantly higher in groups of low socio-economic status residing in low risk areas than in groups of the same status living in high risk neighbourhoods. In addition, the differences between high and low socio-economic groups also followed different patterns in the different residential areas.

Residential density is another aspect of the physical environment which has received researchers' attention [22]. As Evans et al. [23], says, high residential density can weaken social ties and in turn provoke disruption in social support systems. People living in high density situations maintain an excessive number of undesirable social contacts and suffer a lack of privacy [24]. These conditions can interfere with the regulation of interpersonal exchange [25] and cause social seclusion as a coping strategy [26]. In addition, these coping strategies can manifest themselves in other situations, reducing the number of social ties and resources of social support [27].

2.2. *Mental Health and Environment*

Since the 19th century the fields of medicine, sociology and psychology have strived to understand the relationship between social factors and mental health. In 1855, a doctor from New England (US) indicated that a ratio of sixty-four to one existed in cases of dementia in the underprivileged classes in relation to the privileged classes. Durkheim [28] found empirical evidence to confirm that suicide was more frequent among people with few intimate social ties. Durkheim was also concerned about the social changes in industrial societies; according to him industrial development introduced modifications in traditional community links and family ties. He proposed that traditional relationships were dissolved with immigration to urban areas causing a loss in social integration or anomie which was capable of producing serious psychological disorders.

These ideas became the seed of many studies oriented towards the study of the relationship between mental health and the human environment. Among them can be cited Farish and Dunhan [29] who observed that cases of schizophrenia were notably concentrated in the deprived central area of the city as well as in residential areas of minority ethnic groups, in their study on the time-space distribution of certain psychoses in Chicago. It seemed obvious to them that this first observation confirmed Durkheim's idea that psychic disorders followed social marginality and disintegration. As far as minorities were concerned, they proposed the explanation of the appearance of the schizoid component as the rupture in communication within the community.

As time passes, interest is growing in the studies which relate environmental factors with mental health. In Cornell University in the 1950's a systematic study programme was launched to evaluate the extent to which cultural, geographical and interpersonal changes had a negative effect on health. The programme, carried out by Hinkle and Wolf [30], grew from the findings of Selye [31] on the influence of stress on health and they tried to find out which factors could counteract the given effects. They examined diverse factors such as psychological defences, experiences, access to social resources, etc. They specifically observed that a significant difference existed in the personal style of coping with the situations as well as in ways of dealing with social relationships today referred to as 'social support'.

Another important study in the health environment relationship was carried out by Holmes and Rae [32] who proposed a classification of possible environmental stressors; this allowed a scaled evaluation of the environmental conditions to which subjects were exposed and led to the formulation of diverse scales of vital factors and their links with psychological disorders. But all of these relationships although quite significant were not very relevant and provoked much theoretical and methodological criticism.

The same stress factor manifests itself in very different forms in different people due to physiological and psychosocial circumstances. It is here that the role of the interpretation or evaluation of the individual facts relating to the subject might begin. Lazarus [33] proposed an 'appraisal hypothesis' which the subject does and which has been very well received by social scientists as it combines contextual and cognitive factors in coping with stress.

The current recognition that contextual as well as cognitive factors constitute a powerful resource in the prevention of illness, leads to current Community Health programmes in our societies.

2.3. *Mental health and culture*

The authors of the Cyclops project—a USA project for the detection of extraterrestrial life forms—wrote: "Whatever the morphology of other intelligent beings, their microscopes, telescopes, communication systems and energy sources must have been almost the same as ours, until a certain moment in their history. Undoubtedly, there must have been differences in the order of invention and in the appearance and applications of procedures and machines but the technical systems must have been more determined by physical laws of optics, thermodynamics, electrodynamics and atomic reactions than by the peculiarities of the beings handling them. The conclusion is that we should not worry about the problems of communication exchange between biological life-forms of diverse origins. Whatever intelligent life form we might come into contact with, we share with them a broad base of technology and scientific and mathematical knowledge that could be used to build a system for communicating the most subtle concepts" [34].

Our civilization, supposedly so developed and advanced, believes that its own technological practices must essentially be shared by some other intelligent life-form in the Universe. This is obviously false. Our civilization has not been able to communicate with the intelligent animal life-forms which exist on Earth. Our western culture has not even been able to communicate with other cultures without dominating them and placing them in danger of extinction.

The egocentrism of our western civilization is evident. Can we go further? Can we find symptoms that our society is sick? I think we can so. The diagnosis will be difficult because it must be subtle and self-referring, but it will not be less evident even if we do not wish to accept it. We have to point out that finding symptoms of sickness is only the first step towards a cure.

Paul Watzlawick, discussing the catastrophic effects for a culture which comes into contact with a more developed one, writes: "Clinical experience reveals that suddenly coming into contact with new information of overwhelming proportions can have two effects: we either close ourselves to the new facts and try to carry on as though they didn't exist, or we lose grasp of reality. Both enter into the realms of madness." [35]

Examples of this can be found in our western civilization. Due to an unusual development in the mass media and the growth of information from all around the world, a similar phenomenon is taking place in our culture, namely a 'closing off' to our own cultural development and, on the other hand, a 'loss of grasp of reality' option is also in evidence.

(i) The question whether "we close ourselves to new facts and try to live as if they didn't exist when confronted by information of overwhelming proportions", will be analysed first.

The increase of information available to people has made it impossible to process or take in, because of the internationalization and globalisation of the world as our society. One would have to spend twenty-four hours a day listening, watching and reading all the political, economic and general social information available to be well informed; to these would also need to be added the specific information relating to each individual sphere of work. The sheer quantity of information available not only in published studies but also in preprints, informal networks and unofficial electronic networks such as Internet, would guarantee that more than twenty-four hours a day would be necessary to take it all in and this is obviously impossible.

One must resort therefore to the prior selection of the information one will deal with. This normally means that information sources are selected not because of the quality of the information they give us but by how much we believe in the sources, which practically nulls the possibility of contrasting information. To avoid this we take different measures when we can (what happens when we can't or when we are unaware of the problem will be examined later). We resort to 'zapping' when we want to find out different opinions about national and international news on television. Also we read an assortment of newspapers or listen to different radio stations in order to hear the same news items from different points of view. However, this tendency is known and international news services compete at a national and international level, to control the greatest number of news outlets. At the same time anti-monopoly laws are established to guarantee the freedom of information, and so on. But in all this process the individual does not have access to a huge part of the possible information available.

In short, we consciously and necessarily leave a huge part of the potential analysis of reality to one side and in so doing we close ourselves consciously and necessarily to many other interpretations of reality. What happens when this takes place unconsciously will be examined later. It is a 'closing off', a closing ourselves in our own opinion and pre-selected information. This happens among individuals and social groups internationally. What are the opinions and what is the social information we close ourselves off to? The ones which we think or believe either attack or are radically different from our own.

This closing of ourselves in our own opinions and information results in the 'living as if it didn't exist' solution of some individuals and social groups in relation to others, which is becoming a reality in our societies. The greatest part of the inhabitants of urban areas live as if their neighbours didn't exist.

The integral xenophobe version takes it a step further, from 'living as if they didn't exist' to 'trying to stop them existing', 'them' being any group they dislike. The rejection, however, is only a question of degrees. In our 'well educated and developed societies' some social groups live as if others, especially immigrants, conflictive or marginal groups didn't exist. We can see it first-hand in any city. We try to ignore marginal social groups and blame their existence on political powers. Our own sense of irresponsibility reaches the extreme of accepting municipal policies such as 'the banning of certain undesirable social groups' (e.g., the prohibition of begging) or we express a desire for the eradication of socially undesirable groups from our own area through civic demonstrations in a specific district. It isn't important to us that those people are put in prison, move on to another district in the same city or to another city. Each social group seeks to live as if other socially conflictive groups didn't exist.

(ii) We can also observe the existence in our society of the second answer to the dichotomy, that is "losing grasp of reality when faced with overwhelming information".

This second position is evident as above but when:

- We do not know or cannot select information sources, or
- We are unaware of the huge quantity of information we are rejecting and as a result are ignoring other interpretations of reality.

In this case we lose grasp of reality or, to be more precise, we fall victim to the mistake we make, identified by Watzlawick in fooling ourselves into believing that there is only one reality namely the one in which we live. One relevant analysis to our study from the social psychology angle can be found in Kenneth J. Gergen [36].

3. CONCLUSIONS

At this point some conclusions may be reached. In current times, in developed western societies we can see an enormous breakdown in the desire for the rationality of Enlightenment. An enormous fissure has appeared between let's say 'normal' and 'virtual' life. People who are more in contact with Nature have their specific aims and preoccupations, while those living in the western cultural world increasingly face different realities and as a result have different needs. These realities and needs are brought about by new technologies and together create a 'virtual' reality. In this virtual reality the human being feels obliged to practise some processes of reasoning and decision making which are an anomaly to his genetic and cultural knowledge, to his ancestrally acquired knowledge passed down through generations. These new processes of reasoning are what are referred to here as virtual rationality.

The effect of virtual reality has been to separate the human being from his/her old natural reality and transfer her/him to a virtual universe where he/she stops being by definition and becomes a virtuality of himself/herself using virtual rationality in his dealings with others.

This allows us to explain phenomena such as the increase of senseless violence in our societies given that violence is perceived in terms of the virtual reality in much the same way violence is perceived in a cartoon. It can be seen in the ever-more evident use of all types of drug-taking in the widest sense: stimulants to wake up, sedatives to sleep, euphoriants to feel better, etc. Not all is negative however, it also allows us to explain the proliferation of NGOs (No Government Organizations) and the solidarity which can occur in the light of disasters which take place thousands of kilometers away to people we only come across through the mass media.

How has this change taken place?

Let us plot a schematic story. At the beginning of human history the isolated human being was defenseless to face the problems of life; few creatures were born as vulnerable as he/she was. Few were so unprotected by Nature where laws of selection, competitiveness and the survival of the fittest, ruled. But he/she was born into a strong and cohesive social network, the dense social networks of the hunting environment as well as those of the agricultural environment were fundamentally based on strong ties of kinship. Inside those social networks the individual was strongly tied to traditions and norms from which it was difficult to escape, but at the same time provided him/her with the firm security of social support. The solidly established, dense social ties made strong social cohesion possible and guaranteed the survival of the social group. The individual was subordinate to social need.

As already discussed, the transition from rural to urban life meant a high price had to be paid in terms of health. The given transition produced a fundamental change in social relations. The human being lost the social networks to which he/she was strongly tied. Social networks in an urban environment are generally less dense. They are not so strongly based on kinship ties. The cultural level has increased. Social services on offer in big cities have also increased. There is a higher degree of security confronting natural problems. Physical survival is taken for granted. The struggle for life among the species does not exist at a personal level. The struggle among individuals is a moral or legal problem. The struggle among groups or people is still a problem but in general it can be said that the problems of individuals and social groups have changed considerably. Individualism has grown. Anonymity has grown. The associations which a person can belong to explicitly or implicitly have increased astonishingly. So social relationships among urban people are based primarily on friendship, work or on an affinity of very diverse and very specific interests. The individual feels stronger in that he/she belongs to much wider although more unstable networks. Social pressure to belong to these new networks is practically nil. Almost all of them are an option for the individual. In almost all cases the decision to join or leave can be made at any moment.

In social networks of a rural environment, kinship ties, density, intimacy and necessity prevail. In an urban environment friendship, superficiality, instability and voluntary action prevail.

Medical, social and psychological problems are not caused by environment in itself nor by changing relations, but by:

- environmental changes, which cause unexpected or undesirable changes in social relations; as well as
- the perception the individual has of those changes.

It is here that the problem of new technologies appears. By new technologies I am referring to technologies such as those based on computing, laser, digital technologies, genetic technologies, technologies of new materials, etc., and particularly new information technologies which started to revolutionize life towards the middle of the 20th century and which will transform life even more in the 21st century. I highlight 'even more' since, to my understanding, we have not even glimpsed the tip of the iceberg of the revolution which will take place. Making predictions is always risky but certain tendencies continuing from the lines already drawn here can be indicated.

The revolution which will take place in terms of new technologies will be far reaching. New technologies are transforming the human environment more radically than the change from the rural to the urban environment. And this is also happening in the individual's perception of those changes.

I think that the changes which will take place in the human environment will change our environment to a 'virtual' environment. Some approximate characteristics which could help design this environment for better or for worse, without wishing to make evaluations, might be the following:

- The nature of the new environment will bear almost no resemblance to the Nature of the human hunter's physical environment. If the rural and the urban environments are progressively transforming the human habitat, destroying the 'ecological niche' into which the human species is born, the new environment will be of an almost completely artificial nature. To get a quick impression of what I have in mind, a mental comparison can be made between the jungle, the tundra, the savannah or the desert with an urban garden or a metropolis. The change from an urban environment to a virtual environment will create a bigger gap than the one between the jungle and the garden.

- In the new environment there will be complete geographical mobility. The human being will change location much more frequently, because of work or for any other motive, voluntary or otherwise. But, at the same time this fact in itself is unimportant in regard to maintaining social relationships as these will be primarily maintained by technological means, to be more precise, relationships between people will be 'technologically mediated', e.g., most of the time people will use computers for automatic translation, so they will be able to use their own languages. This will result in personal relationships which are more fluent and less binding. There will be almost no physical connection to social groups; person to person social relationships in particular will be sporadic, i.e., the number of components of social groups which maintain vis-à-vis contact will tend to diminish. These social relations will be increasingly instigated by momentary and specific interests, with no need of friendship nor family ties.

- The cinema will join the computer and virtual reality games. We will be able to live and participate in 'stories' built in such a way that our decisions alter the outcome of each of the 'stories' be create. Through new drugs and sensors we will be able to physically feel almost whatever virtual reality. This will bring with it more psychological problems than exist at the moment with television, drugs and role-playing games. However, the benefits are incalculable, such as the possibility of directly 'experiencing' all historical events, recreated specially for each of us, or travelling without moving a muscle; in short there are endless possibilities some of which are already being reflected in science-fiction movies.

- Social networks of the individual with regard to intimacy, time and the proximity of physical relationships will become notably weaker. But on the other hand they will become more dense in number and will surpass all geographical limits. A comprehensive example, which I hope will be taken more as a metaphorical approximation than real, could be the social relationships established in the Internet network. We potentially have 'friends' or 'acquaintances' in all computer terminals, participating in many 'chats', subscribing to many lists in order to receive a wide range of information and we have access to the latest advances being made by each one of our 'Internet colleagues'. Usually the Internet network terminal is situated in the workplace, but we can have it at home simply by adding a modem, and its use is not merely professional on the part of its users. On occasions we must even protect ourselves from distracting messages which appear on screen during work. But proportionately, we have personally met few of these people. With even fewer, we maintain some kind of personal contact and with a very few, we have friendships. In statistical terms no family relationships exist.

To what measure can it be said that those new social relationships are going to provide social support? It is difficult to say but, this social support will be different in the sense that it will be based less and less on vis-à-vis relationships. This will be the case for although we meet more people and travel more, the growth of virtual contacts is increasing much more rapidly than the number of personal contacts. In turn and necessarily so, these personal contacts will become more sporadic.

We will have more and more relationships and friendships but most of them will be virtual and our rationality and behaviour will reflect this, becoming increasingly virtual.

AFFILIATION

Francisco Campos
Department of Philosophy
University of Valencia
Spain

REFERENCES

[1] **Simon, H.**, *Administrative Behaviour*, McMillan Co., NY, 1957.

[2] **Simon, H.**, op. cit. p. xxiv.

[3] **Simon, H.**, op. cit. p. 79.

[4] **March, J. and Simon, H.**, *Organizations*, John Wiley, NY, 1967.

[5] **March, J. and Simon, H.**, op. cit. p. 138.

[6] **Farmer, M. and Matthews, M.**,"Cultural Difference and Subjective Rationality: Where Sociology Connects with Economics of Technological Choice " in: **Hodgson, G. and Screpanti, E.** (eds.), *Rethinking Economics*, Elgar, Cheltenham, 1991, pp. 103-116.

[7] **Miller, G.A.**, "The magical number seven, plus or minus two: Some limits on our capacity for processing information", *Psychological Review*, **63**, 1956, pp. 81-97.

[8] **Slovic, P. and Lichtenstein, S.**,"Comparison of Bayesian and regression approaches to the study of information processing in judgment", *Organizational Behavior and Human Performance*, **6**, 1971, pp. 649-744.

[9] **Björkman, M.**, "Time and risk in the cognitive space " in: **Sjöberg, L.** (ed.) *Risk and Society*, London, Allen and Unwin, 1987.

[10] **Sjöberg, L. and Biel, A.**, "Mood and belief-value correlation", *Acta Psychologica*, **53**, 1983, pp. 253-270.

[11] **Lichtenstein, S., Fischhoff, B. and Phillips, L.**, "Calibration of probabilities. The state of Art to 1980", In: **Kahneman, D., Slovic, P. and Tversky, A.** (eds.) *Judgement under uncertainty: Heuristics and biases* New York, Cambridge Univ. Press, 1982.

[12] **Sjöberg, L.**, "Aided and unaided decision making, Improving intuitive judgment", *Journal of Forecasting*, **1**, 1982, pp. 349-363.

[13] **Wagenaar, W.A. and Timmers, H.**, "The pond-and-duckweed problem: Three experiments on the misperception of exponential growth", *Acta Psychologica*, **43**, 1979, pp. 239-251.

[14] **Simmel, G.**, "The metropolis and mental life", in: **Wolf, K.** (Ed.), *The sociology of George Simmel (1902)*, Glencoe, Ill., The Free Press, 1950.

[15] **Thomas, W. and Znaniecki, F.**, *The polish peasant in Europe and America*, New York, Alfred A. Knopf, 1920.

[16] **McKenzie, Roderick**, "The ecological approach to the study of the human community", in: **Park, R. and Burgess, E.** (Eds.), *The city*, Chicago, University of Chicago Press, 1926.

[17] **Fisher, C.**, *To dwell among friends*, Chicago, Chicago University Press, 1982.

[18] **Crenson, M.A.**, *Neigborhood politics*, Cambridge, Harvard University Press, 1983.

[19] **Harrison, P.**, *Inside the inner city*, Harmondsworth, Penguin, 1983. See also, **Willmott, P.**, *Friendship supports and social support*, London, Policy Studies Institute, 1987.

[20] **Gracia Fuster, Enrique and Gonzalo Musitu Ochoa**, *Informe final de la investigación: Apoyo social: Un análisis empírico en poblaciones de alto riesgo*, Generalitat Valenciana. Consellería de Treball i afers Socials. Direcció General D'Afers Socials, (in press).

[21] **Lin, N., Dean, A. and Ensel, W.**, *Social support, life events and depression*, Orlando, Academic Press, 1986. See also, **Gracia, E. and Musitu, G.**, "Integracion y participación en la comunidad: Una conceptualización empírica del apoyo social comunitario", in: **Musitu, G., Berjano, E. and Bueno, J.R.** (Eds.), *Psicología Comunitaria*, Valencia, NAU Libres, 1990.

[22] **Sundstrom, E.**, "Crowding as a sequential process: Review of research on the effects of population density on humans", in: **Baum, A. and Epstein, Y.** (Eds.), *Human response to crowding*, Hillsdale, NJ: Erlbaum, 1978. Also, **Aiello, J.R., Thompson, D.E. and Baum, A.**, "Children, crowding, and control: Effects of environmental stress on social behavior", in: **Wohlwill, J.F. and von Vliet, W.** (Eds.), *Habitats for children: Impacts of density*, Hillsdale, NJ, Erlbaum, 1984. See also, **Baum, A. and Paulus, P.B.**, "Crowding" In: **Stokols, S. and Altman, I.** (Eds.), *Handbook of environmental psychology*, New York, Wiley, 1987.

[23] **Evans, G.W., Palsane, M.N., Lepore, S.J. and Martin, J.**, "Residential density and psychological health: The mediating effects of social support", *Journal of Personality and Social Psychology*, **57**, 1989, pp. 994-999.

[24] **Baum, A. and Valins, S.**, *Architecture and social behavior: Psychological studies of social density*, Hillsdale, NJ, Erlbaum, 1977.

[25] **Altman, I.**, *The enviroment and social behavior*, Monterey, CA: Brooks/Cole, 1975. See also, **Stokols, D.**, "On the distinction between density and crowding: some implications for future research", *Psychological review*, **79**, 1972, pp. 275-277.

[26] **Evans, G.W.**, "Crowding and the developmental process", In: **Baum, A. and Epstein, Y.** (Eds.), *Human response to crowding*, Hillsdale, NJ, Erlbaum, 1978. See also, **Aiello, J.R., Thompson, D.E. and Baum, A.**, *op. cit.*, and **Baum, A. and Paulus, P.B.**, *op. cit.*, reference [22].

[27] **Evans, G.W., Palsane, M.N., Lepore, S.J. and Martin, J.**, *op. cit.*, reference [23].

[28] **Durkheim, E.**, *Suicide: A study in sociology*, New York, Free Press, 1951.

[29] **Farish, R.E. and Dunham, H.W.**, *Mental disorders in urban areas*, Chicago, University of Chicago Press, 1939.

[30] **Hinkle, L.E., Jr. and Wolff, H.G.**, "Health and social environment: Experimental investigations" In: **Leighton, A.H., Clausen, J.A. and Wilson, R.N.** (Eds.), *Explorations in social psychiatry*, New York, Basic Books, 1957.

[31] **Selye, H.**, *The stress of life*, New York, McGraw Hill, 1956.

[32] **Holmes, T.H. and Rahe, R.H.**, "The social readjustment rating scale", *Journal of Psychosomatic Research*, **11**, 1967, pp. 213-218.

[33] **Lazarus, R.**, *Psychological stress and the coping process*, New York, McGraw Hill, 1966.

[34] Project Cyclops. A Design Study of a System for Detecting Extraterrestrial Life, publication nr. CR 114 445 (revised edition 7/73), NASA/Ames Research Center, Code LT, Moffett Field, California 94035, p. 4.

[35] **Watzlawick, Paul**, *Wie wirklich ist die Wirklichkeit?*, R. Piper & Co., Verlag, München, 1976.

[36] **Gergen, Kenneth J.**, *The Saturated Self. Dilemmas of Identity in Contemporary Life*, Basic Books, Harper Collins Pub. Inc., 1991.

JOHAN DEKLERCK AND ANOUK DEPUYDT

THE CYCLE OF INTEGRATION AND DISINTEGRATION APPLIED TO WESTERN SOCIETY: A FRAMEWORK FOR A FUNDAMENTAL PREVENTION OF DELINQUENCY

INTRODUCTION

In reaction to the current fragmented, controlling and technological approach of crime prevention, we are developing a model for the contextual prevention of crime based on ethical and ecological considerations. Therefore we want to investigate, through the study of religious world views and the ecophilosophical literature, the concept of 'connectedness' as a part of a larger frame of reference of 'integration-disintegration'. Four considerations form the basis of our approach.

1. Since 1993, the university of Leuven, hosts a research project concerning *prevention of delinquency*. This project consists of a theoretical and a practical part and was brought about through a cooperation of four faculties, i.e., the faculty of Law (Department of Criminology), Pedagogical Science (Research Centre for Early Childhood and Primary Education), Medical Science (Department of Health Ecology) and the Higher Institute of Philosophy (Department of Eco-Philosophy). This is not a traditional, specialised project, based on classical criminological theory. The *context of cross-discipline* serves as a basis for developing a new approach to delinquency within its societal framework.

2. Traditional criminological theories have often been based on unspoken ideas of society and of man, and were linked to political ideologies which no longer fit reality. This is the case of legal and law-oriented criminological theories, but also of the labeling theory, Marxist theories, etc.. Since the early eighties, criminology (and other disciplines) tend toward an increasingly methodological and technical approach, linked to a greater need for professionalism. As a result, the content oriented theories shifted to the background. There is a growing need for current analysis of society, contributing to a more fundamental approach to criminological theory. The problem of delinquency can only be answered adequately if it is based on a correct analysis of society, because fluctuations, developments and differentiations of the delinquency problem are linked to an evolving social context [1].

3. There are two premises within this criminological research project. As mentioned above, we start by analyzing new major developments in society. Second, we will use a new definition of delinquency. Before going further into the problem of delinquency itself, we will start by explaining our ideas on these recent developments of society. The major issue we deal with in analyzing society is '*disintegration*'. First we will go further into this concept. Afterwards we will focus on various aspects of a disintegrating society.

4. Western society is ruled by a linear approach of reality (humanity develops from alfa to omega, from a non-developed world to paradise), but complementary to this exists a *cyclic approach* of reality, in which the latter repeats itself cyclically [2]. In spite of the fact that within Western society this approach is rapidly losing ground, we still know it through the hands of a clock, and we experience it in the circle of the seasons, day and night, and life itself.

Together, these four considerations form a dynamic approach of "coming and going" in which two major phases may be distinguished. In the first one, elements tend to organize themselves into a coherent whole, a highly complex system of internal cooperation. We call this phase '*integration*'. In it, we can observe how the system gathers energy and information from its surroundings. The second phase is the one in which a system of high complexity, of different elements cooperating in order to achieve a common goal, disintegrates into its building materials which are no longer coherent or cooperative, but will in turn form the basis of a new integration circle. We call this the *disintegration* phase. In it, we are able to observe centrifugal powers, fleeing from the system towards the surroundings.

The season's circle illustrates this rather well. Every year, the same circle of integration (building) and disintegration (degradation) can be observed in nature.

Through the interaction of the sun and the earth, spring becomes the season of budding green. Blossoms form. The sun gives its warmth longer during the day, birds start brooding and new life forms everywhere. Everything settles into place. This process reaches its culminating point during summer: nature is then being built intensely of partly cooperating and partly competing elements. We call this building process integration. What was being built will start to decay in autumn. Trees lose their leaves, the greenery starts to decay. Rain causes unpicked fruit to rot and saps of plants and flowers withdraw to their roots and tubers, deep under the surface. During winter we are thrown upon ourselves, we remain indoors, while life itself hides under the soil. What remains will rot and compost and will in its turn serve as a nutriment for a new season of spring. Nature demolishes systems to their elementary parts and the forces return to the earth. This is disintegration.

Not only can this cyclical process be determined in nature, but also in man's life span. Youth will be spring, summer is adulthood, aging man will be autumn and during winter one dies and leaves life to the coming generations.

As a matter of fact, the *Einstein meets Magritte* conference shows us the same dynamics. First we gather with certain expectations and a vague idea of what this meeting may mean to us and to the public. As we develop our ideas, the public becomes increasingly involved. A process of collective thought results from the ideas, and one either tries to follow, accept or reject these thoughts, or one may try to add some of one's own ideas. This can be regarded as a process of integration. The totality of it will prove to be more than the sum of separate ideas of those who intervene and the ones observing. Toward the end of the meeting, conclusions can be drawn ("fruit" is picked). Then one begins to think of a next exposition, a coffee break, lunch, leaving, and step by step thought drifts away from this gathering. During this phase of disintegration, centrifugal powers are at work.

These simple examples illustrate the cyclic process in various forms. The hypothesis we scrutinize here is that our society currently finds itself in a state of disintegration. Traditional and sometimes very basic forms of connectedness of man with his social and natural environment are gradually losing their meaning. Four different dimensions can be distinguished: social life is fragmented and individualized, meaning giving systems gradually disappear, man is alienating from nature, and "normal life" is alienating from an institutionalized society.

FRAGMENTATION AND INDIVIDUALIZATION

"The world we are living in is fragmented. Everything we witness and experience is becoming increasingly complex. (...) There is no longer a single culture, but rather subcultures, culture fragments and an uninhibited pluralization of culture and an individualization of human behaviour" [3]. The individual's place and task within society is no longer obvious. In numerous fields of life, the individual can choose almost endlessly. No longer do we have the means to forecast where a child will live its life, what its profession will be, whether it will found a family and what that family will look like, which religion or philosophical thought it will adhere to, which rituals will lead it through the turning points of its existence, where it will find a mate and for how long they will be together, how its culinary and social traditions will be, how it will educate its offspring, which language they will speak, and which technological means will be at their disposal. Only a few decades ago, the answers to these questions were obvious for the vast majority of a population. Nowadays it is rather the opposite. Luc Huyse writes in his book "Beyond Politics": "Well-known screenplays for behaviour gradually lose their power. Next to this, the single individual's opportunities to lead a life of individual ideas, wishes and desires, gain importance." [4] Thus life more and more becomes a set of strictly individual choices, occupying more time and space.

School and professional careers for example turn themselves into more rapid successions of an increasing number of moments on which to choose. The choice in its turn also increases: it is no longer limited to the town or region in which one lives. In the case of higher education, more often than not country boundaries are passed. A professional career is no longer predictable. Having certain capacities or certificates can lead to very different fields, and careers frequently alter. Only very few people work for a single company all their lives.

Next to this, important elements of life such as residing, labouring, kith and kin, and recreation rarely coincide. They alienate, and individuals find it increasingly difficult to compromise. Quite often the place where one lives is not the one where one works, one's colleagues are often not one's friends, and often one's home is far away from one's family. Spare time and holidays are no longer spent at home. These ever-increasing distances are partly met by numerous communication systems and means of transport [5], involving ever-increasing environmental burden and an ever faster rhythm of life. Next to this, we witness how the family, and before it the larger family systems, gradually are falling apart. The number of single parent households increases, as does the number of single adults, the more so in an urban environment. "Distrust, hostility and indifference gradually rule human relations between groups and individuals, especially in urban areas characterized by social disintegration. Less and less people know what social togetherness is in a family, a neighbourhood, a region. Already, a looming loss of material satisfaction and security is bringing about the latent war of all against all" [6].

THE DISAPPEARING OF MEANING GIVING SYSTEMS

This increasing individualization which on the one hand gives people the opportunity to order their pattern of life as they wish, leaves them, on the other hand, with little anchorage. The former centres of ethics and value frameworks seem to gradually lose ground. "The disappearance of social centres round which communities formed in recent history (religions, ideologies, social movements), is striking. Even if the thought that past societies were highly integrated proves historically to be an illusion, it can nevertheless be stated that parts of it really were integrated and that the illusion of integration was widespread. This illusion is no longer there, and even fragments disintegrate. Different nations, different cultures and different social subgroups are all transformed thoroughly. They differ and diverge into a chaotic melee of movements and processes evolving in every which direction" [7].

In this chaos of impressions, ideas, opinions and beliefs, it becomes difficult for the individual to develop a stable identity [8]. Truth and knowledge spread by mass-media remain rather superficial [9]. "In a time of sweeping social breakdown and intellectual fragmentation, it is not surprising to find that the very notion of coherent thinking is being seriously corroded by patchwork eclecticism and ideological faddism. This has happened in earlier periods of social decay and it can be expected to happen today" [10]. From this starting point man is searching for new forms of purpose on different fields. "Exactly because of the fact that traditional systems of meaning giving, religious ones as well as secular ones, are no longer valid, a notion such as cosmology becomes almost religiously important. Man finally would like to know what his place is in this majestic universe" [11].

LOSING THE BOUND WITH NATURE

When discussing our society, the opposition between preserving the environment and scarce natural resources on the one hand and economic and social needs on the other hand, becomes increasingly apparent. We can speak of a new rift in the social landscape, next to the rift between communities, ideological rifts and the opposition between labour and capital [12]. This new rift rapidly plays a more important role, and enters all the different fields of social debate. Society is seeking new ways of dealing with this contrast. "The ecological crisis is rooted in deeply and proves to be more than a mere temporary disruption of a natural balance. It is caused by a certain way of life, by a certain way to attach importance to things and by sacrificing everything to it" [13].

This contradistinction is therefore beyond the discussion of sense and nonsense in preserving nature and how to do so. Recent discussion about man's relation to nature (i.e., the eco-philosophical discussion between anthropocentrism and ecocentrism) is an utterance of a widening gap between nature-bound man and culture-bound man. It points out a fundamental alienation, to the extent that man has alienated from his "naturalness." In this context, Bookchin mentions the link between "first nature" and "second nature." "What it can mean—and what it should mean—is a radical integration of second nature with first nature along far-reaching ecological lines, an integration that would yield new eco-communities, ecotechnologies and an abiding ecological sensibility that embodies nature's thrust toward self-reflexivity" [14]. In this way, the development of culture (second nature) will be formed on life's natural quality, on its opportunities as well as on its boundaries. From this view, culture is a result of nature and forms part of it.

This of course involves a substantial evolution in the organization of social life, man's ideology and his image of himself and of the world. This leaves Guattari pleading for an "ecosophy": "Man passively contributes to the increasing pollution of air and water, the destruction of the forests, the disturbance of the climate, the disappearance of a great number of species. He impoverishes the biosphere's genetic capital, disturbs the natural landscape, suffocates his own cities and gradually neglects cultural values and moral frameworks of human solidarity and togetherness. (...) This circle induces me to bring forward the need for an ecosophy which will link the environmental ecology with a social ecology and a mental ecology" [15].

The opposition between nature and culture comes as the result of the disappearance of "common" life. It is the result of the continuing individualization and privatization of what is common to our lives, and of the growing gap between public life and private life. What is common is reduced to the sum of private property and ideology. Public life is strictly ruled and regulated and quite often the individual citizen no longer considers it to be collective property, responsibility or joy. "In other words, the environmental crisis is also the result of the breakdown of "commons", of what used to be collective and accessible to everyone" [16]. Therefore, "commons" so to speak form a sort of natural midfield, a bond between nature and culture (compare with first and second nature). Here we are dealing with a revaluation of what is collective, "common", economically unpaying, as well as with repairing "ordinary life."

THE BREAKDOWN OF ORDINARY LIFE

Man's alienation from nature often has to do with the distance that has presented itself during these last decades between common life and institutionalized society. The latter is miles away from normal, daily life that can still be found in a considerable part of the planet: sleep where it suits you, drink from the pump, feed a little on your garden, talk to your neighbours, unite on the occasions of birth and death. We still know these elementary processes of life and we may actually go looking for them during our holidays. Complex political decision-making, however, an inflation of regulations and instructions as well as our economy, which is mainly organized around exchange-value, have alienated man from his natural dimension of existence.

Especially the philosopher I. Illich has pointed out this opposition between what he called the heteronomous field of life (ruled by exchange value) versus the vernacular field (which he originally called autonomous), ruled by use-value [17]. "The heteronomous production process has crossed a certain borderline and is now, so to speak by means of a built in system, conquering man's entire life. As a consequence, independence and freedom are lost, specialized and professionalized institutions proclaim what should be considered normal and what is supposed to be good for man" [18]. "According to this philosophy everybody is a consumer and the world a giant supermarket. Life is nothing but a prolonged pursuit of groceries whereby one chooses education from Tesco's and local government from Sainsburies" [19].

In spite of the fact that these fields of life are indissolubly linked to each other, this does definitely not mean that we would be able to transfer the contents of the heteronomous field (which centres around scarcity, material endeavour, financial means, selfishness, an economy based on exchange-value, being productive, and on rights and duties) to the vernacular field. This vernacular field centres around awareness of nature's abundance, feeling to be a member of a group or a community, finding satisfaction in processes that do not seek an immediate result, creativity, an economy based on use-value, a multitude of means, abundance in nature, few goals to achieve (elementary conditions only, such as not being hungry or thirsty, and being happy) and human freedom and autonomy.

Both of these worlds have their own form of existence, their characteristics, their values and their ways of contribution to the whole. The vernacular, non-economised world, however crumpling apart in the West, still forms a considerable and strongly present part of life in the entire world, but it is not always esteemed and respected. We still know it in our Western society, but we put it in the compartment of private life and leisure and we regard it as a banality next to "serious" matters, such as political problems, our dwindling economy, organization of society, unemployment, our unpayable health care, the organized unification of Europe, the continuance of NATO, the battle against the world's civil wars, delinquency, drug addiction and the hormone Mafia. "An obstinate institutionalizing subjects man to the logic of its built in dynamics which can no longer be contained. The super expansion of automatics is an entirely new threat to man" [20].

This threat results in numerous deadlocks that seem to confront our society. Politics loses its message and therefore its credibility. The economy seems to have entered a rapid succession of crises, always resulting in discussion about limits to and opportunities for further growth. Health care becomes unpayable because of the major expansion of what it seeks to do. This brings about the question of what is understood by a citizen's health. Health is said to be the lacking of disease, whilst safety is said to be the lacking of delinquency. In order to permanently confirm this, an ever increasing prevention is necessary [21].

Here we are dealing with a seriously disturbed equilibrium which remains, however, shadowed by social discussions: knowing what is good for oneself, feeling one's needs, having the possibility to decide autonomously, versus dependency on what is socially prestructured or what is ruled by others. It seems as if man no longer dares or wants to live from within and has replaced this urge by a permanent fear of being ill and of becoming a crime victim. The inner ability to judge and decide, to know one's needs, is eroding. Our over-organized society gradually takes initiative from man and continues to empty his inner autonomy. And yet we should not forget the fact that institutions and organizations are only a means to ensure our autonomy, which should enable and support the natural processes as described above, rather than obstructing them.

In describing these four processes we have pointed out our society's major tendencies. Each of them indicates growing distances and a coming apart of things that used to be related and closely interwoven. Hence our choice for the *common notion of disintegration*.

The developments as mentioned above determine the degree, the kind and the shape of the sort of delinquency we are dealing with at present (think, e.g., of growing delinquency against the environment, organized crime, etc.). Against this background, we will now study in depth the problem of delinquency itself. Here as well, on a micro-level, we witness a process of disintegration, of a coming apart of the bond between the offender and the party aggrieved (i.e., the individual victim, the environment, Society).

When analyzing various cases of serious crime, we were able to conclude that the bond between offender and victim was lost every time [22]. This can be illustrated by two examples.

When a train was hijacked in Holland in 1989 the hijackers picked a hostage whom they wanted to kill in order to press their demands. The victim-to-be asked the hijackers permission to write a letter of good-bye to his wife and children. This letter was very personal and created, as the man later put it literally, a "*bond*" between himself and the hijackers. He remained unharmed and somebody else became the victim. When researching the survival of some people in the second world war's concentration camps, the Norwegian criminologist N. Christie came to the same conclusion. Survivors had succeeded to create bonds with their guards. In this way, a Yugoslavian prisoner could give the guard a light for his cigarette after having learned a few words of Norwegian. Others were able to present their family to the guards by means of a photograph. Those who did not succeed in such things were considered objects, dirt, animals.

This brief description leads us toward a more accurate description of delinquency as the *non-existence of a link*, which etymologically can be found in the noun 'de-linq-uency'.

Relinking this missing link is the starting point of this prevention project.

The cornerstone here is formed by the notion of '*linkage*'. The notion of 'linkage' is etymologically present in the word religion, which literally means "re-link" (Latin: "re-ligare"). Linkage is one of the cornerstones of the eight major religions (worshipping and ideology), as we have found in our own research [24]. We will now look for the meaning of "linkage" in Western religious tradition. We will distinguish between religious experience and church structures. When comparing different religions as well as their practice and their theories, we can conclude that religions form a superstructure, a concrete shape of a basically religious feeling of man, determined by culture and time. As de-linking of a religion's contents on the one hand and its structural shape on the other hand gradually increases, a process of alienating is born, which carries the risk that a religion will no longer be based on contents, but will rather maintain itself in the form of external rites, serving other purposes [25].

Therefore linkage cannot be seen as a rule or a regulation. We rather consider it to be a *basic attitude,* a philosophy.

This philosophy can only be acquired if one *experiences* it within oneself. Thus we deal with a religiousness which derives its value not from external and traditional patterns of thought whose very tradition have made them rigid, but rather with one that focuses on daily life and on material fact as such.

In the two examples of delinquency mentioned above there was said to be a linking as being the cause of a delinquent act being prevented. In the first situation, however, someone else became the victim. The second case shows how delinquent behaviour does not take place within the own group (the Norwegian guards), but rather vis-à-vis a third party (the group of prisoners). Both examples illustrate a selective form of connectedness. Historically, this selective linkage has served as a basis of numerous crimes [26]. This leads us to a definition a "linkage" in its broadest sense, i.e., the one of experiencing an implicit, universal link between all living creatures.

Repairing and/or strengthening the basic feeling of linkage with *life as a whole*, with nature, with the ecosystem, is the concrete starting point of the practical part of this research project. Two years ago, the project was activated in 14 elementary schools in Flanders, showing very different characteristics: they differ in size, geographical position, population, methodology, ideology and social status. Furthermore, they belong to different school networks [27].

The idea is that a considerable part of minor delinquency and of behaviour preceding it is connected to a lack of lived linkage to the milieu. From this point of view, developing this attitude of linkage proves to be a basic requirement in avoiding problematic behaviour [28].

Practically, and by means of concrete initiatives, we try to create in these schools a basic attitude of connectedness in daily life of school and class. The aim is developing a basic attitude of linkage in children and in all those who are involved. They should acquire this attitude towards themselves, towards others and towards their material, social and ecological world.

In practice this means that during the course of a schoolyear the entire school team assembles about five times. Knowingly, we do not prescribe a set of activities to be carried out by the teachers. Different school teams, with linkage as a starting point, developed initiatives that were very much their own and were very different.

As far as *the link with oneself* was concerned for example, moments of rest were built in during the day or during the week. A class journal was kept and discussed, there were moments of relaxation and fifteen minute periods of silence with hand-made lanterns. It was decided that healthy fruit juices should be available in the refectories, next to the traditional coca-cola and lemonade. We worked on experiencing and expressing feelings, on art and on children's own initiative.

As to the *link with others*, we taught how to deal with conflict in a different way. We taught children to talk to each other about what happened when there were arguments on the playground, rather than talk to the monitoring teacher. The results are definitely positive. Children are being much more honest with each other, while showing a tendency to exaggerate versus the teacher. Problems tend to be settled more easily. If children prove to be very angry, things are not solved so easily. They need to cool down in order to settle the conflict afterwards. Toddlers often find this difficult: their vocabulary is insufficient. In their case we let them make drawings of the conflict.

As far as the *link with objects and materials* is concerned, in a given school we discussed with the children where objects of daily use in school come from: a bicycle, a pen, a book carrier. Children were given the opportunity to help keeping clean class, playgrounds and refectories.

Linkage to the group or to society was stimulated by paying more attention and devoting more time to togetherness in the form of a break at ten or at four o' clock. With the children's cooperation, in some schools the regulations were rewritten as a code for living together in a simple and positive way. In others schools, the end of the week is celebrated with a singing hour in which all children participate. The link with the children's parents is being discussed and in one school with a mainly north-African population, a biscuit celebration was held in which Moroccan mothers bake biscuits, an activity they find easier to perform than for example taking part in a traditional parents' evening.

The single most important initiative concerning the *linkage with the environment* was the laying-out of small vegetable gardens. In schools that were situated in an urban setting, small crops were sown, so that the children could make contact with the circle of nature.

The most important thing on each of these levels was not to create major projects, but rather to obtain, by means of minor, concrete initiatives, a slow process of awareness. Another important starting point in our prevention project was integrating all these initiatives by linking them to daily life. The very first initiative in the first school project illustrates these points rather well. In a school where the toddler's bread and butter up to then was simply disposed of, it was decided to feed it to the birds. We do not intend to reject the short-lived, intensively focused projects on drugs, vandalism or bullying that are set up in schools. We focus on a school's general climate of life, which is a major preventive agent. We therefore outlined the so-called prevention pyramid [29], which holds various forms of prevention (ranging from general, fundamental prevention, to the specifically curative form) that are supposed to be complementary in a mainstream of measures and having integrated preventive politics as a goal. We especially want to focus on the fundamental level.

It proved to be important for teachers to see that the different circles of linkage we worked on, proved to be very intensively interwoven, rather than separate, self-confined compartments. In one school the lavatories were often treated disrespectably: toilet-paper used to be thrown into the bowls, etc.. This behaviour spontaneously changed, not by teaching the children how they were linked to these objects, but rather by introducing the school's maintenance staff, by inviting these people to school parties, etc.. From the moment the children knew who was responsible for keeping their school clean, they changed their disrespectable behaviour towards such things. In bullying too we found that not only the creation of a bond between victim and tease was important, but that very often the first thing that proved to be necessary was to let them feel the bond with themselves, to approach them separately and supportingly, which tended to reduce the aggression in one party and to strengthen the defensibility in the other.

Because we are convinced of the fact that practicing these ideas of linkage can be meaningful not only in education, but also in numerous other parts of life, we have set up a non-profit organization. It has a broader horizon than the world of education and does not form part of the investigative project. Our organization is called Loung Tak [30], which literally means "horses of the wind." It is the Tibetan name for a "prayer flag." The organization is independent, not connected to any political party, faction or religion and works in a spirit of connectedness and pluralism. "Loung Tak" wants to discover things that unite people and does not care for superficial or specific subdivision, specialization and functional dispersion. Practically, Loung Tak has started the idea of linkage within the world of education, together with The Centre of Experiential Education. We intend to also look for a basic attitude of linkage in other fields and other professions and, possibly cooperating with numerous existing organizations, we will seek linkage between people with ideologically and socially different backgrounds.

AFFILIATION

Johan Deklerck and Anouk Depuydt
Department of criminal law, criminal procedure and criminology
Faculty of Law, Katholieke Universiteit Leuven, Belgium

REFERENCES

[1] See also **Deklerck, J.**, "De preventiepiramide, een orientatiekader voor geïntegreerde preventie", in *Panopticon*, 1996; **Hebberecht, P.**, "Kritische kanttekeningen bij het Belgisch beleid ter preventie van criminaliteit", p. 450, *Panopticon*, Kluwer Rechtswetenschappen, 1988; **Peters, T. and Van Kerckvoorde, J.**, *Onrustgevoelens en de media*, p. 120, Handboek Politiediensten, C.E.D. Samson, June 28, 1989.

[2] **Lenaire,T.**, " Naar een antropokosmologie", pp. 59-96, in: *Rimpels in het water. Milieufilosofie tussen vraag en antwoord*, festal book celebrating the 60th anniversary of Professor E. Vermeersch, Acco, Leuven-Amersfoort, 1994.

[3] **De Moor, B.**, *Wereldbeelden: naar een atlas van modellen*, p. 125, in: **Weiler R. and Holemans, D.** (eds.), *Bevrijding of bedreiging door wetenschap en techniek*, Pelckmans, Kapellen, 1993.

[4] **Huyse, L.**, *De politiek voorbij*, Kritak, Leuven, 1994, p. 34.

[5] In this way it seems as if faster transport coincides with increasing distances te be covered. An hour's walk to the market or to a relative in a nearby village is replaced by an hour's car drive. A journey to Brussels becomes a flight to New York.

[6] **Duintjer, O.**, "Over natuur, vervreemding en heelwording", in: **Achterberg, W. and Zweers, W.** (eds.), *Milieucrisis en filosofie, westers bewustzijn en de vervreemde natuur*, Ecologische Uitgeverij, Amsterdam, 1984, pp. 192–93.

[7] **Apostel, L.**, "De constructie van wereldbeelden", in: **Weiler, R. and Holemans, D.** (eds), *Bevrijding of bedreiging door wetenschap en techniek*, Pelckmans, Kapellen, 1993, p. 114.

[8] In an interview with De Morgen (June 6, 1995), criminologist David Smith (Edinburg University), on the occasion of an international research on "psychological disorders in young people" (with Michael Rutter), indicated that individualization and a free choice of what is morally acceptable, presents young people with a panoramic variety in which they tend to get lost. This major psychic pressure, this form of stress is what is to be paid for more independence, and has as a result a serious increase in crime, depressions, suicide, nutritive disorder and abuse of drugs and alcohol.

[9] In an interview with NRC Handelsblad, following the publishing of two recent books, i.e., "The moral Dimension. Towards new Economics." and "The Spirit of Community. Responsibilities and the Communitarian Agenda (30 December 1994), professor A. Etzioni (Columbia University,

New York) makes a plea for "restoring moral infrastructure" and describes an America that is suffocating in a moral vacuum: "Just as rivers and lakes have deteriorated, society's moral structure is coming apart." He summons Europe to no longer follow the U.S.A.

[10] **Bookchin, M.**, *The Philosophy of Social Ecology*, Black Rose Books, Montreal-New York, 1990, p. 137.

[11] **Van der Veken, J.**, *Naar een geïntegreerd opvoedingsproject*, speech held on September 18, 1992 (unpublished), p.7.

[12] See **Huyse, L.**, *De gewapende vrede: de politiek in Belgie na 1945*, Kritak, Leuven, 1993.

[13] **Van der Veken, J.**, "Is onze wereld nog een kosmos om in te leven" in: **Weiler R. and Holemans, D.** (eds.), *Bevrijding of bedreiging door wetenschap en techniek*, Pelckmans, Kapellen, 1993, p. 102.

[14] **Bookchin, M.**, ibid, p. 177.

[15] **Guattari, F.**, "Naar een herijking van de sociale praktijken", part XXX, *Filosofiemagazine*, 1st year of publication, nr. **3**, December 1992, pp. 38-39.

[16] **Vandevelde, T.**, "Schaarste en overvloed: een sociaalfilosofische reflectie over de milieu crisis", *Tijdschrift voor Filosofie*, Leuven, 1993. He gives the example of land (woodlands, grass verges, etc.) that belonged to no one and to everyone (they were not thought of as possessions) and which were at the disposal of the one who needed them (poor people were allowed to gather dead wood, goats could graze on it, etc.).

[17] **I. Illich** has dedicated most of his writings to these dynamics and its consequences for Western man on different fields such as medicine, education, transport, etc.. I, for one, have applied in my dissertation I. Illich's philosophical points of view to the institution of criminal law, its growth and its contraproductive consequences, see **Deklerck, J.**, *Criminogenese als recht, recht als criminogenese. Een analyse van de strafrechtsinstitutie vanuit Illichiaans perspectief*, K.U.Leuven, 1984 (unpublished).

[18] **Achterhuis, H.**, *De markt van welzijn en geluk*, Ambo Baarn, 1981, p. 56.

[19] Quote by **Sir Ian Gilmore MP**, In: **Hawby, R.I. and Walklate, S.**, *Critical Victimology, International Perspectives*, Sage Publications, London 1994, p.84. This book describes how within the institution of Law, the citizen since the early eighties is being reduced to a legal consumer, a process that is called "consumerism" (p.80 sqq). By offering justice as a kind of service, some crimes are "normalized": "...since women are in receipt of a device which has been tried and tested on other

victims and shown to work successfully. In some ways this constitutes a normalization of rape." (p. 85)

[20] **Illich, I.**, *La Convivialité*, Seuil, Paris, 1973, p. 12 sqq.

[21] See also **Deklerck, J.**, "De preventiepiramide, een orientatiekader voor geïntegreerde preventie", in *Panopticon*, Kluwer Rechtswetenschappen, 1996.

[22] **Depuydt, A.**, *Religie als antwoord op delinquentie: een nieuwe basisfilosofie voor kinderen in het lager onderwijs?*, unpublished criminology dissertation, K.U.Leuven, 1990, pp. 8-14.

[23] **Depuydt, A.**, ibid, pp. 66-69.

[24] **Depuydt, A.**, ibid, pp. 41-42.

[25] **Depuydt, A.**, ibid, pp. 46-64.

[26] Think of the numerous wars fought because of language, belief or race, but also of political tendencies such as Nazism or Stalinism, etc.

[27] Starting from the basic philosophy of "linkage" itself, we explicitly chose starting the project not only in so-called low opportunity schools, but also in elite schools and in middle of the road schools. Low opportunity schools do not simply appear out of the blue: starting from the basic philosophy of "linkage", their existence has everything to do with the existence of elite schools, and vice versa. We tried to establish to what extent the same basic philosophy of linkage could serve as a set of references in different types of schools.

[28] In this context it is interesting to mention the notion of "shame and reintegration", launched by **J. Braithwaite**. He says it is vitally important that, in order to repair wrongdoing, the offender and society should mutually reintegrate, see **Braithwaite, J.**, in: *Crime, Shame and Reintegration*, Cambridge University Press, 1989.

[29] **Deklerck, J.**, ibid.

[30] Contact address: Loung Tak v.z.w., Hondsdonkstraat 51, 3120 Tremelo, Belgium

GPSR Compliance

The European Union's (EU) General Product Safety Regulation (GPSR) is a set of rules that requires consumer products to be safe and our obligations to ensure this.

If you have any concerns about our products, you can contact us on ProductSafety@springernature.com

In case Publisher is established outside the EU, the EU authorized representative is:

Springer Nature Customer Service Center GmbH
Europaplatz 3
69115 Heidelberg, Germany

Batch number: 10739606

Printed by Printforce, the Netherlands